The Man Who Knew Too Much

Dr Stephen Inwood was born in London in 1947, and was educated at Dulwich College and at Balliol and St Antony's College, Oxford. For twenty-six years he was a college and university history lecturer, but he became a professional writer in 1999, after the publication of *A History of London.* He also holds posts at Kingston University and at New York University in London. He lives in Richmond, west London, with his wife and three sons.

Also by Stephen Inwood

A HISTORY OF LONDON

STEPHEN INWOOD

The Man Who Knew Too Much

THE STRANGE AND INVENTIVE LIFE
OF ROBERT HOOKE
1635–1703

PAN BOOKS

First published 2002 by Macmillan

This edition published 2003 by Pan Books
an imprint of Pan Macmillan Ltd
Pan Macmillan, 20 New Wharf Road, London N1 9RR
Basingstoke and Oxford
Associated companies throughout the world
www.panmacmillan.com

ISBN 0 330 48829 5

1 3 5 7 9 8 6 4 2

A CIP catalogue record for this book is available from
the British Library.

Typeset by SetSystems Ltd, Saffron Walden, Essex
Printed and bound in Great Britain by
Mackays of Chatham plc, Chatham, Kent

For William and Jessie Inwood

Acknowledgements

As Isaac Newton said in a famous letter to Robert Hooke, most of what each of us achieves is built upon the work of others. Without the research of many scholars on the history of science and the early years of the Royal Society, it would have been quite impossible for me to have written this book. I have acknowledged my debts to them in my notes and bibliography. I am grateful to the Royal Society for allowing me to use their library, and to the inter-library loans service at Thames Valley University, my ex-employer, for finding so many books and articles for me. Ursula Carlyle, the archivist of the Mercers' Company, allowed me to use the material on Hooke and Gresham College which is held in the Company's archives. I have received helpful advice from Michael Wright, Curator of Mechanical Engineering at the Science Museum, on Hooke's wheel-cutting engine, and from Dr Anthony Geraghty of the Glasgow School of Art, on Hooke's role in building the City churches. My old friend Nigel Greenwood spotted some errors in the hardback edition, and helped me correct them. My brother, Michael Inwood, gave me advice on the meaning of some of Hooke's Latin phrases, my sister, Jacqueline Inwood, helped me to explore his birthplace on the Isle of Wight, and my mother, Jessie Inwood, found me a picture of Freshwater.

After the publication of *A History of London* in 1998 I was lucky enough to find – or to be found by – a sympathetic and resourceful literary agent, David Godwin, who showed me how to leap painlessly from the academic to the literary world when events forced that choice upon me. At Macmillan, I would like to thank Tanya Stobbs, for her critical and intelligent reading of the original manuscript; Nicholas Blake, for discovering the ambiguities and inconsistencies in my text however well I tried to hide them, and for giving me the benefit of his own knowledge, especially of maritime matters; Jane Henderson, for the index; and Jeremy Trevathan and his assistant, Stuart Evers.

My wife and sons were tolerant about sharing their house with a seventeenth-century scientist for the past three years, though they are probably relieved that he has at last started packing his bags. Tom, Joe and Benji put up with my Hooke anecdotes, and at least pretended to be interested when I asked them, for the umpteenth time, 'Guess who invented that?' Anne-Marie, my wife, has been unfailingly supportive, a pleasure to live with in every way. She never showed the slightest doubt about my ability to make a success of what I had undertaken, and never told me that I should get a proper job.

Preface

In the 1990s, when I was working on *A History of London*, I came across an obscure and fascinating character who was so inventive, able and energetic that he had exerted a profound influence over three distinct aspects of London's development in the later seventeenth century. He led the way in replanning and rebuilding the City after the fire of 1666, he developed technology that promoted the growth of the important London watch and scientific instrument industry, and helped to turn the Royal Society from an aristocratic club into a permanent feature of London's intellectual life. The fact that this man, Robert Hooke, was depicted (when he was mentioned at all) as an embittered and troublesome individual who made it his business to quarrel with greater and nobler scientists (especially Isaac Newton), and to claim credit for their work, drove me to find out more about him. Isaac Asimov summarizes the standard opinion of Hooke, a 'nasty, argumentative individual, antisocial, miserly, and quarrelsome', who took a 'malignant pleasure in controversy', and drove Newton to a nervous breakdown. Could Hooke have been the malicious pantomime character, with a twisted back and a personality to match, that appeared in popular reference books and general works on the history of science?

As I looked a little further into Hooke's life I discovered that he was not really a forgotten figure. Many scholars, attracted by Hooke's immense versatility, and by the sense that he has not been dealt with fairly by standard historical accounts, have been working on Hooke's scientific and mechanical achievements, producing scholarly articles, conference papers, doctoral theses and annotated editions of his writings. Some of them, perhaps taking their mission to resurrect Hooke's name and reputation too literally, were even trying to find Hooke's body, which was reburied in an unmarked grave in the nineteenth century, and to reinter it with appropriate ceremony in a more fitting place in time for the tercentenary of his death in 2003. But specialists do not often write biographies, and there had been no general account of Hooke's life, encompassing his science, his building, and his personal life, since Margaret 'Espinasse's book in 1956.

Hooke was a man of ideas, a thinker who was at the forefront of the scientific revolution of the later seventeenth century. His life cannot be properly appreciated if it is presented only as a series of incidents, scandals, or disputes with rival scientists. We have to look at and understand his ideas, his inventions, the principles that guided his work. Sometimes this takes a little concentration, but Hooke's science was largely intuitive, experimental, and non-mathematical, and it is not so difficult for non-scientists to understand. I hope that readers will find that the pleasure of watching Hooke and his colleagues grappling with many of the great questions that interested scientists for the next 200 years will repay a little effort. For Hooke stood right in the middle of the wonderful world of Restoration science, in which scholars were rebuilding their understanding of the universe and the natural world after the collapse of the old Greek and medieval certainties. He worked alongside (and sometimes in conflict with) some of the greatest figures in the modern history of science – Huygens, Boyle, Wren, Halley, Flamsteed, Newton, Leibniz, Hevelius – and left his mark on almost every scientific and mechanical project of his day. Nothing, from the rotation of planets and the nature of light to the origins of fossils and the life cycle of the gnat, escaped his attention.

Restoration England had many multi-talented citizens, but the diversity of Hooke's accomplishments was impressive even in his own time, and would be unthinkable today. As well as making an important contribution in almost every scientific field, Hooke was a notable scientific artist, a pugnacious controversialist, a brilliant designer of watches, telescopes, quadrants and scientific instruments of all sorts, a surveyor and urban developer of the first rank, and one of the most important designers and builders of country mansions, town houses, churches, hospitals and monuments of his time. Such a variety of interests and activities makes for a fascinating life, but not a simple one. It might have been easier to tackle each area of Hooke's career in turn – Hooke the mechanic, Hooke the architect, Hooke the experimental scientist, Hooke the coffee-house conversationalist, and so on. But in

Hooke's real life all these activities were crowded together into every week, and I thought it was important to capture Hooke's frenetically busy life as he lived it day by day, or at least month by month. I also wanted to follow Hooke's life through its full extent, not stopping when he ceased to be a central figure in the scientific world. What happens to us in our old age is as important as anything else in our lives, so I have done my best to go with Hooke into his final years, and to be with him at his death bed.

Hooke lived his final years with the growing fear that his scientific work would be forgotten, and as far as the non-academic world is concerned this is exactly what happened. Now, three hundred years after his death, it is time for this difficult, ugly, tireless and brilliant man to be remembered again.

A Note on Money

In the seventeenth century the pound sterling was divided into twenty shillings (shortened to s.), and a shilling was divided into twelve pennies, or pence (shortened to d.). So an amount might be expressed as £2 10s 6d, or £2/10/6d.

The value of money in Hooke's lifetime cannot be converted into 2002 values, because prices and wages have grown at different rates, and the prices of various goods and services have changed in different ways. Coffee, tea, tobacco and sugar, which were exotic and expensive in Hooke's day, have become cheap everyday commodities, but a large house in St James's Square, which Hooke could have bought for £5,000 in the 1670s, would cost several millions today. Bearing these difficulties in mind, multiplying seventeenth-century sums of money by a hundred will give a general indication of their purchasing power in 2002.

A Note on Dates

Until 1752 England used the Julian calendar, which was then ten days behind the Gregorian calendar adopted in other West European countries. Hooke and all his English contemporaries used the 'old style' Julian calendar dates, and so have I. In official usage the new year began on 25 March, but most people treated 1 January as the first day of the year. Sometimes dates between 1 January and 24 March were written with both years (1665/6), but I have given all dates in the modern manner, with the new year starting on 1 January.

'Science' and 'Natural Philosophy'

Hooke and his contemporaries used the phrase 'natural philosophy' where we might say 'science', to denote the study of the material universe and its laws, and 'natural philosopher' instead of the nineteenth-century term 'scientist'. I have used the contemporary and modern terms interchangeably.

Contents

List of Illustrations

Section One

Section Two

A Chronology of the Life of Robert Hooke

18 July 1635 Born in Freshwater, Isle of Wight.

October 1648 Death of John Hooke, Hooke's father.

October 1648 Hooke travels to London, and begins an apprenticeship with the painter Peter Lely.

c. 1649–1653 A student at Westminster School, under the headship of Dr Busby.

c. 1650 Onset of Hooke's curvature of the spine.

1652 London's first coffee house, Pasqua Rosee's Head, opens.

1653 Hooke goes to Christ Church, Oxford.

c. 1655–6 Hooke joins John Wilkins' scientific group at Wadham College as a paid assistant. He works with Christopher Wren, Seth Ward, and others.

c. 1657 Hooke starts work as Robert Boyle's paid assistant.

c. 1658 Hooke makes a working air pump, enabling Boyle and Hooke to conduct many experiments on vacuums, air pressure and combustion.

1658 Hooke later claimed to have invented the spring-regulated watch at this time.

May 1660 Restoration of Charles II.

28 November 1660 Royal Society founded, but without a royal charter until 1662.

1661 Hooke's first surviving publication, a tract on capillary action.

November 1662 Hooke starts working unpaid as a curator of experiments for the Royal Society. Elected a Fellow in June **1663.**

December 1662 Experiments on weight at the top of Westminster Abbey.

c. **1663–4** Hooke tells Boyle, Moray and Brouncker of his spring-regulated 'longitude' clock, but rejects their proposals and decides to keep its details secret.

May 1664 Hooke sees a spot on Jupiter, proving that it rotates.

June 1664 Sir John Cutler offers to fund a £50 pa lectureship on science and trades for Hooke.

September 1664 Hooke moves into Gresham College.

1664 Hooke uses the freezing point of water as the standard zero on glass thermometers.

1664 Hooke conducts unsuccessful trials of the Reeve–Gregory reflecting telescope.

November 1664 Hooke carries out an experiment to show that a dog could be kept alive by blowing air directly into its lungs.

December 1664 to **April 1665** Hooke and Wren study the paths of two comets.

11 January 1665 Hooke is formally elected Curator to the Royal Society, at £30 a year.

January 1665 *Micrographia* published.

March 1665 Hooke becomes Gresham Professor of Geometry.

April 1665 Hooke shows his new seventeen-inch quadrant.

June 1665 Bubonic plague epidemic begins in London.

July 1665 Hooke, Wilkins and Petty move to Durdans, near Epsom.

January 1666 Plague subsides, and Hooke returns to London.

March 1666 Hooke establishes the rotation period of Mars.

23 May **1666** Hooke's seminal paper on the mechanics of planetary motion.

2–6 September **1666** Great Fire of London. Wren, Evelyn and Hooke prepare plans for a rebuilt city.

12 September **1666** Hooke's reflecting quadrant with micrometer adjustment is ready for demonstration to the Royal Society.

October **1666** Hooke nominated as one of the City's three representatives on the Commission to survey the ruined City. Formal appointment as City Surveyor, March **1667**.

February **1667** First City Rebuilding Act lays down rules for new streets and houses.

10 October **1667**, May **1668** Hooke, Lower and King show that fresh air, rather than the motion of its lungs, keeps a dog alive.

c. **1667–68** Hooke's lectures on earthquakes develop his ideas on fossils and the changing shape of the Earth.

February **1668** Hooke demonstrates a watch regulated by 'a little spring of tempered wire'.

May to June **1668** Hooke designs a new Royal Society college (not built).

June to December **1668** Hooke's correspondence with Hevelius over telescopic sights for quadrants.

July to October **1669** Hooke observes Draco, in an attempt to prove that the Earth moves, and gives a Cutler lecture on this in **1670**.

1670 Hooke is appointed one of Wren's two assistants in rebuilding fifty-one City churches. Payments start in **1671**.

November **1670** The Royal Society Council decides to rebuke Hooke for neglecting his duties.

December **1670** Royal College of Physicians employs Hooke to design and build a new college. The work lasts from **1671** to **1679**.

February to March **1671** Hooke tests his ability to survive in a sealed and depressurized cask.

March 1671 Hooke experiments on the effects of vibration on flour.

March 1671 Wren and Hooke start work on the Fleet Canal and Thames quay.

December 1671 7,000 (of 8,394) City houses have now been rebuilt, and City life is getting back to normal.

11 January 1672 Royal Society sees Newton's reflecting telescope, and Hooke starts work on a better one.

February to June 1672 Hooke's dispute with Newton over the nature of light and colour.

10 March 1672 Hooke's diary begins.

October 1672 Hooke's work on the Monument begins, and continues until 1677.

January 1673 Harry Hunt arrives at Gresham College as Hooke's assistant.

February to March 1673 Hooke produces a calculating machine to beat Leibniz's.

18 March 1673 After four months' work, Hooke shows that air loses 5 per cent of its volume in combustion.

April to June 1673 Hooke and Boyle create gases that do not sustain combustion.

August 1673 Nell Young leaves Hooke's employment.

November 1673 Royal Society meetings return to Gresham College after six years in Arundel House.

11 December 1673 Hooke lectures against Hevelius's observational methods.

February 1674 Hooke demonstrates the first practical Gregorian telescope.

April 1674 Hooke starts work on the Bethlem Royal Hospital (Bedlam). Work continues until 1676.

March 1674 Hooke's first published Cutler Lecture, *An Attempt to Prove the Motion of the Earth.*

May 1674 Hooke's collaboration with watch and instrument maker Thomas Tompion begins.

September 1674 Hooke starts designs for Montagu House, Bloomsbury. Building work continues until 1680.

December 1674 Hooke publishes *Animadversions on the First Part of the Machina Coelestis of Johannes Hevelius*, urging the use of telescopic sights and describing his new equatorial quadrant and universal joint. His dispute with Hevelius lasts until 1679.

February 1675 Hooke's priority dispute with Huygens and Oldenburg over the spring-regulated watch begins. For the rest of the year Hooke and Tompion race to produce a good spring watch.

July 1675 Hooke helps design the Greenwich Observatory, which was later (in 1676–7) equipped with his quadrants.

October 1675 Hooke publishes *A Description of Helioscopes*, an account of many of his mechanical achievements. Its postscript gives his side of the spring watch dispute.

December 1675 to January 1676 Newton's letters to Oldenburg rekindle the dispute with Hooke over light and colours.

January 1676 Hooke forms a New Philosophical Club.

January to February 1676 Hooke and Newton exchange conciliatory letters, including Newton's 'shoulders of giants' compliment.

February to March 1676 Royal Society Repository moves out of Hooke's rooms, into its own gallery.

May 1676 Shadwell's play *The Virtuoso* mocks Hooke and the Royal Society.

June 1676 Hooke's sexual relationship with Grace Hooke begins.

September 1676 *Lampas* published, with a postscript attacking Oldenburg.

January 1677 Hooke starts working as an agent and architect for John Hervey in St James's Square and the Strand.

February 1677 Hooke designs Escot House, in Devon, for Sir Walter Yonge.

8 July 1677 Thomas Crawley becomes Hooke's assistant (until January 1680).

9 July 1677 Jonathan's coffee house opens.

August 1677 Grace Hooke goes to the Isle of Wight for ten months.

5 September 1677 Oldenburg dies of malaria.

12 September 1677 Tom Giles, Hooke's cousin, dies of smallpox.

November 1677 Hooke sees micro-organisms in water, and shows them to the Royal Society.

30 November 1677 Hooke and Grew elected joint secretaries of the Royal Society; Lord Brouncker ousted as President.

27 February 1678 Hooke's brother John hangs himself in Newport, Isle of Wight.

April 1678 *Lectures and Collections* published, containing *Cometa* and *Microscopium. Cometa* shows that Hooke does not know the inverse square law.

June 1678 Hooke sees 'tadpoles' in semen, using a single-lens microscope.

November 1678 *Lectures of Spring* published, developing Hooke's theory of matter and elasticity.

May 1679 Hooke's weather clock completed.

24 November 1679 to 17 January 1680 Hooke and Newton correspond on planetary motion, the path taken by a falling body and the inverse square law.

16 January 1680 Hooke's falling-bullet experiment in Garraway's.

December 1679 to March 1680 Hooke's series of congruity experiments, exploring the nature of matter.

February 1680 Hooke's series of lectures on light begins. He explains the inverse square law.

May to June 1680 Hooke visits Willen Church, Buckinghamshire, and the site of Ragley Hall, Warwickshire.

September 1680 Hooke finishes work on Moses Pitt's Atlas.

September 1680 Hooke meets Robert Knox, recently returned from seventeen years' captivity in Ceylon.

1681 Hooke designs Ramsbury Manor, Wiltshire.

27 July 1681 Hooke demonstrates an iris diaphragm and wheels for finding the frequency of musical notes (Savart's wheel).

November 1681 Hooke and Flamsteed argue over lenses and rhumb lines.

December 1681 to April 1682 Five editions of *Philosophical Collections* published.

June 1682 Hooke's lecture on the mechanical nature of memory.

August 1682 Hooke visits Ramsbury Manor, Wiltshire, to supervise building work.

October 1682 Hooke's legal dispute with Cutler over unpaid arrears begins.

30 November 1682 Hooke loses his position as Secretary of the Royal Society, and his place on the Council.

February 1683 The Royal Society appoints two new curators, alongside Hooke.

February to March 1683 Hooke's experiments to investigate the relationship between vibrations and gravity.

6 June 1683 Hooke's salary as Curator is replaced by a gratuity based on performance.

January 1684 Wren challenges Hooke to explain the laws of celestial motion from the inverse square law. Hooke fails to do so.

February 1684 Hooke wins his case against Cutler and receives £475.

November 1684 Newton sends *De Motu* to Halley.

1 December **1684** Hooke is re-elected to the Royal Society Council, on which he sat for thirteen of the next seventeen years.

February to March **1685** Hooke's paper on friction, road surfaces and carriages.

28 April **1686** The first book of Newton's *Principia Mathematica* is presented to the Royal Society. Hooke claims credit for two of Newton's key ideas, but wins no support.

20 June **1686** Newton's letter to Halley denounces Hooke for his errors and pretensions.

December **1686** to March **1687** Hooke gives a series of lectures on fossils and the changing shape of the Earth.

February **1687** Grace Hooke dies, aged twenty-six.

March to June **1687** Hooke argues with John Wallis over the compatibility of Hooke's ideas on the history of the Earth with biblical texts.

1688 Hooke receives his last payment from the Royal Society.

1 November **1688** Hooke's last diary begins. Final entry, 8 August **1693.**

June **1689** Hooke designs Shenfield Place, Essex.

15 September **1689** Aubrey's letter to Anthony Wood, on Hooke's achievements.

18 December **1689** Hooke lectures on the uses and effects of cannabis.

January **1690** Hooke appointed Surveyor to the Dean and Chapter of Westminster.

January **1690** Hooke presents his design for Haberdashers' Almshouses. Building work continues until **1695.**

15 April **1693** Sir John Cutler dies.

December **1693** Hooke's last known 'view' of a City property.

18 July 1696 Hooke wins his two-year case against Cutler's estate for arrears and future payment.

April 1697 Hooke begins to write the history of his life.

August 1699 The Gresham Joint Committee warns that the College has to be rebuilt.

10 June 1702 Hooke's last recorded contribution to a Royal Society meeting.

3 March 1703 Hooke dies in his Gresham College rooms.

6 March 1703 Hooke is buried in the church of St Helen Bishopsgate.

'An Exact Surveigh' of the City of London in 1667, after the Great Fire. Surveyed by John Leake, engraved by Wenceslaus Hollar. Gresham College is on Broad

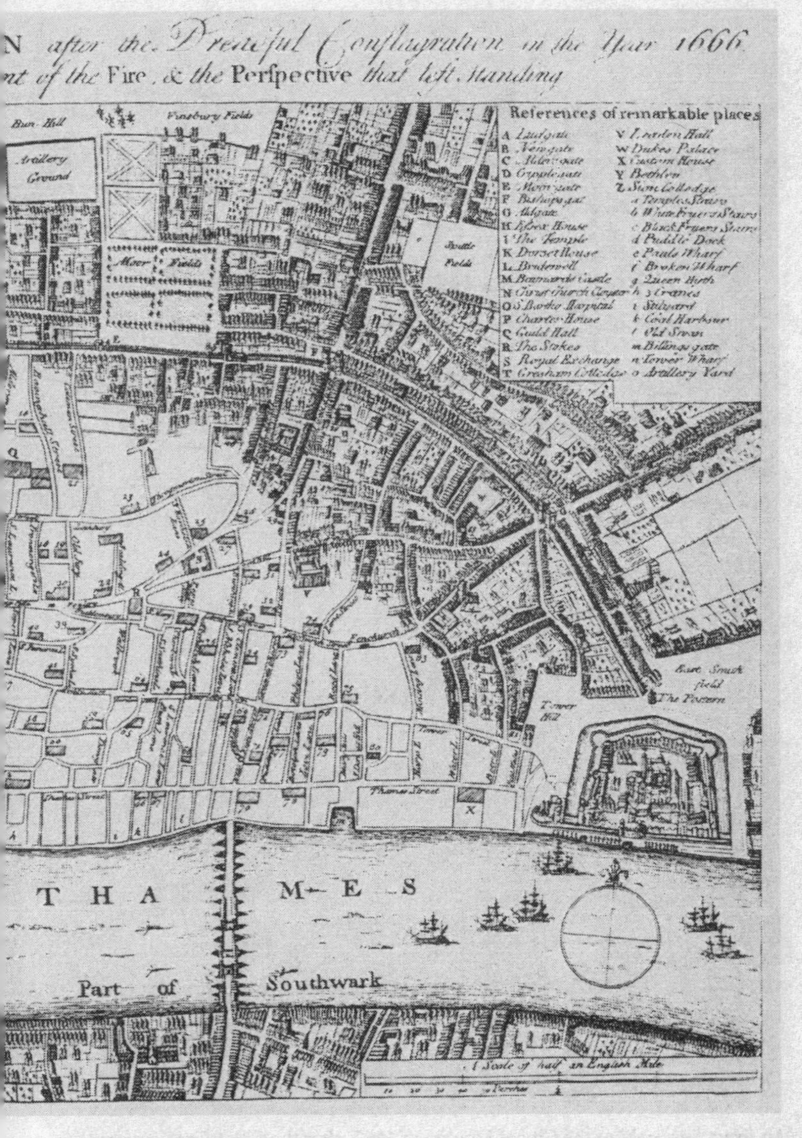

Street, in the undamaged north-east corner, with Moorfields nearby. (W. Besant, *London in the Time of the Stuarts* (1903).)

A section of John Ogilby's 1 inch to 100 feet map of London in the early 1670s, published in 1676. This shows the small area in which Hooke spent most of his

life, between Gresham College, the coffee houses and bookshops off Cornhill, the Royal Exchange and the Guildhall. (Guidhall Library, London)

The Man Who Knew Too Much

The truth is, the Science of Nature has been already too long made only a work of the Brain and the Fancy: It is now high time that it should return to the plainness and soundness of Observations on material and obvious things. It is said of great Empires, That the best way to preserve them from decay, is to bring them back to the first Principles, and Arts, on which they did begin. The same is undoubtedly true in Philosophy, that by wandring far away into invisible Notions, it has almost quite destroy'd itself, and it can never be recovered, or continued, but by returning into the same sensible paths, in which it did at first proceed.

Micrographia, 1665

These I mention, that I may excite the World to enquire a little farther into the improvement of Sciences, and not think that either they or their predecessors have attained the utmost perfections of any one part of knowledge, and to throw off that lazy and pernicious principle, of being contented to know as much as their Fathers, Grandfathers, or great Grandfathers ever did, and to think they know enough, because they know somewhat more than the generality of the World besides: . . . Let us see what the improvement of Instruments can produce.

Animadversions on the Machina Coelestis of Johannes Hevelius, 1674

I have had the misfortune either not to be understood by some who have asserted I have done nothing, or to be misunderstood or misconstrued (for what ends I now enquire not) by others who have secretly suggested that their expectations – how unreasonable soever – were not answered . . . And though many of the things I have first Discovered could not find acceptance yet I finde there are not wanting some who pride themselves on arrogating of them for their own – But I let that passe for the present.

Hooke in the Royal Society, 26 June 1689

Certainly there are many *Species* of Nature that we have never seen, and there may have been also many such *Species* in former Ages of the World that may not be in being at present, and many variations of those *Species* now, which may not have had a Being in former Times: . . . when we consider how great a part of the preceding Time has been . . . unrecorded, one may easily believe that many Changes may have happened to the Earth, of which we can have no written History or Accounts. And to me it seems very absurd to conclude, that from the beginning things have continued in the same state that we now find them, since we find everything to change and vary in our own remembrance; certainly 'tis a vain thing to make Experiments and collect Observations, if when we have them, we may not make use of them; if we must not believe our Senses, if we may not judge of things by Trials and sensible Proofs, . . . but must remain tied up to Opinions we have received from others, and disbelieve every thing, tho' never so rational, if our received Histories doth not confirm them.

Hooke in the Royal Society, 25 July 1694

1. 'The History of My Own Life'

(1635–1653)

It was 3 March 1703, in Gresham College, in the City of London. Dr Robert Hooke, Gresham Professor of Geometry and Curator of Experiments for the Royal Society, lay dead on his bed. In death, as in life, he was not an attractive sight. His ragged clothes were twisted about his emaciated body like a winding sheet, and the lice were so thick on his corpse that 'there was no coming near him'. Those who searched his rooms after his death found a profusion of papers, letters, notebooks, calculations and lectures, but not the one document that they were looking for. Hooke had often talked of leaving a generous bequest to the Royal Society, his lifelong employer, so that a library, a laboratory and a lecture could be endowed in his name. But the will was never written, and Hooke's property passed to his next of kin (probably his cousin), Elizabeth Stephens (née Hooke), an illiterate woman whose signature was a pirate's hook.

In any case, Hooke's household possessions were not those of a wealthy man. An inventory made shortly after his death listed an unimpressive collection of beds, chests, chairs and tables, some china and pewter kitchenware, and an assortment of old clothes. There were three pairs of breeches, nine waistcoats, two velvet caps, a black hood, five pairs of stockings, two nightgowns and (for Hooke, though a bachelor, had not always lived a celibate life) three pairs of stays, a yellow necklace and a petticoat. He had accumulated a large hoard of pieces of cloth – Colchester baize, yellow silk, purple and yellow paragon, blue linen, coarse Silesian linen – and in the cellar were stored several hundredweight of scrap brass, iron and lead. The life of leisure was represented by a draughts board, some pots for drinking coffee and chocolate, a collection of prints and paintings and two broken harpsichords. There were many reminders of the fact that old

Dr Hooke had been one of the greatest scientists and mechanics of his age. In his parlour, where the Royal Society had often met, there were three lodestones (magnetic oxide of iron), two large globes, three telescopes and three pairs of brass scales, and in the cellars below it there were a pestle and mortar, two bench vices and a pair of smith's bellows. Hooke's magnificent library, collected during a lifetime spent in auction rooms and bookshops, contained many bound pamphlets and over 3,000 books, half in Latin and hundreds in French and Italian. These were valued by Edward Millington (an auctioneer who had sold many of the books to him in the first place) at just over £200, but when auctioned later they raised a much greater sum.[1]

So far, everything the searchers found was as they might have expected in the home of an unmarried and untidy old scholar whose final years had been darkened by disease, blindness and disappointment, and who had lived out his old age in apparent penury. But when they opened his great iron chest they found a fortune beyond all expectation for a man whose salary from the Royal Society and Gresham College, when it was paid, had been £80 a year. The chest held nearly £8,000 in old and new money, and another £300 in gold and silver. This was almost £1m at today's values, a fortune a successful London merchant would have been happy to leave.[2] Rumour increased Hooke's hoard to £12,000, and his enduring image as a twisted old miser was established.

*

REPUTATION IS CAPRICIOUS. Robert Hooke, who was regarded in his day as one of the greatest masters of the new science (or natural philosophy, as they would have called it), is now a forgotten figure. If his name is remembered at all outside specialist circles it is as the originator of a simple law of springs, Hooke's Law, which states that the extension of a spring will be proportional to the force applied to it (or 'strain is proportional to stress'). The memory of his brilliance as an inventor of scientific and mechanical instruments has faded as his devices have been superseded by more modern inventions. Yet in the index of an authoritative encyclopaedia of the instruments of science Hooke's name is mentioned more times than that of any other inventor: eighteen entries each for Lord Kelvin and Carl Zeiss, eleven for Jesse Ramsden, eight or nine for Isaac Newton, James Clerk Maxwell and Ptolemy, and twenty for Robert Hooke.[3] His second career as an architect and surveyor, in which he was Christopher Wren's colleague and assistant in rebuilding the City churches after the Great Fire of 1666, and one of the two men most responsible for the rapid reconstruction of the ruined City as a place of work and residence for nearly 100,000 people, is almost unknown. How many London guidebooks tell their readers that several of the twenty-three surviving 'Wren' churches in the City were probably designed by Hooke? Versatility, Hooke's outstanding characteristic, is not a quality much valued by posterity. The process of historical simplification usually determines that the greatest names are remembered while the confusing cast of secondary characters is ignored. Two great figures, Isaac Newton and Christopher Wren, dominate the histories of science and architecture in the late seventeenth century. We assume that every building constructed in London after the Great Fire was designed by Wren, and every step on the way to the understanding of universal gravitation and the nature of light was taken by Newton. Newton himself, in a letter to Robert Hooke, made the famous observation that every great scientist builds on the work of those who preceded him: 'If I have seen farther it is by standing on the shoulders of giants'. But it is the man on the top of the pile, not those under his feet, who wins the glory.

Pope's famous couplet, written in 1730, was an early contribution to the reduction of the story of late seventeenth-century science:

> Nature and Nature's laws lay hid in night:
> God said, *Let Newton Be!* And all was light.

In this popular view of the origins of modern science, Robert Hooke is simply a shadowy creature of the pre-Newtonian night. For twenty years after Hooke's death Newton ruled the English scientific world, and used that position to establish his towering reputation, belittling the work of Hooke and others in the process. A similar task was performed for Sir Christopher Wren by his son, whose memoir of his father's life, *Parentalia*, virtually ignored Hooke's contribution to the rebuilding of London and claimed the entire enormous enterprise for Wren.

Hooke, who was far from indifferent to questions of reputation, anticipated the danger that his life's work would be forgotten, and in his final years decided to make a record of his own achievements. But blindness and pain overtook him before his task was finished. The comprehensive list of all his discoveries and inventions he had planned to make was never compiled, and the autobiography he started to write stopped when he reached his teens. The small pocket diary containing this fragment has been lost, but his friend and biographer Richard Waller saw it, and quoted or summarized its contents. This is how it began:

> *Saturday April* the *10th* 1697. I began this Day to write the History of my own Life, wherein I will comprize as many remarkable Passages, as I can now remember or collect out of such Memorials as I have kept in Writing, or are in the Registers of the ROYAL SOCIETY; together with all my Inventions, Experiments, Discoveries, Discourses, &c. which I have made, the time when, the manner how, and means by which, with the success and effect of them, together with the state of my Health, my Employments and Studies, my good or bad Fortune, my Friends and Enemies, &c. all which shall be the truth of Matter of Fact, so far as I can be inform'd by my Memorials or my own Memory, which Rule I resolve not to transgress.[4]

Robert Hooke was born on 18 July 1635 in the small town of Freshwater, near the western tip of the Isle of Wight. His father, John Hooke, was the curate of All Saints', the parish church of Freshwater, and his mother (John Hooke's second wife) was Cecelie, née Gyles. John Hooke's house disappeared long ago, but we know from the inventory taken when he died that it was a modestly furnished cottage, with a parlour, hall, study, kitchen and buttery downstairs and three attic bedrooms above them. Robert Hooke would have shared one of these low-ceilinged rooms with his brother John until the older boy went off to Newport to become a grocer's apprentice in 1644.[5]

Robert Hooke was a sickly infant, and his parents nursed him at home, feeding him a diet of milk and fruit, instead of sending him out to a nurse as they had with their three older children, Anne, Katherine and John. For seven years the child was expected to die, but his grasp on life gradually grew stronger, and he became agile and energetic, though not robust. The young Hooke was a fast learner, and his father briefly hoped that he might be trained for a career in the Church. But study gave the boy headaches, and John Hooke, whose own health was failing, abandoned his son's education. Left to follow his own inclinations, Hooke imitated the skills of local craftsmen, and developed an aptitude for building mechanical devices. His lifelong interest in clocks and navigation began in these years. As he wrote in his notebook: 'Seeing an old Brass Clock taken to pieces, he attempted to imitate it, and made a wooden one that would go: Much about the same time he made a small ship about a Yard long, fitly shaping it, adding its Rigging of Ropes, Pullies, Masts, &c. with a contrivance to make it fire off some small Guns, as it was sailing cross a Haven of a pretty breadth.'[6] The sluggish River Yar, where Hooke would have sailed his little warship, is just a few yards from All Saints' Church. Another adult preoccupation no doubt had its origins in his Freshwater childhood. The western end of the Isle of Wight is rich in fascinating geological formations, which hide (we now know) the bones of many dinosaurs. If Hooke wandered along the fossil-rich beaches of Totland or Freshwater Bay, or dug petrified shells from the huge chalk cliffs that run from Compton Bay to the Needles, or collected coloured sand from the famous stratified cliffs of Alum Bay, the seeds of his later speculations about the extinction of species, the violent formation of mountains and dramatic changes in sea level, might have been sown. If he walked a mile or two northwards from All Saints' Church to the sloping clay cliffs of Bouldnor, and watched them cracking in summer, oozing in winter, and sliding steadily into the Solent, it might have occurred to him then that the world was not as it was when God first created it nearly six thousand years earlier.

Hooke's mechanical genius developed in his early years, and so did his confidence in his ability to master other men's skills without formal teaching. When John Hoskins, a famous painter of miniatures, visited the Isle of Wight in the 1640s, Hooke watched him at work and decided that the techniques of the artist were well within his reach. His friend John Aubrey tells the story: 'Mr Hooke observed what he did, and, thought he, why cannot I doe so too? So he getts him chalke, and ruddle [red ochre], and coale, and grinds them, and putts them on a trencher, gott a pencill, and to worke he went, and made a picture: then he copied (as they hung up in the parlour) the pictures there.'[7] As usual, Hooke's perhaps irritating faith in his own talents was not entirely unjustified. His skill with the pen and pencil was invaluable later on, when he needed to communicate his microscopical and mechanical discoveries to a wider audience, and in his career as an architect and surveyor.

The later 1640s were a time of unusual excitement in the sleepy history of the Isle of Wight. In November 1647 King Charles I, after being defeated first by Parliament and then by the Scots, fled from Hampton Court to Carisbrooke Castle, in the centre of the island, passing within a mile or two of the Hooke family's house. From the Castle, where he was held in benign imprisonment, Charles encouraged the Scots to launch an invasion of England. After this was defeated by Cromwell in August 1648 the King entered into futile negotiations with a Parliamentary delegation which had been sent to the island. It is possible that young Hooke travelled to Newport to watch the delegates coming and going in September and October, or that he walked to Freshwater harbour early on 1 December 1648 to see the King taken under armed guard to the mainland. This, the King's last journey, ended on an executioner's block in Whitehall two months later.

By the time of King Charles' execution, Robert Hooke was probably in London. In October 1648 his father had died, dragged down by those familiar seventeenth-century assailants, 'a Cough, a Palsy, Jaundice and Dropsy'. Hooke was by then thirteen, old enough, by the standards of the time, to make his own way in the world. Though his mother Cecelie was still alive (and remained so until 1665) he took his inheritance, £40 from his father and £10 from his grandmother, and set off. Later in his life he told John Aubrey that he had brought £100 with him to London. Perhaps his memory misled him, but it is also possible that he had raised more money by selling his father's books, which had also been left to him.[8]

Hooke's journey took him from an isolated island community to one of the biggest cities in Christendom, a monster of nearly 400,000 people. London had spread beyond its Roman and medieval walls long since, but in the first decades of the seventeenth century its growth, especially in the west, had been unusually rapid. In the 1620s and 1630s the first West End squares had been laid out at Covent

Garden and Lincoln's Inn Fields, and the town was beginning to take on its sprawling eighteenth-century shape. By the 1640s the population living in the suburbs of Westminster, Stepney, Southwark, Clerkenwell and Shoreditch easily outnumbered those living in the old City, within or adjacent to the ancient Roman walls.[9] The City itself was immensely busy and crowded, and its overhanging wooden houses and narrow medieval streets were waiting for the disaster that engulfed them in 1666.

At the end of the 1640s daily life in London was getting back to normal after the dangers and privations of the Civil War. Oliver Cromwell and his army had established their control of the capital in December 1648, and in the 1650s Lord Protector Cromwell showed his more conservative colours and suppressed the movements of political and religious radicalism that had threatened London's equilibrium in the 1640s. In some respects Puritan London was a duller place than it had been under the Stuarts. The Bankside and Clerkenwell theatres had been closed down and the royal court and aristocracy had abandoned Westminster in 1642. But a new leisure institution which was to be of enormous importance in Robert Hooke's life arrived in London just after he did. In 1652 Pasqua Rosee, a Turkish Greek, opened London's first coffee house, Pasqua Rosee's Head, in St Michael's Alley off Cornhill, in the middle of the City. By the end of the decade there were more than eighty coffee houses in the City, and for many Londoners they rivalled taverns as centres of social life.

Robert Hooke's plans took him first to Westminster, London's growing western suburb, rather than into the City. A well-placed friend or patron must surely have given him advice and introductions, enabling him to meet influential Londoners who would not have taken much interest in a friendless thirteen-year-old migrant. Perhaps inspired by the ease with which he had picked up John Hoskins' skills, Hooke intended to take up an apprenticeship with the great Dutch portraitist, Peter Lely. Lely had been a court painter, but in the 1640s circumstances had driven him into private practice. It was Lely who later painted the famous picture of Cromwell, the Lord Protector, 'warts and all'. Hooke's stay with Lely did not last very long, however, perhaps only a few months. The smell of paint, like the study of religion, made his head ache, and he thought that the skill could be acquired (he explained to Aubrey later) without formal training. 'Mr Hooke quickly perceived what was to be donne, so, thought he, why cannot I doe this by myselfe and keepe my hundred pounds?'[10]

Instead, Hooke took his money to the great Dr Richard Busby, the famous (and famously severe) headmaster of Westminster School, the oldest and most prestigious school in London. Dr Busby specialized in the production of clergymen, and left the mark of his rod on the backsides of at least sixteen future bishops. Hooke was at Westminster, living in Dr Busby's house, probably from 1649 to 1653, though he was not often seen around the school, and his fellow pupils included the future poet and playwright John Dryden, the philosopher John Locke, Robert South, who later became one of the strongest theological critics of the new science, and several others whose lives touched his later in the century. Under Dr Busby's care Hooke learned to play the organ, quickly mastered the first six books of Euclid (the basic mathematical text), became proficient in Latin and Greek, dabbled in Hebrew and other Eastern languages and (he told Aubrey) 'invented thirty several ways of flying'.

Perhaps his interest in the possibility of human flight was stimulated by his growing bodily deformity. At Westminster he was 'very mechanicall', and it was at this time, he said, that 'he first grew awry, by frequent practicing, turning with a Turn-Lath, and the like incurvating Exercises'.[11] We need not accept his word on the origins of his twisted back. There is a condition known since its identification in 1921 as Scheuermann's kyphosis, in which a severe and inflexible stooping of the thoracic (central) spine is brought about by the development of wedge-shaped spaces between the vertebrae. Sometimes small growths of disc material, known now as Schmorl's nodes, push into the end-plates of the vertebrae, opening an angle at each joint of five degrees or more. The condition is more common in boys than girls, and generally becomes obvious in early adolescence. Hooke first noticed his problem when he was sixteen. The causes of Scheuermann's kyphosis are still unknown, but genetic or nutritional factors are probably involved, and trauma or heavy work might contribute to it. The condition is not necessarily painful, but it is likely to create stresses in other parts of the spine. Present treatments include physiotherapy, bracing and, as a last resort, surgery, but of course none of these was available to Robert Hooke. In severe cases the forward stooping continues to develop after adolescence, and it is clear that Hooke's problem grew worse as he got older. Friends who described him in his middle and later years agreed that he was a sorry sight, with his thin and crooked body, his over-large head, sharp facial features and protruding eyes.

Nevertheless, Robert Hooke had come a long way since 1635. He had survived his very dangerous early years and developed into an intelligent and agile, if not very well-formed, young man. He had demonstrated great mechanical, mathematical and artistic talents, he had been educated at London's finest school and he had discovered how to fly. Now his apprenticeship as a natural philosopher was about to begin.

2. The Revolution in Science

ROBERT HOOKE GREW UP at a time when European understanding of the universe was going through a fundamental revolution. In 1543 Nicholas Copernicus' *De Revolutionibus Orbium Cœlestium* (*On the Revolutions of the Celestial Spheres*) made the extraordinary proposal that the Earth revolved around the Sun along with all the other planets, rather than remaining stationary at the centre of the universe, as Aristotle and Ptolemy had taught and men had previously believed. It was nearly fifty years before the Copernican system, which was based on argument rather than evidence, received confirmation from astronomical observations, and even longer before it was accepted by the majority of scientifically minded men. Between 1574 and 1601 the Danish astronomer Tycho Brahe made very accurate naked eye observations of the movement of the Moon, planets and comets. On the strength of these, he dismissed the long-accepted idea that the stars and planets were held in place by a succession of crystalline celestial spheres, and replaced it with a set of planets in orbit around the Sun, held in place by forces which were not yet understood. Brahe did not accept the Copernican system (he believed that the planets revolved around the Sun, but the Sun revolved around the Earth), but after his death his assistant and successor as Imperial Mathematician in Prague, Johannes Kepler (1571–1630), used the records of his observations to confirm that the universe was heliocentric, to propose that planetary orbits were elliptical rather than circular, and to propound three laws which governed the motion of the planets in their orbits around the Sun. Kepler's work took thirty or forty years to win general acceptance, but eventually, in the 1670s and 1680s, his laws provided an essential foundation for Isaac Newton's work on universal gravitation.

It was the work of Galileo that made the Copernican universe familiar and credible to educated Europeans. In 1609 he made himself a telescope and used it to observe the heavens. His observations showed that the Earth was not the unique celestial body that traditional astronomers, the followers of Ptolemy and Aristotle, had assumed. The Moon's surface was rough and mountainous like the Earth's, and Jupiter had satellites or moons of its own. Moreover the universe was far more extensive, and stars far more numerous, than had ever been suspected. Galileo first published these results in 1610, but his two most influential works, the *Dialogue on the Two Chief Systems of the World, Ptolemaic and Copernican* and *Discourses on Two New Sciences*, were published in 1633 and 1638, around the time of Hooke's birth. These works showed that the weight of evidence and argument was now on the side of the Copernican universe, and introduced a new science of mechanics, the study of motion and force, which began to make sense not only of the motion of the planets, but of the forces governing motion on the Earth itself. The same rules applied to both.

The errors of Aristotle and the medieval scholars who had for centuries accepted his authority were exposed. The absolute distinction between levity and gravity was dismissed: all bodies, however weightless they seemed, were heavy, and were drawn towards the centre of the Earth. The idea that a body could not be affected by more than one force at a time was replaced with the principle that different forces could act upon a body simultaneously. This made possible the realistic study of the paths taken by cannon balls and other projectiles, and enabled Galileo to explain why falling bodies were not left behind by the motion of the Earth. Galileo showed that the speed at which bodies fell towards the Earth was not directly proportional to the weight or density of the body, but that all falling bodies fell with the same accelerating velocity, or would do so in a vacuum, where they would not be affected by atmospheric resistance. The effect of the medium through which a body passed was not (as Aristotelians believed) to assist or prolong motion, but to retard it. Without such resistance, bodies in motion would never stop moving. Just as Aristotle's laws of physics made some sense in a stationary world, so Galileo's made sense in a world that was in constant motion. The universe, including the Earth and everything on it, was subject to one set of mechanical laws. These laws were later refined and restated as Newton's laws of motion, but the basic principles were available to anyone able to read the two books of dialogues written by Galileo in the 1630s.

Galileo's last two books made available to all educated men ideas which only a few had grasped earlier in the century. At almost the same time the English physician William Harvey demonstrated the circular motion of the blood around the body and the function of the heart in promoting this process. The errors of ancient and medieval science were revealed, and a great scholarly edifice was broken down. All the old certainties, derived from the teachings of Aristotle, Ptolemy and Galen, were abandoned, and the whole of Creation had to be investigated and understood afresh. Copernicus, Kepler and Galileo had led the way, but there was an enormous amount of work still to be done. Galileo had reached his conclusions by reasoning, mathematics and observation, rather than by experiment or precise measurement. He had probably not even dropped lead and iron balls from the top of the Leaning Tower of Pisa to test their acceleration, as later biographers claimed. How the laws of motion operated in the real world of friction, resistance and imperfect shapes had yet to be established, and the accurate observations made possible by the telescope, the microscope, the thermometer and the pendulum (all invented since 1590) had yet to be made.

In London in the 1640s and 1650s, as in Paris, Florence, Oxford and other centres of learning, there were groups of scientists and thinkers who were trying to grapple with these fundamental changes in their picture of the universe, and working on ways to apply the techniques and principles of the new science to a wide range of practical and theoretical problems, from the satellites of Jupiter and the fall of heavy bodies to the grinding of telescope lenses and the circulation of the blood. Some followed the methods advocated by Francis Bacon, Lord Verulam, in the 1620s, and focused on the collection of observations which would eventually enable them to discover fundamental laws of nature, while others inclined towards the more deductive approach of René Descartes (1596–1650), who favoured the development of general propositions which could be afterwards tested (and in Descartes's case generally disproved) by practical observation.

Seventeenth-century scientists took their positions somewhere on the spectrum between these two approaches, the Baconian (inductive, empirical, arguing from fact to theory) and the Cartesian (deductive, hypothetical, arguing from theory to fact). Hooke paid homage to Bacon, 'the incomparable Verulam', and he was always a proponent of experiment, observation and accurate measurement. But temperamentally he was drawn towards the intuitive approach of Descartes, in which bold hypotheses were proposed long before substantial evidence for them had been collected. The difference between Descartes and Hooke was that Hooke, the empiricist, tested his hypotheses by experiment and observation, and that Hooke's intuitions were very often right, while Descartes's were nearly always wrong. Hooke tested and rejected many of Descartes's theories, but like most later seventeenth-century scientists he accepted his mechanical philosophy, in which the universe was composed of matter in motion, and all natural phenomena could be explained as collisions between particles of matter. The universe and all it contained, including living things, were machines, obeying the laws of mechanics.

Two London scientific groups formed in the mid-1640s are of particular interest, because the Royal Society traces its ancestry from them. One, often called the '1645 group', included the mathematicians and astronomers John Wallis, John Wilkins, Samuel Foster and Theodore Haak, and various London physicians, including George Ent and Jonathan Goddard. This group met once a week in Gresham College or in Goddard's rooms in Wood Street to discuss the 'New Philosophy' in all its ramifications. Many years later John Wallis described their wide-ranging interests:

> the circulation of the blood, the valves in the veins, . . . the lymphatick vessels, the Copernical hypothesis, the nature of comets and new stars, the satellites of Jupiter, the oval shape (as it then appeared) of Saturn, the spots in the sun, and its turning on its own axis, the inequalities and selenography of the Moon, the several phases of Venus and Mercury, the improvement of telescopes, the grinding of glasses for that purpose, the weight of air, the possibility or impossibility of vacuities, and nature's abhorrance thereof, the Torricellian experiment in quicksilver, the descent of heavy bodies, and the degree of acceleration therein, and divers other things of the like nature.[1]

The second group was led by the influential German Protestant Samuel Hartlib, and included Henry Oldenburg and Robert Boyle, who called it the 'Invisible College'. Not much is known about its activities, but its interests seem to have been more social and theoretical than those of the '1645 group'.

At the end of the 1640s, when Robert Hooke was settling down to his studies in Westminster, some of the leading figures in London science decamped to Oxford, where Cromwell's ejection of obstinately Royalist professors and college heads in 1648 had made room for more politically compliant academics. John Wilkins became Warden of Wadham College in 1648, John Wallis and Seth Ward became Savilian Professors of Geometry and Astronomy a year later, and Jonathan Goddard (Cromwell's physician) was appointed Warden of Merton College in 1651. At Wadham, John Wilkins established a circle of promising scientists, drawn there by his reputation as a leading exponent of the New Philosophy, and by his genial tolerance of men of all political and religious beliefs. By 1652 there were around thirty men meeting in Wilkins' rooms at Wadham, including Seth Ward, John Wallis, Jonathan Goddard, Thomas Willis, Laurence Rooke and Matthew and Christopher Wren. Membership of Wilkins' 'philosophical club' was formal but fluid. One of its early leaders, William Petty, left Oxford for Ireland in 1651, Robert Boyle arrived in 1655 and Henry Oldenburg, tutor to Boyle's nephew, came to Oxford for about a year in 1656. At least twenty members of this group became Fellows of the Royal Society in 1663, and the Wadham philosophical club, itself descended in part from the London clubs of the 1640s, can be regarded as the prototype and originator of the Royal Society of London.

The Oxford scientists were ready to turn their minds and hands to any branch of theoretical or practical science. Wilkins was a keen botanist, an inventor of improved ploughs, glass beehives and a wonderful fountain for his college garden. His many projects included sundials, waywisers (to measure the distance travelled by coaches), fortifications and ways of attacking them, improved sailing ships, navigational techniques, building methods and the development of a universal language. With his brilliant young protégé Christopher Wren he worked on thermometers, magnets, perpetual motion machines, new balances and wagons, and model submarines and flying machines. Wren developed a double writing instrument and studied tiny objects through a microscope, and joined Ward and Rooke in stocking a mobile observatory in Wadham College tower with the latest astronomical instruments.[2] The study of anatomy at Oxford had suffered when the Royalist William Harvey resigned the wardenship of Merton in 1646, but members of the philosophical club did their best to revive it. William Petty made himself and his 'patient' famous in December 1650 when he brought an apparently dead woman, Ann Greene, back to life on the dissecting table after she had been hanged, and then successfully protected her against those who wanted to hang her again.[3] Thomas Willis worked on the anatomy of the brain, and Wren began experimenting on the effects of injecting various liquids into the bloodstreams of dogs.

This was the rich and eccentric world of English science that Robert Hooke entered in the 1650s, and in which he spent the rest of his life. The rejection of Aristotelian science meant that natural philosophy had returned to a second childhood, and every question had to be asked and answered afresh. The world and the universe were full of wonders and puzzles which would, these men were certain, yield their secrets to rational thinkers equipped with the latest experimental techniques and observational instruments. Almost nothing was beyond their understanding, but hardly anything was as yet understood.

3. Hooke at Oxford

(1653–1662)

ROBERT HOOKE REMAINED IN Dr Busby's care until 1653, when he followed Robert South and John Locke to Christ Church, an Oxford college which had a special relationship with Westminster School. By this time, it appears, Hooke's money had run out, and he had to pay his way at Oxford by acting as a chorister and as a servitor to a richer student named Goodman. A glossary of 1656 defined the term: 'Wee use the word Servitor in our Universities, where the poor or meaner sort of Schollars . . . execute the office of a Servitor or attendant to those of greater wealth and quality.'[1] Perhaps he found this a humiliating position, but he did not say so, and it was not an unusual way for poorer scholars to make ends meet. The room he occupied, he said later, was the one in which Robert Burton, author of *The Anatomy of Melancholy*, had hanged himself in 1640.

Robert Hooke had been studying at Christ Church for about two years before he made contact with John Wilkins' scientific group. His mechanical talents were quickly recognized, and he was taken on as a paid assistant. His first job in the group was with the physician Thomas Willis, whose particular interest was in the brain and nervous system. Through Willis and Wilkins he was introduced to the wealthy amateur scientist Robert Boyle, who employed him (probably starting in 1656 or 1657) to work in his new laboratory next to University College in Oxford High Street. A plaque marks the site today. Hooke lived under Boyle's roof, and received an income from him at least until 1662, and possibly until 1664. In July 1663, when he was working unpaid for the Royal Society, Hooke still saw himself as 'belonging' to Boyle.[2]

Boyle was a great advocate of the problem-solving power of the experimental method. We take it for granted that scientists conduct experiments, but it is an assumption that does not hold true before the middle of the seventeenth century. Some earlier scientists had performed experiments, or written as if they had performed them. The English physician William Gilbert (1540–1603) had made many impressive experiments on the nature of magnetism and electricity, and Francis Bacon (1562–1626) had repeatedly emphasized their importance in the accumulation of scientific knowledge. But on one of the rare occasions that Bacon actually followed his own teachings and carried out a practical trial – he stuffed a chicken with snow to see if it could be preserved – he caught a chill and died. Many earlier experiments were imagined rather than performed, with results that were predicted or assumed, not discovered. Galileo had contradicted the ancient belief in the impossibility of vacuums and described the behaviour of objects in airless conditions, but he had never created a vacuum or seen what happened in one. Boyle was one of a new generation of scientists who believed that real experiments, rather than 'thought experiments', offered a promising way forward in the accumulation and demonstration of scientific knowledge. As a son of the immensely wealthy Earl of Cork (who is buried in staggering splendour in St Patrick's Cathedral, Dublin), he had the money to put his belief into practice.

The problem that fascinated Boyle in the 1650s was suggested by a demonstration carried out in 1644 by Galileo's pupil Evangelista Torricelli. Torricelli had taken a long tube with one end sealed, filled it with mercury, placed his finger over the open end of the tube, inverted it into a dish of mercury and removed his finger, ensuring that no air bubbles entered the tube. Instead of flowing away entirely, a column of mercury almost thirty inches high remained in the tube, leaving a gap at the top. This apparatus, the first barometer, inspired scientists all over Europe. What was in the space in the tube? Was the column of mercury supported (as Torricelli supposed) by the weight of the 'sea of air' surrounding the Earth, and what determined its height? In Paris, Blaise Pascal tried the device at the top of Notre Dame, and others took it to the tops of mountains to see whether the reduced atmospheric pressure there would shorten the column. Boyle's intention was to try the Torricellian experiment inside a vacuum, where atmospheric pressure would be removed. His confidence that a vacuum could be created was based on the work of the

Saxon scientist Otto von Guericke, who had made a pump and used it to draw the air from two large copper hemispheres fitted together tightly as a sphere – the famous 'Magdeburg Hemispheres'. When he showed the experiment to the Emperor Ferdinand III in Regensberg in 1654, thirty horses could not pull the hemispheres apart.

Boyle had to make an air pump which was easier to operate than Guericke's cumbersome two-man machine, and a glass receiver or vacuum chamber in which experimental devices could be placed and observed. This presented great practical difficulties, especially in preventing leaks from around the piston and from the sealed lid of the vessel, which the famous pumping specialist Ralph Greatorex (later Samuel Pepys' favourite scientific instrument maker) could not overcome. Boyle recognized Hooke's unusual mechanical ingenuity, and decided to employ him on the project. According to Hooke's own account, he 'contrived and perfected the air pump for Mr Boyle' in 1658 or 1659, since Greatorex's version was 'too gross to perform any great matter'. Hooke tended to claim credit for almost everything and his version of events might be disbelieved, but Boyle, unusually for his time, openly acknowledged that his assistant had made him a pump that worked.[3]

The machine sounds unimpressive now, but in the 1650s and 1660s it was regarded as one of the most important, sophisticated and costly scientific instruments of the age. Even ten years after its construction there were probably only six or seven pumps in Europe, of which three had been made in Boyle's laboratory. A fifteen-inch glass globe or 'receiver' with a brass lid four inches in diameter stood above a brass cylinder which contained a tight-fitting wood and leather piston or sucker. The brass lid contained a second smaller stopper, which was shaped and lubricated to allow it to be turned without letting air into the globe. By turning it, the experimenter could pull a string to activate an experiment, perhaps by pulling the trigger of a pistol to test the ignition of gunpowder in a vacuum. The pipe connecting the globe and the brass pump could be opened or closed with a stopcock, and the piston was drawn down the cylinder (thus pulling air out of the globe) with an iron rack and pinion device, a notched bar moved by the turning of a cog wheel, much as we would find now on a car jack or an adjustable spanner. The mixture used for making the lids and the piston airtight depended heavily on the lubricant and sealant qualities of 'sallad oil'.[4]

Using this apparatus, which could create high as well as low pressure in the globe, Boyle and his assistants carried out forty-three experiments on the qualities of air. Among other things, this helped Boyle to establish the law of gases that bears his name: at constant temperature the volume of a gas is inversely proportional to the pressure exerted on it. This was a specific law of elasticity, a subject that fascinated Hooke all his life. Boyle and Hooke also conducted experiments on burning coals, charcoal, candles and gunpowder in a vacuum, which led them towards the conclusion that fire was not one of the four 'elements' (along with earth, air and water), as Aristotelians had believed for nearly two thousand years, but a chemical process. Boyle, the ideal 'gentleman philosopher', set new standards of clarity and objectivity in the conduct of scientific experiments, and gave his eager and talented employee a valuable grounding in the new experimental philosophy. For lessons in the orderly conduct of scientific disputes, and the careful adjudication of matters of credit and originality in research and discovery, a young scientist could not have had a better teacher than Robert Boyle. Not everyone would agree that Hooke learned these lessons well.

In the later 1650s, while Hooke was working with Boyle, he was also drawn into the exciting world of magnets, pendulums, waywisers, telescopes and microscopes inhabited by Seth Ward, John Wilkins, Christopher Wren and the other members of the Wadham group. Their range of interests suited Hooke's omnivorous delight in every branch of science. Seth Ward, the Savilean Professor of Astronomy, introduced him to the study of planets and pendulums, and his schoolboy fascination with flight was encouraged by Wilkins, who had published in 1640 a book on the possibility of flying to the Moon. Hooke showed his plans for a flying machine to Wilkins, and made a contraption of springs and wings which he managed to keep in the air for a while. Among Hooke's papers his friend Richard Waller found some sketches of artificial bat-like wings for a man's arms and legs, and 'a Contrivance to raise him up by means of Horizontal Vanes plac'd a little aslope to the Wind, which being blown round, turn'd an endless Screw in the Center, which helped to move the Wings, to be manag'd by the Person by this means rais'd aloft'.[5] Several failures taught Hooke that man's natural muscles were not strong enough for flight, and he set to work on designing artificial ones. The results of several trials of these were disappointing.

Another interest Hooke shared with Ward and Wilkins was the possibility of devising an international language or 'Universal Character', rather like the language of music or mathematics, for the easy communication of scientific ideas. Wilkins had written on this in the 1640s, and later (in 1668) completed an *Essay Towards a Real Character and Philosophical Language.* Hooke proposed at that time that all scientific work should be reported in Wilkins' universal language, and he remained fascinated by the idea of a universal symbolic language and other ways of communicating across language barriers or over long distances to the end of his life. This sounds eccentric, but it was a project that had previously attracted Francis Bacon and René Descartes, and which would interest the great mathematician Leibniz in the 1660s and 1670s. As in the case of mathematical algebra, symbols could be an aid to reasoning as well as communication. Hooke was especially interested, in the 1650s and later, in the possibility of devising a symbolic language which would ease the task of recording and memorizing the mass of facts discovered by past and future scientists, and of drawing general principles (or 'raising axioms') from them. His own weak memory, he always said, was in particular need of assistance. He often spoke of a symbolic language, a 'philosophical algebra', which would allow scientists to solve problems with the speed and certainty of mathematicians, and in 1676 (when Leibniz was working on a similar language) Hooke claimed that he had mastered a system of 'mechanical algebra' in the late 1650s, while he was at Oxford. But he never produced a clear written account of his philosophical algebra, and its meaning has to be pieced together from scattered and ambiguous references. Like the secret of flight, it was an accomplishment he chose to keep to himself.

Unlike some of his rivals Hooke was incapable, by inclination and circumstances, of concentrating single-mindedly on one problem or project at a time. While he was operating and perfecting Boyle's air pump, he was also working with Christopher Wren on the barometer and microscope, making astronomical instruments under the direction of Seth Ward, and trying to devise a clock that would keep accurate time at sea. In 1661 his first publication, a little tract on capillary action, appeared. In this, he considered a phenomenon he had observed in his experiments with Boyle. When a narrow glass tube, open at both ends, stood vertically with one end in a pot of water, water in the tube rose above the surrounding water level. The surface of the water in the tube was concave, in what we would call a meniscus, and the narrower the tube, the higher the water rose. To explain these observations, which had not been seriously discussed before, Hooke began to develop his own theory of matter. The tendency of some fluids to mix with other fluids or to cling to a solid arose from a quality which he called 'congruity', and their resistance to such mixing he called 'incongruity'. Water was more congruous with glass than air was, so it found it easier to enter the tube than the air did, and atmospheric pressure on the water was reduced. Hooke did not claim to know the causes of congruity, but he suggested that it was a mechanical quality of matter, perhaps connected with the motion of its constituent particles. Four years later, in his amazing book *Micrographia*, he provided a much bolder and more comprehensive theory of matter, in which congruity still played a central part.

In 1660, the year of the Restoration of Charles II, Robert Hooke was twenty-five. He was already on close but not equal terms with England's leading scientists, and he had been Boyle's indispensable assistant in one of the most significant experimental programmes of the time, and (in Boyle's words) 'made him understand Descartes'.[6] His unique skill with the air pump was well known, and his work on capillary action would demonstrate that his talents were intellectual as well as mechanical. In addition, his childhood interest in clocks and his urge to win himself financial independence had drawn him into the race to solve the lucrative problem of longitude, and thus into a collision course with one of the greatest scientists of the age, Christiaan Huygens. His character, like that of the fluids he described in his little tract, was a combination of congruity and incongruity. He mixed freely with his fellow scientists, learning all he could from them, and impressing them with his intellectual and mechanical ingenuity. He was an active and useful member of the Oxford scientific community. Yet his personality also had a strong element of incongruity, a degree of secretiveness, a reluctance to share insights and inventions with others, especially where profit or glory might be involved. This was a difficult mixture, and we will see how Robert Hooke handled it when he entered a much more public world of science in the 1660s.

4. The Royal Society

(1660–1664)

BETWEEN BROAD STREET and Bishopsgate in the old City of London, on a plot now filled by the NatWest Tower (now known as Tower 42), Thomas Gresham's great Elizabethan mansion once stood. The mansion consisted of brick and timber buildings around a large courtyard, about a hundred yards square. Its near neighbour was Crosby Place, the fifteenth-century mansion whose great hall survives today in Chelsea, where it was reconstructed piece by piece in the 1920s. Gresham, an extremely wealthy merchant who was also financial agent and adviser to the last three Tudor monarchs, was the founder of the Royal Exchange, London's new business centre. When he died in 1579 he left a substantial income from the revenues of the Royal Exchange to the City of London and the Mercers' Company so that they could convert his mansion into a college. They were to employ seven lecturers (in divinity, rhetoric, music, geometry, astronomy, civil law and physic) to live there in scholarly celibacy and give weekly lectures in English and Latin for the education and practical benefit of the citizens of London. The first lecturers (or professors, as they were known) were appointed after the death of Gresham's wife in 1596, and, as lecturers do, they at once set about reducing the number and length of the lectures they had to deliver. By the 1630s the College's reputation for teaching was not very high, though it was respected as a centre of research. Many of its professors in recent years had been nonentities or absentees, drawing their annual salary of £50 without performing the simple duties that the founder had imposed on them. But in the 1640s the College had been the focal point for the new science in London, and now, with the drift of scientists back from Oxford to London, it assumed this role again.

When the Wadham group began to break up at the end of the

1650s several of its most active members, including William Petty, Laurence Rooke and Christopher Wren, went back to London and took professorships at Gresham College. Petty went to Ireland to conduct a government land survey, but Wren, Rooke and other ex-members of the Wadham club started meeting informally with other London scientists, including several who had been members of the London groups of the 1640s, in the College or in one of London's many taverns or new coffee houses. By November 1658 these meetings were a regular fixture, a focal point for men who took an interest in natural philosophy. The collapse of the Cromwellian regime and the restoration of Charles II in May 1660 brought new recruits – Wilkins, ejected after a year from the mastership of Trinity College, Cambridge; Sir Robert Moray, who had been in exile with the King; and William Petty, dismissed from his post in Ireland. After Wren's regular Gresham lecture on 28 November 1660, a group of twelve, chaired by Wilkins, decided to turn their informal debating club into a permanent association for the promotion of 'Physico-Mathematicall Experimentall Learning'. What these men wanted was a formal scientific institution like the Accademia del Cimento (Academy of Experiments) in Florence or the Montmor Academy in Paris, but one whose survival would not be dependent, as the Continental ones were, on the whim of an individual royal or noble patron. They had another model in Francis Bacon's *New Atlantis* (published in 1627), which had put forward the idea of a publicly funded research institute, a 'Solomon's House', which would conduct collaborative investigation into 'the secret motions of things'. Despite – or perhaps because of – the bitter divisions of the past twenty years they were all determined that their common interest in scientific enquiry and experiment should unite them, and that political and religious differences should be put aside. They hoped that their Baconian approach to science, which emphasized its practical benefits to citizens and the state, and which welcomed the contributions of interested amateurs as well as scholarly specialists, would attract the support of rich and aristocratic Londoners, and especially the patronage of Charles II, who had shown a flattering (but, as it turned out, superficial) interest in science. Charles II's support, they believed, would bring the society the permanence of a chartered corporation, and the possibility of a generous royal endowment to pay for a college staffed with curators, operators and other salaried officials.

These principles and aspirations, along with the King's repeated failure to come up with the money the society hoped for, shaped the Royal Society for the next forty years. The royal charters which it won in 1662 and 1663 ratified its permanent constitution: a President, a Treasurer, two elected Secretaries and an elected decision-making Council of twenty-one fellows. They also gave the Royal Society of London its impressive name, a coat of arms with a Horatian motto, '*Nullius in Verba*' (Bound to no man's words) and the privileges of a chartered corporation, but not the expected royal endowment. As a result, although the Royal Society had the legal right to buy property and to employ staff it did not have the money to do so. It could never afford the large scientific staff it had envisaged, it was unwilling or unable to pay its few employees a regular salary, and it was forced to hold its weekly meetings in rooms in Gresham College. These took place every Wednesday afternoon from three to six o'clock, until January 1667, when they were moved to Thursdays. Thursday was the usual meeting day until the end of 1680, except for a year of Wednesdays from May 1672 to June 1673. From January 1681 onwards, the Society met on Wednesdays again.

Nobody suffered from the Royal Society's lack of funds more than its first and best secretary, Henry Oldenburg. Oldenburg came to London from Bremen in 1640, when he was in his mid-twenties, and in the 1650s he made his living as a diplomat and a tutor. His command of European languages and his long-standing membership of the London and Oxford scientific groups made him the obvious choice as one of the Society's secretaries in 1662. His duties included taking minutes at Society and Council meetings, keeping the Journal and Register Book, conducting a vast and very important correspondence with foreign scientists, and publishing the *Philosophical Transactions*, a prestigious periodical that largely reflected the Society's work. Oldenburg's devotion and hard work were vital to the success of the Royal Society in its first fifteen years, but until the Society started paying him a salary in 1669 he and his family struggled to live on a little property from his two marriages (his first wife died in 1666), some money from Boyle and the *Philosophical Transactions*, and two £40 gifts from the Royal Society. Oldenburg was the central figure in a vast European exchange of information, ideas and opinions, spreading news of experiments and publications, and stimulating the competition and debate through which he thought natural philosophy would advance. Those who liked to work in secret sometimes took his zeal as a correspondent for treachery or trouble-making. He was locked up in the Tower of London by Charles II for two months in 1667, and in the 1670s Hooke came increasingly to regard him as a spy and a mischief-maker.[1]

The Society was dependent for its income on the joining fees and annual subscriptions of its fellows, and it was therefore keen to attract men of wealth and high status, even when their scientific accomplishments were small. So the core of under twenty active and competent scientists in the Society had to carry a much larger number of well-off dabblers without substantial scientific or mathematical backgrounds. Social distinction was a surer route to a Royal Society fellowship than scientific worthiness in the 1660s. Aristocrats, privy councillors and members of the royal family had virtually automatic membership if they wanted it, and at least in the 1660s it was as fashionable in courtly circles to patronize (in both senses of the word) scientists as it was to show favour to painters or musicians. In that decade well over half the 261 men who joined the Royal Society as fellows were aristocrats, courtiers, politicians or independent gentlemen, while less than a quarter (57 of 261) were scholars or doctors.[2] The Society was a willing victim to this fashion. Its literature boasted of its aristocratic membership, and its £2 joining charge and £2 12s. annual fee excluded many scholarly schoolteachers and craftsmen who would have brought intellectual distinction to its meetings. The Royal Society's motives were mixed. In part, it simply needed the subscriptions and gifts its richer fellows might (but often did not) contribute to its funds. It saw aristocratic and especially royal approval as the best way of winning public and political support for natural philosophy and thus guaranteeing its future. But it is also true that most of the leaders of the Royal Society were themselves well-off gentlemen who enjoyed the company of others of equal or higher social status, whether or not they were active scientists. And even if some aristocrats, gentlemen and courtiers treated the Royal Society as a rather inconveniently located club and came to its meetings to be amused or intrigued rather than to advance the frontiers of science, they played their part as trustworthy observers, as witnesses to the success of an experiment or the demonstration of a new phenomenon. For in those days the evidence of a gentleman was as gold compared with the base currency of a craftsman's testimony.

The preponderance of amateurs among its fellows had important effects on the work of the Royal Society. To hold the interest of gentlemen scientists the Society had to focus its weekly meetings on the more accessible branches of science, placing an emphasis on experiments with predictable or spectacular results, the demonstration of interesting gadgets or natural curiosities, and the collection and compilation of information and observations, rather than on the more demanding mathematical sciences. Thomas Sprat's *History of the Royal Society*, which was written in the 1660s under the guidance of the leaders of the Royal Society, is a manifesto for the Society's empirical, experimental, utilitarian and inclusive approach to science. Sprat emphasized the valuable part that could be played by 'plain, diligent and laborious observers: such, who, though they bring not much knowledg, yet bring their hands, and their eyes uncorrupted . . . and can honestly assist in the *examining* and *Registring* what the others represent to their view'.[3]

In theory, if not in practice, the Royal Society was a community of equals, in which every fellow was capable of taking part in the collective pursuit of natural philosophy. One of the Royal Society's earliest projects was to compile a comprehensive History of Trades, in which the methods and secret skills of all known crafts would be collected and revealed by fellows working in committees. This was work to which even the dullest dilettante could contribute without making a fool of himself, and it was also part of the process of assembling all existing knowledge that Bacon had recommended his followers to undertake. Fellows were also expected to take their turn in entertaining the weekly meetings with demonstrations or experiments. In its second meeting, in December 1660, the Society made regulations for the conduct of experiments and coined a new word for those who would supervise them – 'curators'. A committee was created to collect ideas for suitable experiments, and a programme based on the talents of the more scientifically competent fellows – the 'virtuosi' – was initiated. The system of voluntary curators was not a great success. Only a handful of fellows had serious work to demonstrate, there was a lack of continuity between one meeting and the next, and no coherent research programme developed.

Towards the end of 1662 the leaders of the Royal Society decided that its meetings would benefit from the skills of a full-time professional experimental scientist. Robert Hooke was the obvious man to fill this entirely new position. He was a protégé of two leading lights in the Royal Society, Boyle and Wilkins, and he was already known as a skilful designer of experiments and instruments, and as a competent scientist with a treatise on the behaviour of liquids in glass tubes to his name. He had done promising work with the microscope and telescope, and he was Europe's greatest master of the most important experimental apparatus of the moment, the air pump. Above all, he would be able to design experiments which would impress King Charles II and stimulate his generosity to the Society. Since he had no independent income Hooke was ready to take employment, and even to do so without the immediate prospect of payment. Robert Boyle was prepared to release his servant for the Society's benefit, and on 12 November 1662 the Royal Society accepted Sir Robert Moray's proposal that Robert Hooke should be employed as a curator of experiments, 'to furnish them every day when they met, with three or four considerable Experiments', and to undertake other tasks suggested by the fellows. Hooke's formal appointment was delayed until 1665, but his work for the Society began straight away, and he was soon entertaining the weekly meetings with experiments with compressed and rarefied air, falling bodies, refracted light, exploding glass balls, microscopical observations and machines for testing the strength of gunpowder. After six months, in June 1663, he was elected a Fellow of the Society, exempt from the usual fees, and in December 1663 he was asked to live four days a week in Gresham College to prepare experiments to show the King, though for a while he still kept lodgings of his own.

The enormous benefits of employing Hooke were gained at a very low cost to the Royal Society. He agreed to work unpaid and without formal appointment until the Society could afford to pay him, and the leaders of the Society did their best to ensure that others would supply the bulk of the £80 annual salary that he expected from them. He cooperated in this search for an alternative income. In May 1664

Isaac Barrow resigned as Gresham Professor of Geometry, and Hooke was a competitor for the position, probably with the Royal Society's support. When the Gresham committee met on 20 May Hooke was beaten by Arthur Dacres, a physician at St Bartholomew's Hospital, who had also applied for the position two years earlier. A few days later Hooke was drinking with John Graunt, statistician, draper and Fellow of the Royal Society, in a public house near Graunt's home in Birchin Lane, by the Royal Exchange. The two men fell into conversation with the wealthy financier Sir John Cutler, a friend and regular coffee-house companion of Graunt's. Hooke told Cutler of his failure in the Gresham College election, and Cutler offered to match the £50 annual Gresham salary to enable Hooke to continue his work for the Royal Society. In 1683, when he was involved in a legal dispute with Cutler, Hooke claimed that this £50 had been offered as an unsolicited and unconditional gift, and perhaps this was how he recollected the conversation after twenty years. But the Council of the Royal Society, which discussed Cutler's offer on 22 June, knew that they would have to accept Cutler's terms before the gift was agreed. In effect, Cutler wanted Hooke to be given a new Gresham professorship in the History of Trades, with the obligation to lecture once a week in vacation time (sixteen weeks a year) on trades, crafts and the mechanical aspects of science. Despite his later claims that the gift was given without specific conditions, Hooke clearly understood that he was obliged to deliver sixteen lectures a year 'to the advancement of art and nature', on subjects determined by the Royal Society. He acknowledged this duty, and Cutler's generosity to him, in his preface to *Micrographia*, and signed a formal document in January 1667 which described the timing and general subject matter of the lectures he had undertaken to give.[4]

On 27 July 1664 the Council of the Royal Society rushed to confirm its intention to elect Hooke as its Curator of Experiments, but kept his appointment and his £80 salary secret until Cutler's £50 was safely in the bag. When Hooke's appointment was discussed in November, and when he was formally elected Curator in perpetuity on 11 January 1665, it was made clear that he would only receive £30 a year from the Royal Society, with Cutler's £50 making up the difference. In effect the Royal Society had diverted Cutler's money into its own funds and left Hooke with a tiresome lecturing duty and dependent for much of his income on an increasingly unreliable patron. Cutler stopped paying the stipend in 1670 and after years of reminders and complaints Hooke was twice (in 1682–4 and 1694–6) forced to go to court to get the money that was due to him. The Royal Society, which had negotiated the terms of the Cutler deal and saved £50 a year on the strength of it for twenty-four years, left Hooke to fight most of this battle for himself.[5]

At the beginning of September 1664 Hooke moved into rooms in Gresham College, which was to be his home and laboratory for the rest of his life. He did not yet hold one of the seven Gresham professorships, but this position was regularized in March 1665, when he petitioned for a reconsideration of the previous year's election of Dacres to the geometry post. The Royal Society had been told in June 1664 that Hooke had been defeated by the casting vote of the Lord Mayor, Sir Anthony Bateman, who had no right to sit on the committee. In fact, as Professor Michael Cooper has discovered, the new committee found out that Dacres' election had been even more irregular than that. The original committee of nine men had voted 5:4 for Hooke, but Sir Anthony Bateman, who had no proper position on the committee, had declared Dacres the winner. Two other members of the Bateman family, Thomas and Sir William, had voted for Dacres, and it looks as though this powerful City clique had tried to deliver the lectureship to their favoured candidate. The new Lord Mayor, Sir John Lawrence, who chaired the committee which uncovered all this, became 'a good and sure freind' to Hooke, and they often met. Thanks to the Royal Society's secret informant (perhaps one of Hooke's five supporters on the committee) and to his own persistence, Hooke was installed as Gresham Professor of Geometry in March 1665.[6] His rooms were in the south-east corner of the College, near its Bishopsgate entrance, and consisted (according to the 1703 inventory) of a library and parlour and two smaller rooms on the first floor, three cellar or workshop rooms below them, and a garret in the loft. By this time his appointment as the Royal Society's Curator of Experiments had also been confirmed, and thus two of the main institutional components of Hooke's future career were secured. He was no longer dependent on the patronage of Robert Boyle, though he always treated his ex-master with the utmost deference and reported regularly to him by letter on his scientific work. Robert Hooke, Gresham Professor of Geometry, Cutler Lecturer, Fellow and Curator of Experiments of the Royal Society, was established as England's first professional research scientist in a world occupied almost entirely by 'virtuosi', physicians, aristocratic dilettantes and gentleman scholars of independent means.

*

HOOKE HAD MADE a start, but it was his great hope that his unusual talents would bring him more than a modest salary and a few rooms in Gresham College. Ever since the later 1650s he had been working on an important project which might, if carefully handled, make his name and his fortune. The measurement of longitude at sea was one of the great scientific and practical problems of the age. Until sailors could do this, navigation would remain a hit or miss affair, and long-distance sailing for commercial or military purposes, upon which England's prosperity and power increasingly depended, would be unsafe and uncertain. Hooke had considered four ways of solving this problem, and rejected two of them as impractical. He dismissed the old idea that navigators could establish their position by using a lodestone or compass to measure the difference between magnetic and true north at different places, and the possibility that the speed and direction of a ship's path could be accurately tracked by loglines and water fans. The third method, establishing a ship's position by accurate observations of the Moon or of Jupiter's satellites, depended on high-quality telescopes and astronomical skills which sailors did not possess, though it was possible that new instruments of celestial measurement might one day make this easier. Hooke's lifelong work on accurate sextants and quadrants was driven by his belief in this possibility. This left the idea that the time at noon or sunset could be compared with the time at a location of known longitude, such as London. For this, a clock or watch that could keep good time during a long sea voyage would be necessary. This was the solution he decided to pursue in Oxford and London in the late 1650s and early 1660s.

Since the fourteenth century there had been mechanical clocks in which the driving power was supplied by the pull of weights, and in which the motion of the mechanism was slowed and regulated by a device known as a verge escapement. The verge was a rotating vertical shaft on which there were two flag-like metal projections called pallets. As a toothed ratchet wheel (the crown wheel) was turned by the pull of the weights, it was alternately caught and released by the pallets. The swinging of the verge, and thus the movement of the pallets, was caused by the pressure of the crown wheel's teeth on the pallets. There was nothing in the mechanism to ensure that the movement of the verge was regular, and as a result these clocks were very inaccurate, and measured hours rather than minutes. Galileo's discovery that a swinging pendulum kept almost regular time opened the way for great improvements in the accuracy of clocks. In 1656 the great Dutch scientist Christiaan Huygens applied the pendulum to the movement of a traditional verge escapement mechanism, and produced a clock which was far more accurate than any then existing. There were problems with the accuracy of the pendulum clock, especially when the arc of the pendulum's swing was wide, as it had to be when the traditional verge escapement was used. These could be solved by altering the path of the pendulum's swing, using longer pendulums which could swing in narrower arcs, or devising a more reliable and accurate escapement than the verge. Hooke, Huygens and others worked on these improvements in the 1660s. A more obvious difficulty was that the pendulum could never be used in a pocket watch, and was too sensitive to rough motion to keep accurate time at sea. The eventual solution to both these problems was to replace the pendulum with a straight or spiral spring, whose oscillations are as regular (or isochronous) as a pendulum's, but which is less sensitive to external motion.

There was nothing new about using springs in the place of weights as the driving force in a clock or watch. Coiled springs had been used as a power source for timepieces since the fifteenth century, but they had not been used before to give a clock a regular beat. Hooke's innovation, he claimed later, was to use the regular vibrations of a spring to keep time in place of Huygens's pendulum, enabling him (so he thought) to make a portable clock or watch which would remain accurate at sea. Some time after the restoration of King Charles II to the throne, probably in 1663 or 1664, Boyle arranged a meeting between Hooke and two of the most influential figures in English science, Sir Robert Moray and Lord Brouncker. Hooke showed them a watch regulated by a spring, though he seems to have hidden the details of its workings from them, 'concealing the way I had for finding the Longitude'. A contract was drawn up by which Hooke agreed to reveal the secrets of his watch to the three men, on condition that he received with their assistance a fourteen-year royal patent and a generous allocation of the profits that would be made from charging the masters of ships for using the longitude timekeeper. If the deal had been completed (and if the watch had worked) Hooke would have received several thousand pounds, and all his financial troubles would have been over. But he balked at one clause that the others insisted on including in the agreement. He could not accept that anyone who improved on his watch in any way should be given the patent and the profits in his place. It was easy enough to improve on someone else's invention, he said. So he broke off the negotiations, and resolved to keep the details of his longitude watch secret until he could think of a way of ensuring that he would profit from it. Hooke told this story in 1676 to strengthen his case against Huygens over the invention of the spring-regulated watch. But its outlines were true, and were repeated by Moray in a letter to Huygens in 1665. Various drafts of the contract, and a draft of a law compelling ships' masters to pay for using the timekeeper, were found among Hooke's papers after his death.[7]

Since the dispute with Huygens over the invention of a working spring-regulated watch was one of the critical events in Hooke's career, calling into question his honesty and his scientific abilities, it is important to establish, as far as it can be done, what Hooke had achieved in watchmaking by the mid-1660s. One undated bundle of papers now in Trinity College, Cambridge, eleven much-amended and confusing sheets, offers the best chance of doing this. The papers have been analysed and transcribed by Michael Wright, of London's Science Museum, who has also tried to reconstruct the sort of watch Hooke apparently had in mind. The earliest papers, six of the eleven sheets, were probably written in 1664 for presentation to Charles II shortly after Hooke's negotiations with Boyle, Moray and Brouncker. They describe his reasoning in rejecting existing ways of finding longitude, including Huygens's pendulum clock, and explain what he had done to develop an alternative method. First, he invented a new escapement mechanism to replace the irregular and unreliable verge escapement: 'I made notches or teeth in the crown wheel and fixt a certain Stay or catch with a small spring to it by which meanes the motion of the whole fabrick was stopt untill such time as the catch or Stay being lift up the whole fabrick would move untill the Next tooth was met by the Stay and Stopt.' In other words the catch which repeatedly stopped and released the notched crown wheel would be acted upon by the constant force of a straight or leaf spring, creating the first 'constant-force escapement' clock.[8]

This clock would be fine, he said, for making accurate scientific observations (timing bullets, falling bodies and musical vibrations, measuring astronomical distances, and so on), but it became obvious to him (though not to Huygens) that when the clock or watch was subjected to the irregular movement one expected in a ship or a pocket a pendulum would not give his constant-force escapement the regular lift it needed to allow the crown wheel to move forward in perfect time. To replace the pendulum, he invented a spring-driven balance wheel. He 'poised a wheel soe exactly upon its two poles as sharp as needles, that the centre of its motion & that of its gravity were both the same . . . And because natural gravity could take noe hold of it as to its motion about its centre; I contrivd an artificiall one . . . by applying two springs soe that their motion drawing one against the other should not receive any irregularity from the shog [shake] of the instrument.' Finding that a ship's rolling motion would disturb the regular oscillation of this balance wheel, he introduced a third original idea, a set of paired wheels on the same axis, one above the other like two stacked plates. These two identical balance wheels were linked by two smaller gear-wheels, so that they would move in exact opposition to each other. In his arrangement, any circular motion that tended to speed one balance wheel would tend to slow down the other, and the effect of a ship's rocking motion would be nullified.[9]

Hooke was sure that he had designed a practical longitude clock, and ended his draft letter to Charles II on an optimistic note: 'This is that Invention which has been soe long and by soe many sought, though to the best of my knowledg not found or known to any yet but my self and two freinds to whom I very lately Reveald it, which I have had perfect as it now is, by me these three yeares . . . I doe humbly therefore throw both my self & it at Your M feet'.[10]

In 1675, during his dispute with Huygens over the invention of the spring watch, Hooke claimed that he had invented it in 1658. This draft letter to the King, if it was written in 1664, places the claimed invention in 1661, and suggests that Hooke later pushed the date back about three years in order to strengthen his case against Huygens. His assertion in his lecture *Of Spring*, published in November 1678, that he had invented the watch eighteen years earlier – in 1660 – is more plausible. Whether the watch worked, or what kind of springs (straight or spiral) were applied to the balance wheels, we do not know. Hooke was still working on this and other clocks in the later 1660s, which makes it plain that the clock described in his 1664 papers did not work as well as he hoped it would. It is certain, in any case, that the watch would not have been accurate on long sea voyages, because Hooke had not overcome (and had barely considered) the effects of temperature changes on the size and flexibility of the spring. This problem was solved by John Harrison a hundred years later. Nevertheless, Hooke was convinced that the solution to the great longitude puzzle was within his grasp, and several times in the 1660s he repeated his claim to have invented a spring-driven marine timekeeper, revealing further tantalizing details of its operation, but not, as far as we know, the nature of the spring. His great mistake was to forget that other men, at least as clever and resourceful as he was, were working on the same problem. There were such people even in his own city. William Clement, a London watchmaker, devised a better and much more influential mechanism than Hooke's constant-force escapement in 1671. This was the anchor escapement, which consisted of a curved arm shaped like an inverted anchor, with triangular points or pallets at each end, which rocked back and forth on its central pivot to hold and release the teeth of the crown wheel, allowing it to move one notch at a time and producing the regular ticking motion of a modern pre-electronic clock. The often-repeated claim that Hooke invented the anchor escapement originated in William Derham's *The Artificial Clock-Maker* (1696), not with Hooke, and is now regarded as untrue.

5. 'Full of Employment'

(1662–1664)

HOOKE WAS FIRST NAMED as one of the Royal Society's Curators of Experiments in November 1662, when he was twenty-seven. By the time he was thirty he had taken control of most of the Society's experimental programme, and the system of voluntary and unpaid curators had withered away. The Society's immensely wide-ranging scientific and technical programme in the 1660s was driven by Hooke's energetic and amazingly fertile mind, and its meetings in the later 1660s relied almost entirely on his performances. Christopher Wren, a friend but not a flatterer, wrote to him on 20 April 1665: 'I know you are full of employment for the Society wch you allmost wholy preserve together by your own constant paines'.[1] Among Hooke's many advantages over other curators were his ability to turn his hand to almost any branch of natural philosophy from anatomy to ballistics, his full-time commitment to the Royal Society's service, his genius as a designer of experimental apparatus and scientific demonstrations and his unrivalled skill with the air pump.

Hooke's enormous energy, ambition and appetite for knowledge in the 1660s is suggested by this undated list of all the work he intended to do:

Theory of Motion	Improving shipping
of Light	— watches
of Gravity	— Opticks
of Magneticks	— Engines for trade
of Gunpowder	— Engines for carriage
of the Heavens	

Inquiry into the figures of Bodys
— qualitys of Bodys[2]

Some men might have been daunted or overwhelmed by this huge agenda, and others would have chosen a single specialism, but Hooke set to work on it like a beggar at a banquet. And the extraordinary fact is that within ten years he had made a significant contribution in nearly all these fields. He was not, as some of his critics claimed, an idle braggart, but an extremely industrious one.

In 1661 Boyle was working on an improved version of the air pump, and in May that year he presented the original Boyle–Hooke pump to the Royal Society. Christiaan Huygens witnessed experiments with this air pump at Gresham College in the spring of 1661, and returned to the Hague to make a better one of his own. Lacking an expert in the maintenance and use of its pump, the Royal Society watched the experimental initiative slip into Huygens's hands. It was Huygens, for instance, who discovered a mysterious quality known as 'anomalous suspension', which attracted much attention in the 1660s. For a while this phenomenon, the failure of the column of water in a barometer inside a vacuum chamber to fall as expected, was believed to challenge Boyle's Law and to indicate some special quality in the air that emerged from water during the experiment. It was actually an effect produced by surface tension, the attractive forces between liquid and glass and leaks in the air pump.[3]

It was only when Hooke became a curator that the Gresham College pump started to provide the succession of experiments that had been hoped for. He brought the Royal Society's pump up to the new standards set by Boyle and Huygens, and produced a series of experiments which demonstrated the qualities of air at high, low and normal pressure. Some of these had great entertainment value, but they also raised issues of scientific or practical importance. His first demonstration before the Royal Society, on 19 November 1662, is a good example. The explosion of glass bubbles as they cooled down, or the pleasing inrush of air or water when the bubbles were opened up, amused the audience, but also suggested questions about the expansion and contraction of gases and other bodies by heat and cold, and of the strength of arched or curved structures under external pressure. One of Hooke's outstanding skills was his ability to draw many theoretical, useful and amusing lessons or suggestions from a simple experiment. In December 1662 experiments on the tendency of a hollow glass ball to rise to the top of cold water, but to remain suspended halfway up a vessel of water that was gently heated, led him to a series of speculations. The effect of heat on water was to dissipate or loosen it, he said, reducing its ability to sustain the glass ball, and promoting a flow of warmer and lighter water from the bottom to the top of the vessel. The colder water at the bottom of the vessel supported the ball, but the warmer water could not. On the strength of this, he suggested a type of water boiler for brewers or dyers, in which a small heated copper device at the bottom of a wooden tub would create a circulation of hot and cold currents which would eventually heat the whole vessel. The same principles, he said, could be used in making a perpetual motion device, in which warmed liquid might circulate around endless pipes or through another liquid (the distant ancestor of the lava lamp?), or a novelty weather glass in which the heat or cold of the seasons would bring appropriate creatures floating into view. He also wondered whether the planets were floating, each according to their density, in a vast fluid (the aether) which was thinner (more rarefied) the nearer it was to the heat of the sun. Finally, he speculated that ships that set off from the polar seas had to be loaded at less than their capacity, to take account of the reduced buoyancy they would experience in warmer waters. Within weeks, a questionnaire had been prepared for a fleet bound for Greenland, to satisfy the Society's curiosity on this and other matters.[4]

The drive to turn scientific knowledge to practical effect led Hooke and the Royal Society repeatedly towards matters of navigation and seamanship. England was already an important naval power, with the beginnings of a Caribbean and North American empire, and its ships traded in all the world's oceans. Hooke's adult years saw a great expansion in England's mercantile interests, transforming her from a north European commercial power into a global one. The tonnage of shipping involved in England's foreign trade had risen by around 60 per cent between 1630 and 1660, and probably rose by 80 per cent or more between 1660 and 1688. Most English ships sailed on well-established coastal and north European routes, but the number venturing into less familiar Atlantic and Pacific waters was growing fast. In the early years of Hooke's scientific career England was engaged in a struggle with the Dutch for maritime supremacy, which led to three naval wars. In the second of these the two rivals fought each other in America, West Africa, the Caribbean and the East Indies as well as in European waters. In the peace of 1667 England gained the port of New Amsterdam (renamed New York), but in general the second Dutch War was a humiliating demonstration of the superiority of Dutch ships and naval skills. If the scientists of the Royal Society wanted to prove their value to the nation and the King they had to apply their science to correcting this maritime imbalance.

Scientists knew very little about the sea. The causes of its tides and currents were still a mystery, and nobody knew how deep the oceans were, whether their deeper waters were salty or fresh, what the sea pressure was at depth or what creatures lived on the seabed. In the early 1660s Hooke produced several devices which might (if they worked) enable curious sea captains to discover some of the qualities of the mysterious deep-sea world. In September 1663 he described a depth sounder which could measure depths greater than those that could be measured with a weighted line. A hollow ball would be linked to a heavy weight by a spring clip, and thrown overboard. When the weight hit the seabed the clip opened, and the hollow ball floated back to the surface, where it would be recovered. The time taken by the ball's journey to the seabed and back would indicate the depth of the sea when it was compared with the time taken at a known depth. The depth sounder seemed to work in ideal conditions, and it was included in the set of demonstrations the Royal Society ordered Hooke to prepare for Charles II in 1663. Hooke had difficulty in persuading practical sailors, who were not really interested in the depth of waters deeper than they could already measure with their lead lines, to try his device, and when it was tested in the Straits of Gibraltar in 1678 it took so long to find the balls once they had surfaced in choppy seas that it was impossible to measure the time their journey had taken.[5]

Hooke's apparatus for taking a sample of deep-sea water was a square bucket with upper and lower hinged lids which opened upwards as it was lowered through the water on a weighted bracket. When the device was pulled up the movement would force the two lids to close, trapping the water which had entered the bucket at its lowest point. This deep-sea water could then be examined for saltiness and marine life. This simple machine was much used by oceanographers in the eighteenth and nineteenth centuries. In related experiments Hooke was also working on the question of fluid resistance and the speed at which objects sank. Among other things, he found that a body with a blunt rear end like a cone, weighted to sink point-first, sank faster than one shaped like a double cone, pointed at front and rear. This confirmed what practical shipwrights already knew: a ship with a flat stern would travel through the water faster than one with a pointed stern.[6]

In February 1664 Hooke was included in a committee on diving. He first proposed a chain of inverted buckets to supply a diver with air, and then made two lead boxes which could be supplied with air from the surface, and from which the diver working on the river or seabed would occasionally breathe through a pipe. A diver was found to try the equipment out, but in May a test in the Thames failed, apparently because the man had not had enough practice. In June a demonstration in front of the Royal Society, in which the diver and his lead boxes were submerged in a large tub, was more successful, and further trials in the Thames were conducted. Hooke also experimented with a trick which divers were said to use, and tried breathing underwater with a sponge dipped in 'very good Sallet-oil' in his mouth. It was a complete failure: 'I was as soon out of breath, as if I had no Sponge, nor could I fetch my breath without taking in water at my mouth'. He had hopes, though, that an oily sponge might clean used air and make a small supply last much longer.[7] He worked on other marine ideas – diving goggles, a lifejacket, a submarine for the Thames, a whale harpoon – but interest in all these seemed to wane in the later 1660s. He continued to refine his undersea sampling and measuring devices, and many years later, in 1691, produced a comprehensive account of his life's work in the field. His ideas aroused little interest at the time, but in the long run he was proved right. Sea depths were measured with very long weighted lines until the 1850s, when Hooke's depth-sounder was reinvented by an American midshipman, J. M. Brooke. So Hooke's invention, a buoyant cylinder carried to the seabed by a weight that uncoupled itself when it hit the ocean floor, was used to measure the Atlantic seabed when the first transatlantic telegraph cables were laid in the 1850s and 1860s.[8]

Though sailors were unimpressed by his deep-sea gadgetry, Hooke had other schemes for improving the art of navigation. The Royal Society's Council granted him up to £10 to spend on improvements to his secret marine timekeeper device, and he had also started work on improving the accuracy of the various astronomical instruments used by navigators to establish their latitude by measuring the angle between the horizon and the Sun or a star, the quadrant, sextant, cross-staff and back-staff. One way to do this was to make much bigger instruments, but another, which had more value for seafarers, was to increase their precision by introducing better sights and finer adjustments. In February 1665 Hooke showed the Royal Society a quadrant with a radius of seventeen inches 'contrived by himself' which would make 'celestial and terrestrial observations with more exactness than by the largest instruments, that have been hitherto publicly known'. He had achieved this unprecedented accuracy, he declared, by 'the contrivance of a small roller, that moved upon the limb of it', making it possible to adjust the moving arm of the quadrant in divisions of ten seconds, or ⅟₃₆₀ of a degree. He had also given the quadrant new sights to match this level of accuracy, but he did not say whether this was done by using cross hairs or telescopic lenses.[9]

Unlike the majority of Royal Society members Hooke was not a collector of disconnected observations or random curiosities. If he tackled a subject, whether it was the atmosphere, the sea, gravity, the microscopic world or the weather, he liked to do the job properly. In September 1663, at the suggestion of Dr Wilkins and the Royal Society, he began collecting daily records of the weather, in the hope that he could eventually understand and predict weather patterns. Almost anyone else would have compiled long lists of descriptions or figures, but within a few weeks Hooke had produced a comprehensive programme for 'Making a History of the Weather', to teach others (and perhaps himself) how to do the work systematically. The observer should record changes in the strength and direction of the wind, the air temperature, humidity, air pressure, and the appearance of the sky, all in the same location. He should note peculiar effects of the weather on human society ('what Diseases are most rife, as Colds, Fevours, Agues, &c') and crops, animals and insects, the timing and effects of thunderstorms ('souring Beer or Ale, turning Milk, killing Silk-worms'), and unusual tides or comets. Hooke had a particular interest in recording these observations on a standardized form, so that a month's information could be seen in one view. In this way, he said, trends and relationships could be spotted, and the 'Laws of

Weather may be found out'. It was important, too, that observers should speak the same scientific language, record quantifiable information according to agreed scales, and describe the sky in standardized phrases. For the first time, he tried to establish a universal vocabulary for recording cloud conditions – clear, checkered, hazy, thick, overcast, hairy ('a Sky that hath many small, thin and high exhalations'), watered (a mackerel sky), waved, cloudy and lowering. Hooke hoped that these standards would be adopted by observers all over the world, so that worldwide weather patterns could be compared and understood, 'the benefit of which way is easily conceivable'.[10]

As in so many of his enterprises, Hooke was pursuing an idea originated by Christopher Wren, who had demonstrated a self-recording rain gauge in January 1661 and had proposed an automatic weather clock which would record wind strength and direction, temperature, air pressure, rainfall and sunshine without the need for constant observation. In 1664 the Royal Society told Hooke to improve on Wren's weather clock and produce a working machine. It took him fifteen years to complete the task, but in the meantime he set to work on refining the individual components of the clock. In the winters of 1663–4 and 1664–5 he was the first to establish the freezing point of distilled water as the standard zero degrees mark on sealed glass thermometers, an important step towards the standardization of thermometers in the next century. The idea was widely adopted (but not generally credited to Hooke) in the eighteenth century, when the boiling point of water was introduced as the second fixed point. It was Hooke, too, who demonstrated the expansion and contraction of glass in heat and cold, and the effects of this on the accuracy of thermometers.[11]

After various experiments, he discovered that the beards (ears) of the wild oat and the wild geranium bent according to the degree of humidity in the air, and could be used as a hygroscope. Hooke made a portable siphon barometer, in which a U-bend in the lower (open) end of the glass tube, giving it the appearance of an inverted walking stick, removed the need to stand the tube in a reservoir of mercury. He also invented, probably at the end of 1663, the wheel barometer, in which the rise and fall of the mercury was translated into the rotation of a pointer on a dial. He achieved this by curving round the axis of the rotating pointer a thread with a small weight hanging down on either end. One of the weights floated on the surface of the mercury at the curved end of the tube, so when the mercury level changed the weight and the thread moved, turning the axis and the pointer. This made small barometric changes easier to see, and could also be used as the basis of a barograph, an automatic recorder of changes in air pressure, when Hooke finally produced his weather clock. The friction which stopped the barometer recording small changes in pressure could be overcome, as the future owners of domestic versions of Hooke's wheel barometer were to discover, by tapping.[12]

These were the first of many important improvements to the barometer proposed by Hooke in the course of his career. He installed one of his new barometers in his Gresham College rooms as soon as he had moved into them in September 1664, and began collating its movements with his daily weather observations. He wrote excitedly to Boyle five weeks later:

> I have also, since my settling at *Gresham* college ... constantly observed the baroscopical index ... and have found it most certainly to predict rainy and cloudy weather, when it falls very low; and dry and clear weather, when it riseth very high, which if it continues to do, as I have hitherto observed it, I hope it will help us one step towards the raising a theorical pillar, or pyramid, from the top of which, when raised and ascended, we may be able to see the mutations of the weather at some distance before they approach us, and thereby being able to predict, and forewarn, many dangers may be prevented, and the good of mankind very much promoted.[13]

Thus Hooke foresaw the possibility of scientific weather-forecasting. Later, in December 1664, he reported to Boyle that the relationship between the weather and the height of mercury was more complicated than he had thought at first, but his characteristic belief that his work would eventually bring great practical benefits was not shaken.

Hooke's meteorological observations were not confined to his laboratory. On the morning of 7 June 1664 he observed strange air and cloud movements as a powerful thunderstorm approached Westminster, and that afternoon he went over to Piccadilly, one of the newest West End streets, to see a house that had been struck by lightning. Inside it he found an old bird-catcher unconscious and bleeding from the mouth, and questioned several carters and builders who had been thrown down by the force of the blast. They spoke of the smell of brimstone – sulphur – as the lightning struck, but to Hooke the damage to the house looked like the impact of a huge bullet or bolt, rather than the work of fire.[14]

In Hooke's view, to achieve a thorough knowledge of the nature and qualities of the air was one of the most important tasks in natural philosophy. His plan, which he presented to the Royal Society on 25 February 1663, was to undertake a programme of experiments and observations to discover the composition of air, the extent and weight of the atmosphere, the part air played in promoting fire and sustaining the life of plants, animals and fish (did fish breathe?), how the air supported clouds and birds, its resistance to pendulums and falling bodies, and the possibility of human flight. His programme would cover the part played by air in transmitting sound and light, its refractive and reflective qualities, and ways of sending messages through the air, as well as the explanation of winds and air currents, and their role in changing the weather. It was typical of Hooke that his thoughts combined abstract and practical matters: predicting the weather, improving navigation, regulating pendulums, making wings, sending messages, supplying air to divers, and using air pressure to make water pumps, air-guns and weather gauges. In January 1664 he demonstrated a gun powered by compressed air which, to judge from the dent it made in a Gresham College door, would have killed a man at twenty yards.[15]

Hooke pursued his interest in respiration in various ways. In January 1663 he tried to investigate the qualities of air by enclosing a burning lamp and a chick in a sealed container, to discover which would expire first. The chick survived. This was similar to the slightly later experiment of John Mayow, Hooke's Wadham friend, by which the life-sustaining element in air, which Mayow called 'nitro-aerial spirit', was identified. There was general uncertainty in the early 1660s about the function of breathing in sustaining animal life. Harvey had suggested that the motion of the lungs was necessary to agitate or mix the blood, and this simple mechanical view, in which the vital process was the pumping movement of the lungs rather than the entry of air into the bloodstream, was widely held. Various attempts to solve the problem failed, but on 2 November 1664 Hooke proposed the dissection of a living dog 'displaying his whole thorax, to see how long, by blowing into his lungs, life might be preserved, and whether anything could be discovered concerning the mixture of the air with the blood in the lungs'. Perhaps this idea came from Boyle in Oxford, but it was Hooke, watched by Oldenburg and Dr Goddard, who carried out this unpleasant but revealing experiment in Gresham College on 7 November. In a world in which the torture of animals was still regarded as a pleasant sport, their use in scientific experiments was rarely questioned. In 1663 the Society had asked Hooke to graft a cock's spur and feathers onto its head, and to remove a section of a dog's skin and graft it back on again. But the November 1664 experiment, which involved cutting the dog open so that all its major organs were exposed and blowing into its windpipe through a cane attached to a pair of bellows to see how long its heart could be kept beating, was almost too much for Hooke. Keeping the dog alive was easy enough, and it seemed to Hooke that the heart could be kept beating 'as long, almost, as there was any blood left in the vessels of the Dog'. But he wrote afterwards to Boyle:

> though I made some considerable discovery of the necessity of fresh air, and the motion of the lungs for the continuance of animal life, yet I could not make the least discovery in this of what I longed for, which was, to see if I could by any means discover a passage of the air out of the lungs into either of the vessels of the heart; and I shall hardly be induced to make any further trials of this kind, because of the torture of the creature.[16]

A little squeamish by the standards of his day, Hooke was happy to return to the dissection of vipers, and the more humane investigations of animal life that he could pursue through his microscope. But this was a pioneering experiment, and was repeatedly referred to by those (including Hooke) who were trying to discover the secrets of respiration in the later 1660s.

Hooke also made significant progress in discovering the part played by air in combustion. It was well known that the presence of air was usually necessary for combustion, but nobody understood what its contribution to the process was. Hooke seized upon the observation that gunpowder burned without air, and reasoned that one of the ingredients of gunpowder contained the fire-promoting agent that was also present in air. In January and February 1665 he conducted several experiments before the Royal Society to show that this ingredient was 'a nitrous substance inherent and mixed up with the air'. He demonstrated that two of the three ingredients of gunpowder, charcoal and sulphur, would not burn in an airless container, but that both rekindled when the third ingredient, nitre or saltpetre (in our terms potassium nitrate or KNO_3) was added. Samuel Pepys saw this demonstration on his first visit to the Royal Society as a member on 15 February, and though he seemed to miss the particular purpose of the experiment he found the whole experience 'a most acceptable thing'. It was plain to Hooke, as he put it in *Micrographia*, that combustion 'is made by a substance inherent, and mixt with the Air, that is like, if not the very same, with that which is fixt in Salt-peter'. This 'nitrous substance', though Hooke did not name it, was oxygen, which makes up 48 per cent of the weight of saltpetre. It took more than a century for this idea to be rediscovered by Lavoisier and Priestley, who gave oxygen (meaning 'producer of acid') its present name, and provided a chemical account of combustion which would have been quite beyond Hooke. In the century between the work of Hooke and Lavoisier the accepted explanation of combustion was the entirely misleading idea that all combustible matter contained a hypothetical substance known as phlogiston, which was liberated by burning.[17]

Like most natural philosophers since Galileo, Hooke was interested in the behaviour of falling objects, acceleration and air resistance, the weight of objects at different heights, the impact of one body upon another, and the light such things might shed on the great mysteries of gravity and the nature of matter. In this area as in most others, it was Hooke's inclination to reach conclusions by observation and measurement rather than mathematics or abstract theorizing. Early in 1663, he produced a machine which would, he hoped, measure the impact of bodies of different weights falling from various heights. His device was a set of scales with a weight at one end. If the falling steel ball hit the other end of the scales with enough force to lift the weight and move the scales a spring would be released. Hooke quickly saw that his experiment was unsatisfactory. It did not take account of the loss of force involved in the slight flattening of the ball and denting of the plate it landed on, or of the overall weight of the structure that had to be moved, or the fact that the scale might be moved a little, but not enough to release the spring. He used the occasion to consider the difficulties of creating practical tests for the laws of motion, and to declare that one of those laws, Descartes's fourth rule of impact (that a heavier body at rest can never be moved by the impact of a lighter body of whatever speed), had been proved wrong. Hooke's device for measuring the velocity of falling bodies was a little more convincing. In August 1664 he produced a machine in which a pendulum beating half seconds would be halted when a string was pulled taut. An object attached to the string was dropped at the same moment as the pendulum was set swinging, and the number of swings before the string was tightened (plus the position of the pendulum when it was stopped) indicated the time taken for the object to fall a measured distance. Hooke's empirical demonstration of Galileo's assertion that falling bodies accelerate at a uniform rate and that bodies of different weights fall with almost the same velocity was of great interest to Huygens, though he wondered whether Hooke had enough knowledge of pendulums to make his measurements accurate. A few weeks later Hooke used a similar device to measure the speed of a bullet shot from a carbine (the bullet started a pendulum when it left the gun and stopped it when it hit the target), but it was hard to measure such speeds accurately.[18]

Some experiments could best be performed at altitude. In December 1662 Hooke went almost to the top of Westminster Abbey to find out whether objects weighed less at height, as Dr Henry Power of Halifax had recently reported. Using accurate scales and a weight on a very long thread (so that only *one* side of the scales would be affected by the altitude) he was unable to detect a difference between a body weighed at seventy feet and near ground level. Hooke had an excellent eye for the circumstances that might distort an experiment and produce misleading results, and he was convinced that condensation of moisture on the thread or weight, or perhaps air currents and differences in air pressure, had affected Power's results.[19] He proposed further tests to establish whether differences in weight (if they could be detected) resulted from variations in air density at different heights or from changes in the Earth's 'magnetical attraction'.

These experiments, as he knew, had real practical significance, as well as importance in understanding the laws of gravity. If gravity varied with height then so would the timing of a pendulum's swing, and Huygens's and the Society's search for a standard one-second pendulum as a universal measure of length was futile.

Westminster Abbey, which did not get its great West Towers until 1745, was not high enough for Hooke's purposes. In August 1664 he took his first trip to the top of the tower of St Paul's Cathedral, as a junior member of an experimental committee led by Sir Robert Moray and John Wilkins. Although its mighty steeple had been destroyed by lightning in 1561, the old Cathedral's 204-foot tower was still the highest accessible point in London. That summer Hooke and his colleagues timed the swing of a 200-foot pendulum, tested the air pressure at height with Hooke's portable barometer, and found a slight difference in weight between an object at 200 feet and suspended near ground level on a long brass wire, though vibrations in the wire and air movements made the trials inconclusive. The magnetical experiments Hooke had planned were all spoiled by the fact that the decaying stonework of the Norman tower was held together with iron bars, and experiments on the velocity of falling objects had to be abandoned when Hooke's pendulum measuring device broke.[20] Two years later the old Cathedral burned down in London's Great Fire, and future experiments at such heights had to wait until Hooke's own 202-foot pillar in Fish Street (the Monument to the Great Fire) was finished in 1676.

Nevertheless, Hooke had learned enough to produce a paper for the Royal Society in December 1664 which cast doubt on the search for a universal standard pendulum, on the grounds that the rotation of the Earth was slowing down (thus altering the length of a second year by year), that gravity probably varied with time, height and location, and that its force (if the Earth could be compared with a giant lodestone or magnet) might be stronger near the poles and weaker near the equator. 'All bodies and motions in the world seem to be subject to change, of which we may find instances, even in the very sun itself.' These were bold and original ideas, and their presentation, challenging one of the Royal Society's most cherished projects, signified Hooke's emergence as a scientist of confidence and authority.[21]

Boyle had shown that the twenty-nine-inch column of mercury in a barometer at ground level would need a column of air 35,000 feet (nearly seven miles) high at normal pressure to support it. Since it was known that the atmosphere became lighter and more rarefied at height, the actual column of air required to sustain the mercury would have to be much higher than this. In December 1662 Hooke tried to calculate the height of the atmosphere by dividing Boyle's 35,000 feet into a thousand segments of equal weight, each of which would be slightly taller and less dense than the one below it, because the pressure of the air above it would be less. Discounting such variables as air temperature and water vapour, he calculated that the twenty segments just below the highest segment extended for twenty-five miles, and that the highest segment, expanding freely with no weight bearing down upon it, might stretch for many hundreds of miles, or even (he added later) indefinitely.[22] The fact that the atmosphere became steadily thinner with height had important implications for astronomical observations, he argued. The light from stars or planets passing through the atmosphere would be refracted not once but an infinite number of times as it passed from one density to another, creating a continuous curving which he called inflection. Inflection, he argued in *Micrographia* (1665), cast doubt on previous estimates of the distance of the Sun and Moon from the Earth and explained some familiar sights in the heavens. The effects of atmospheric refraction, rather than the qualities of heavenly bodies themselves, accounted for the redness of the Sun or Moon when they were near the horizon, and explained why stars twinkled and sometimes seemed to disappear.[23]

In astronomy and microscopy Hooke had an irrepressible belief in the importance of better lenses and more powerful optical instruments, and could hardly control his irritation with those who stuck to the old methods. In late June 1663 he was in the shop in Long Acre (near Covent Garden, 'over against the Foot and Leg') owned by Richard Reeve, London's finest optical instrument maker, when he met the famous philosopher Thomas Hobbes. Hobbes, whose defence of strong government in his great work, *Leviathan*, had made him a royal favourite, was the most persistent and formidable critic of Boyle's air pump and of experimental science in general. As they stood in the shop the old man (who had been born in 1588, the year of the Spanish Armada) inspected Hooke closely, intrigued perhaps by his stoop, or because he recognized him as Boyle's assistant and protégé. Hooke's appearance certainly made him someone to stare at. John Aubrey knew him in the 1660s, and gave us this description of him: 'He is but of middling stature, something crooked, pale faced, and his face but little belowe, but his head is lardge; his eie full and popping, and not quick; a grey eie. He has a delicate head of haire, browne, and of an excellent moist curle.'[24] Richard Waller, who knew him in his old age, adds some detail:

> As to his person he was but despicable, being very crooked ... This made him but low of Stature, tho' by his Limbs he should have been moderately tall ... He was always very pale and lean, and laterly nothing but Skin and Bone, with a meagre Aspect, his Eyes grey and full, with a sharp ingenious Look whilst younger; his Nose but thin, of a moderate height and length; his Mouth meanly wide, and his upper Lip thin; his Chin sharp, and Forehead large; his Head of a middle size ... He went stooping and very fast (till his weakness a few Years before his Death hindred him) having but a light Body to carry, and a great deal of Spirits and Activity, especially in his Youth.[25]

Hooke, naturally enough, was disconcerted by the attention, but what really upset him (he told Boyle) was Hobbes' obstinate rejection of the latest scientific advances.

> I found him to lard and seal every asseveration with a rounded oath, and to undervalue all other men's opinions and judgements, to defend to the utmost what he asserted though never so absurd, to have a high conceit of his own abilities and performances, though never so absurd and pitiful, &c. He would not be persuaded, but that a common spectacle-glass was as good an eye-glass for a thirty six foot glass as the best in the world, and pretended to see better than all the rest, by holding his spectacle in his hand, which shook as fast one way as his head did the other; which I confess made me bite my tongue.[26]

Nothing irritated Hooke more than to see a scientist rejecting the benefits of modern optics, and (as we shall see) he did not always bite his tongue. Reeve's long career came to a sudden and dramatic end. In October 1664, just over a year after that awkward meeting between Hooke and Hobbes, he lost his temper with his wife and threw a knife that killed her. 'The jury found it manslaughter, and he and all his goods are seized on', Hooke wrote to Boyle. Reeve, who had made telescopes for Charles II, managed to get a royal pardon, but he died soon after the case. His less skilful son Richard took over the business, but leadership in London's optical world passed to the illiterate Christopher Cock. Cock, whose shop was also at Long Acre ('at the two twisted posts'), became a trusted friend and colleague of Hooke's, and thus the Royal Society's favourite optical instrument maker.

Hooke, Wren and others interested in astronomy or microscopy were dependent on the skills of men like Richard Reeve and Christopher Cock. Hooke had first been introduced to astronomy in Seth Ward's observatory at Wadham, using Reeve's telescopes, and Wren's pioneering work with the microscope, which Hooke continued with brilliant success, was done on Reeve's instruments. The telescopes used until the 1670s were refracting telescopes, with an object lens at the front to collect and focus the light and an eyepiece to magnify the image to the desired size. All object lenses suffered from chromatic and spherical aberration: the images they produced were fringed with colour, because light broke into different colours when refracted, and partly out of focus because of the curve of the lens. The best solution to the problem of chromatic aberration was the development of a practical reflecting telescope, in which a concave mirror took the place of the object lens. James Gregory, a Scot, designed one of these in 1663 and came to London in 1664 to have Richard Reeve make him a mirror with a six-foot focus. Hooke was apparently involved in the trials of the reflecting telescope that was made using this mirror, but the results were so poor that he returned to refracting telescopes, and Gregory, 'much discouraged with the disappointment', went off to Padua, where optical craftsmen were, he hoped, more skilful.[27] The problem of spherical aberration was reduced by the use of larger and flatter lenses with a longer focal length, in bigger and more inconvenient telescopes. Wren used a huge thirty-six-foot Reeve telescope for his observations of Saturn at Oxford in 1657, and installed one of the same size (perhaps the same one) on a sturdy pole in the courtyard of Gresham College in 1658, after his appointment as

Gresham Professor of Astronomy. Hooke watched Wren's telescope being hoisted up the pole on a pulley (his sketch of the scene still exists), and used it himself until it was dismantled (perhaps to be moved to Whitehall for the King) in the early 1660s. It is possible that Hooke managed to retain the most important and expensive part of the Wren telescope, its huge object lens. At least, he was in possession of such a lens in September 1664, when the Royal Society offered to pay for a new thirty-six-foot tube for it to be erected as before in Gresham College.

Hooke used this telescope to make further observations of the Moon in the autumn of 1664, and described the results in *Micrographia*. His imagination added what his eye could not see. He concluded from the quality of light reflected from the crater Hipparchus that 'the Vale may have Vegetables *analogus* to our Grass, Shrubs and Trees; and most of these encompassing Hills may be covered with a thin vegetable Coat . . . such as the short Sheep pasture which covers the Hills of *Salisbury* Plains'. Leaving these speculations aside, Hooke returned to his workshop to find out how the Moon came to be covered in saucer-shaped pits. He found that he could reproduce the walled crater shapes by dropping a bullet into a soft mixture of water and pipeclay, or by boiling a pot of alabaster on the fire. Since it was difficult to imagine where so many external impacts might have come from, Hooke decided that craters were the product of earthquakes and volcanic eruptions generated by underground fires and escaping vapours, as in his boiling alabaster. This experimental work on the formation of craters anticipated the investigations of the geologist Sir James Hall by more than a hundred years.[28] Hooke tried to persuade the readers of *Micrographia* to accept that similar changes had taken place on Earth, too: 'I am apt to think, that could we look upon the Earth from the Moon, with a good Telescope, we might easily enough perceive its surface to be very much like that of the Moon'. This idea, that the Earth had been shaped by centuries of underground explosions, fascinated him for the rest of his life. It also struck him that the shape of the Moon, both in its spherical figure and in its surface details, suggested that it was subject to a gravitational force emanating from its centre, just as the Earth was. 'For I could never observe, among all the mountainous or prominent parts of the Moon, . . . that any one part of it was plac'd in such a manner, that if there should be a gravitating, or attracting principle in the body of the Moon, it would make that part to fall, or be mov'd out of its visible posture'. Since the Moon did not spin on its axis as the Earth did, then the explanation for its gravitation (and therefore that of the Earth) had to be sought in some other cause. Hooke must have felt that he was moving, by this process of elimination, towards an answer to the biggest question of all.[29]

Hooke had an unfailing belief in the power of bigger and finer lenses to reveal the secrets of the universe. He enjoyed dealing with the mechanical problems of erecting and manipulating telescopes as high as a house, and later he gave lessons to the great Hevelius, the outstanding astronomer of the age, in building structures to support long telescopes. He encouraged Reeve to make object lenses with a focal length of sixty feet or more, and he used them when he could. In 1666 Boyle bought a sixty-foot telescope for Hooke's use at Gresham College. But even with a good twelve-foot telescope he had been able to see seventy-eight stars in the Pleiades cluster, more than twice as many as Galileo had counted, and far more than the seven that could be seen on a good night with the naked eye. And it was with a twelve-foot telescope that Hooke discovered, on the evening of 9 May 1664, a dark spot in the southern hemisphere of Jupiter, and watched it moving for two hours, proving for the first time that the planet rotated on its axis. When Hooke published his discovery in March 1665 the Italian astronomer Giovanni Cassini used it to establish the rotation period of Jupiter as nine hours fifty-five minutes.

Just as in his boyhood days on the Isle of Wight, Hooke was convinced that no craft, however specialized, was beyond him. Instead of giving instructions to lens grinders whose lenses usually turned out to be faulty, he believed he could master their craft for himself. In 1663 he decided to grind microscope lenses of his own, hoping to make a single-lens microscope of clarity and power, and perhaps to reduce chromatic aberration. His first effort, a tiny lens he had to examine through a magnifying glass, was disappointing: 'the tool was not very true, nor my hand well-habituated to such an employment,' he wrote to Boyle in July 1663. 'And therefore I despair not of better success in my next attempt'. Another one, which he had hoped to send to Boyle in October 1664, was so small that it got brushed away and lost. At the same time, news of Huygens's latest observations of Jupiter (he saw the shadows thrown by Jupiter's moons) spurred Hooke to look for a way of grinding better telescope object lenses than Reeve could make, using a lens-grinding machine of his own design. 'What my success therein shall be, I shall be sure to acquaint you with', he told Boyle. The machine, described in *Micrographia*, was a type of lathe in which two rods topped with curved grinding heads could be adjusted to hold a piece of glass between them at any chosen angle, and rotated by a foot-driven pulley-wheel. Adjustment of the distance and angle between the two rods and replacement of the two grinding heads meant that lenses for small microscopes and for telescopes with tubes of 1,000 feet and more (Hooke claimed) could be made on the same machine. 'It seems the most easie, because with one and the same Tool may be with care ground an Object Glass, of any length or breadth requisite, and that with very little or no trouble in fitting the Engine, and without much skill in the Grinder.' Not everyone was convinced by his extravagant claims. The Parisian astronomer Adrien Auzout wrote questioning the capacity of the grinder to make lenses of the vast size mentioned in *Micrographia*, and Hooke was involved in one of the many disputes which enlivened his professional career.

One of Hooke's most promising ideas was to make compound lenses filled with water, oil of turpentine, salad oil, alcohol or some other clear liquid, to increase their clarity and power. He gathered a great deal of information about the refraction of different liquids during 1664, when he designed and made a 'refracting engine' to measure the angle of refraction of various liquids and transparent solids (ice, glass, diamonds, resins). This engine, a cunning arrangement of long wooden rulers fitted with sights and hinged together on a long upright pole (in an X shape) to follow and measure the path of a beam of light as it passed into a liquid, was illustrated and described in *Micrographia*. As the rulers were bent to follow the beam, their ends moved along two wooden scales which showed the angles of incidence and refraction (before and after the ray entered the liquid) and enabled Hooke to test and confirm Huygens's mathematical rules of refraction.

The variety of Hooke's work in these frenetic early years at the Royal Society was extraordinary. Along with his work on astronomical and nautical instruments, the weight and composition of air, gravity, hydrostatics (the study of pressure and equilibrium in liquids at rest), the study of weather, the behaviour of light, the longitude clock and the series of microscopical observations that were to be published as *Micrographia*, Hooke was also carrying out work on a huge number of projects suggested to him by his masters at the Royal Society, or by his own restless curiosity. Most of the Royal Society's ideas were inspired by the need to prove the usefulness of science to the public at large, and especially to people of wealth and power. For the Royal Society, he had to design a series of experiments on the strength of gunpowder and the breaking point of different types of wood; he was asked to take plaster casts of stones found in the Earl of Balcarres' heart, to investigate the cleanliness and cost of various fuels, and to test several inventions submitted to the Society by Prince Rupert, Charles II's cousin. He produced a whale-killing crossbow (made from whalebone), a treble writer ('to write three copies as one') and an amphibious wheel. Hooke also had to gather material on manufacturing techniques for his Cutler lectures and for the Society's ambitious and comprehensive History of Trades. This was a tiresome enterprise, he confided to Boyle: 'I am now engaged in a very great design, which I fear I shall find a very hard, difficult, and tedious task.'[30]

One of his most time-consuming tasks was to test new designs for coaches and carriages, and to produce new vehicles of his own. In March 1663 he was asked to report at length on a cart with legs instead of wheels, and a month later he reported on a Chinese three-seater one-wheeled cart that worked 'like a wheel-barrow'. That November the Society asked him for a pasteboard model of his own one-wheeled vehicle ('to travel with ease and speed') and his engine for travelling across land and water, and in September 1664 patents were prepared for 'several new-fashioned chariots' that Hooke had designed. Public coach services within London and between the capital and other towns were growing rapidly in the 1660s and 1670s, and there were many attempts to improve their comfort and efficiency. Coaches, unlike carts, had their passenger compartments suspended comfortably on a cradle of long straps or springs, and Hooke, with his knowledge of the qualities of wood under strain and of the science of springiness, was naturally included in the Royal Society committee on coach design which met in April 1665.

Samuel Pepys happened to meet the research group – Hooke, Lord Brouncker, Wilkins and Moray – near the Royal Exchange on May Day, and took a trip by river with them from Tower Wharf to Greenwich. After an indifferent dinner they rode in some experimental coaches, which was the purpose of their journey. 'And several we tried, but one did prove mighty easy (not here for me to describe, but the whole body of that coach lies upon one long spring) and we all, one after another, rid in it; and it is very fine and likely to take.' The Society soon took out a patent on a new omnibus, but if Hooke made an important contribution to the development of coaches he did not claim credit for it (as he surely would have done) when he lectured on carriages in 1685.[31]

Robert Hooke was a sort of scientific showman, producing new demonstrations, gadgets or microscopical observations week after week, trundling his air pump or velocity-measuring engine down the corridors of Gresham College for the entertainment of the Fellows of the Royal Society. In October 1663 he was given the additional job of looking after the Royal Society's new collection of specimens in the west gallery of Gresham College, the Repository. The exhibits were gradually accumulated: petrified wood, strange bones, metal ores, unnatural deformities of one kind or another. In May 1665 Boyle presented the Society with a monstrous colt's head with a double eye in the centre of its forehead, and Hooke was told to dissect it, describe it and label it for display.

The biggest and most important show of all did not take place. In July 1663 the Royal Society started preparing for the promised visit of Charles II. The Society's hopes of impressing the King depended heavily on Hooke's work:

> Mr Hooke was charged to show his microscopical observations in a handsome book to be provided for him for that purpose: to weigh the air, both in the engine and abroad: to break empty glass balls; as also to let the water ascend into them after they have been emptied; to provide the instrument for finding the different pressure of the atmosphere in the same place, as likewise the hygroscope made of the beard of a wild oat.[32]

In the event, Charles II decided against the visit. Far from being impressed by the work of the Royal Society, the King found their efforts – and especially Hooke's work on atmospheric pressure – laughable. On 1 February 1664 Samuel Pepys recorded this telling scene, in which Sir William Petty, one of the leading figures in the Society, tried to explain his plans for a double-bottomed (twin-hulled) boat to the King:

> To Whitehall, where in the Duke's chamber the King came and stayed an hour or two, laughing at Sir W. Petty, who was there about his boat, and at Gresham College [the Royal Society] in general. At which poor Petty was I perceive at some loss, but did argue discreetly and bear the unreasonable follies of the King's objections and other bystanders with great discretion – and offered to take oddes against the King's best boats; but the King would not lay, but cried him down with words only. Gresham College he mightily laughed at for spending time only in weighing of ayre, and doing nothing else since they sat.[33]

It was easy enough to laugh. Hooke must have seemed a comical figure, and the work he was doing was remote from the concerns of a worldly pleasure-seeker like Charles II. The fact that he was advancing the frontiers of scientific understanding in so many fields impressed hardly anyone outside his own scholarly circle. His contraptions looked outlandish and often failed to work, and their promised practical benefits generally did not appear. To those who could not understand the new science his preoccupation with the atmosphere, the weather, the behaviour of liquids, the speed of falling objects and the process of respiration and combustion must have seemed eccentric and remote from the business of the real world. Nevertheless, by the autumn of 1664 Hooke's great work on the microscope was finished and printed, and its publication was only delayed by the Society's insistence on having the work vetted by several fellows. There were high hopes in the Royal Society that the book would make the great public impact that they had been striving for. Hooke, on the other hand, was not so sure. In false or real modesty, he wrote to Boyle on 24 November 1664:

> As for the microscopical observations, they have been printed off above this month; and the stay, that has retarded the publishing of them, has been the examination of them by several of the members of the Society; and the preface, which will be large, and

> has been stayed very long in the hands of some, who were to read it. I am very much troubled there is so great an expectation raised of that pamphlet, being very conscious, that there is nothing in it, that can answer that expectation.[34]

By the following spring Hooke and the Royal Society would know whether Hooke's masterpiece had put the reputation of the Society and of natural philosophy itself on a stronger footing. Would there still be laughter in Whitehall when they had seen *Micrographia*?

6. A Secret World Discovered

(1665)

ON A FINE FROSTY MORNING at the start of 1665 Samuel Pepys, clerk of the King's ships, spent some hours over naval business with the Duke of York in Whitehall, visited his barber, 'sported a good while' with a young woman at the Swan tavern, and had sexual intercourse with Mrs Martin of Bow Street at a cost of 2s. in wine and cake. After that, 'sick of her impudence', he had a fine French dinner in Covent Garden piazza with Lord Brouncker, President of the Royal Society, which he was about to join. Before returning home, where his wife 'industriously and maliciously' made him read aloud a letter by Sir Philip Sidney on jealousy, and where he had to witness the vexing but arousing sight of his pet bitch being mated by a hired dog, he visited his bookseller. In the bindery there he saw 'Hookes book of the Microscope' in preparation, and found it so pretty that he ordered a copy. He collected it a few weeks later (on 20 January) and took it home with him, 'a most excellent piece, and of which I am very proud'. Pepys already had a microscope, and he had read some of Henry Power's 1664 volume on microscopic observations, *Experimental Philosophy*, but Hooke's work was something altogether new. The following day, after working at his office till past midnight, he sat up till two o'clock reading *Micrographia*, 'the most ingenious book that ever I read in my life'.[1]

Hooke got involved in microscopical work while he was at Oxford. In 1656 or 1657 he had observed and drawn the tiny globules of metal which flew out as sparks when flint was struck against steel. He included the work in *Micrographia*, adding some thoughts on the folly of traditional beliefs in the 'element' of fire and on Descartes's mistaken explanation of what happened when sparks flew. In 1661 Christopher Wren was asked to produce a series of microscopical studies of insects for Charles II. He completed several studies and presented them to the King before deciding, in September, that he had no more time for the project. At the suggestion of John Wilkins, the man whose patronage had first drawn him into natural philosophy at Oxford, Hooke was asked to take over the job. Once he was convinced that Wren had given up the study Hooke agreed to pursue it, and as soon as he started working for the Royal Society he began bringing his microscopical illustrations to the weekly meetings. He began rather unpromisingly with frozen urine, but between April and December 1663 he presented more than forty studies of insects and other tiny creatures, plants and man-made objects to the Society. Of course, the fellows insisted on Hooke providing pictures of all their favourite objects – sage leaves, dogs' blood, deer hairs, 'petrified snow', viper powder and (repeatedly) tiny eels in vinegar (nematodes). Hooke was happy enough to cast his net wide, because his aim was to demonstrate the power and utility of the microscope, and to appeal to a wide readership of inquisitive gentlemen not unlike those who came to Gresham College every Wednesday afternoon.[2]

Although the Fellows of the Royal Society could set Hooke's experimental agenda they could not put limits on the immense speculative power of his mind. The best they could do was to insist that Hooke should insert in his dedication to the Society a statement that the more hypothetical parts of *Micrographia*, its 'Conjectures and Quæries', were 'not done by YOUR Directions'. With this condition satisfied, the Society licensed the work (as they were empowered to do) and had it printed and published by their printers, John Martyn and James Allestry of St Paul's Churchyard. By this compromise, Hooke was able to include in a popular illustrated book of 'Physiological Descriptions of Minute Bodies made by Magnifying Glasses' (the book's subtitle) a series of bold and brilliant observations and hypotheses on light, colours, refraction, combustion, heat, gravity, the mechanical nature of animals and plants, the congruity of matter, the origins of fossils, the changing shape of the Earth, the stages of life, the medicinal value of plants, the instinctive behaviour of animals, the atmospheric refraction of light, the formation of clouds and the origins of lunar craters. Several of the devices he had developed in the early 1660s, including the lens grinder, the wild-oat hygrometer, the wheel barometer, the refraction engine and a thermometer with standardized calibration based on the freezing point of water, were described, and his part in establishing Boyle's Law was recounted. Thus Hooke summarized his work of the previous five years, exhibited his ideas and inventions to a wider audience, and publicly staked his claim to their authorship. So wide a field of scientific investigation was covered in the book that for the rest of his career Hooke was inclined to react to other people's ideas – especially Newton's – by claiming that he had already advanced them in *Micrographia.*

Micrographia began (after an ingratiating dedication to the King and a briefer one to the Royal Society) with an extraordinary twenty-eight-page preface which explained some of the guiding principles of Hooke's scientific thinking. Mankind's powers of observation, memory and reason, he argued, had been undermined over the centuries by 'negligence and intemperance'. Our senses were weaker than those of many other creatures, our memory allowed important things to be 'overwhelmed and buried under more frothy notions', and our understanding, fed by our inadequate senses, inclined either to 'gross ignorance and stupidity' or to 'confident dogmatizing on matters, whereof there is no assurance to be given'. Hooke's proposal – and one of his fundamental beliefs – was that sense, memory and reason could be restored to their original perfection by 'the real, the mechanical, the experimental philosophy'. Our weak senses could be immeasurably strengthened by the use of artificial organs of sight, hearing, smell, touch and taste. Using better telescopes 'we may be able to discover living Creatures in the Moon, or other Planets', and with the microscope (as *Micrographia* would prove) 'this the Earth it self, which lyes so neer us, under our feet, shews quite a new thing to us, and in every little particle of its matter, we now behold almost as great a variety of Creatures, as we were able before to reckon up in the whole universe it self'.

Hooke, along with Boyle and Wren and many of his contemporaries, accepted the proposition advanced in Descartes's *Principles of Philosophy* (1644) that the universe was composed of minute particles in a state of constant motion and arranged (by the hand of God) into perfect mechanical forms. The bodies of animals, plants and insects, like the universe itself, were complicated machines whose workings could be understood by close and careful observation. Hooke did not accept Descartes's speculations uncritically, and had in fact disproved some of his laws of collision and refraction in his early experiments for the Royal Society. He was always interested in putting theories to the test of actual observation, and he hoped that when stronger microscopes had been developed atoms or corpuscles, 'the compounding Particles of matter', would at last be visible. In the meantime, even his own faulty instruments might confirm that plants and insects were indeed beautifully designed machines, inhabitants of what he, like many scientists of his time, believed was a mechanical universe.

Though *Micrographia* concentrated on the value of improving the power of the eye, Hooke did not forget to mention the ways in which the other senses could be helped by mechanical aids. Hearing could easily be improved by techniques he had already been working on: 'I know a way, by which tis easie enough to hear one speak through a wall a yard thick'. And air was not the only medium through which sound could pass: 'I have, by the help of a distended wire, propagated the sound to a very considerable distance in an instant, . . . and this not only in a straight line, or direct, but in one bended in many angles'. He had also made trials (he said) of a way of increasing the flow of air 'through the grisly meanders of the Nose' to improve the sense of smell, and had high hopes that a method might be found of smelling or sensing the fumes emitted by underground minerals, so that they could be found 'without the trouble to dig for them'. His own wheel barometer, thermometer and hygrometer (all described and illustrated in *Micrographia*) had sharpened men's perception of the pressure, heat and dampness of the air, and the microscope, by revealing the true surface of things, had in a way improved on the sense of touch.

Hooke's argument was that philosophers should not claim to understand the world until they had properly examined it. Reason, which stood at the top of the pyramid of understanding, should not be a tyrant over the lower senses but their wise master, supervising a fruitful cycle of observing, recording and understanding.

> So many are the links, upon which the true Philosophy depends, of which, if any one be loose, or weak, the whole chain is in danger of being dissolved; it is to begin with the Hands and Eyes, and to proceed on through the Memory, to be continued by the

> Reason; nor is it to stop there, but to come about to the Hands and Eyes again, and so, by a continuall passage round from one Faculty to another, it is to be maintained in life and strength.

This was Hooke's answer to the choice between Baconian fact and Cartesian theory – that there should be a constant and fruitful interaction between the two. He promised that he would one day explain his method for overcoming the weakness of reason and memory, and that he would soon reveal the great advances he had already made using this new method. He promised to publish his 'new and not inconsiderable' work on human flight, and to demonstrate his solution to the greatest navigational puzzle of the age, the problem of longitude. For the moment his modest purpose was to improve the performance of the senses: 'all my ambition is, that I may serve to the great Philosophers of this Age, as the makers and the grinders of my Glasses did to me; that I may prepare and furnish them with some Materials, which they may afterwards order and manage with better skill, and to far greater advantage.'

Although Hooke was keen to emphasize the significance of his own achievements, he also wanted to persuade his readers that anyone with resolution and integrity, 'a sincere hand and a faithful eye', could do what he had done. Unusual imaginative or speculative powers were not necessary, and could even become the enemy of good observational work. To encourage gentlemen to discover for themselves the 'high rapture and delight of the mind' that could be found in experimental philosophy, as well as to demonstrate his own ingenuity, Hooke made a point of explaining the methods he had used, and the ways he had found of overcoming the practical difficulties of his work. This also had the advantage of adding authenticity to his observations, and enrolling his readers as 'witnesses' of the work he had carried out alone in his laboratory.[3] None of the other great microscopists of the seventeenth century wrote so openly about the methods they used as Robert Hooke. So when we see him castigated later on as a hoarder of secrets and a dealer in codes, we should remember that it was Hooke, not Malpighi, Swammerdam, Leeuwenhoek or Nehemiah Grew, who revealed the mysteries of the microscope.

The most powerful type of microscope until the early nineteenth century was the single lens, which had the magnifying power to reveal bacteria and other unknown organisms. Hooke was probably the first to make one of these, and was certainly the first to explain how it was made. He described how he had melted a piece of very clear Venice glass so that it ran down in a thread, then caught the globule that formed at the bottom of the thread and ground and polished it with a whetstone and fine earth to make a lens of less than a tenth of an inch across. This tiny glass bead held in a brass or pewter mount, the single-lens microscope, gave a clearer, brighter and far more magnified view than larger microscopes with two or three lenses, but Hooke found it awkward to hold the lens and the object viewed so close to his eye, and he feared that it would damage his sight. He used it when he wanted to see the smallest objects in the greatest detail – a nettle sting, or the tiny shell-shape of a minute grain of sand. For most of his observations he used a compound microscope with a six- or seven-inch tube, probably made by Richard Reeve, with an eyeglass, an object glass, and a middle (field) lens which he usually removed to brighten the image. Hooke introduced several improvements to make his observations better and easier than those of earlier microscopists. Instead of the cumbersome tripod, he made an ingenious single pillar stand which enabled him to alter the height of the microscope, and to swivel it on a ball and socket joint. He told a correspondent that this allowed him to follow the movements of a living insect: 'the best way is to accustom oneself to direct the microscope wth ones hand and then there is nothing so easy as to keep pace with any kind of wandring insect'. He overcame the problem of the darkness of highly magnified images by using a convex lens or a globe of water to focus the light from his south-facing window onto the object being viewed. To allow his work to continue when it was gloomy or dark he used a lighting device – a condenser – of his own invention. A lamp, a brine-filled glass globe and a convex lens were mounted on movable arms on a short pillar so that the rays from the lamp were focused through the globe and directed by the lens onto the object being viewed. Several of these innovations later became common features of commercially produced microscopes, especially those made by John Marshall towards the end of the century.[4]

Hooke's compound microscopes probably had a maximum magnification of around 50x, though his drawings were often more than twice that scale. This meant that he was limited to seeing unfamiliar detail on familiar objects: fleas, gnats, moss, threads of flax, nettle stings and so on. Hooke's huge picture of a flea, one of the finest illustrations in *Micrographia*, was a wonderful celebration of the beauty and strength of a hitherto despised creature. But to advance understanding of the causes of the bubonic plague, which made its final disastrous attack on London in the year *Micrographia* was published, he would have needed to see not only the rat fleas that carried the disease, but also the tiny bacteria they transmitted in their bite. By deciding against using the inconvenient but much more powerful single-lens microscope which he had made, Hooke left it to the great Dutch microscopist Anton van Leeuwenhoek to exploit its much greater power to see and describe previously invisible things – red blood corpuscles, bacteria and spermatozoa – for the first time in the 1670s.[5]

Hooke's great advantage over Leeuwenhoek was his ability to draw what he saw, and to do so with great beauty and accuracy. This was not a straightforward matter, as he explained in his preface. He had to view the same object many times and in many lights before he was sure of its true form and colours: 'The Eyes of a Fly in one kind of light appear almost like a Lattice . . . In the Sunshine they look like a Surface cover'd wth golden Nails; in another posture, like a Surface cover'd with Pyramids; in another with Cones; and in other postures of quite other shapes'. The study of insects and other tiny living creatures presented particular difficulties. They could not be killed, because when they died they quickly shrivelled and lost their natural shape. Most could be immobilized by sticking their feet in wax or glue, but ants had to be rendered senseless by being dipped in alcohol. When an ant awoke from its drunken stupor and tried to stagger off, Hooke gave it another soaking.[6]

Although *Micrographia* is regarded as one of the founding works of modern entomology, and its most spectacular engravings were of insects, only fifty of its 246 pages dealt with living creatures. Hooke began with a hundred pages on man-made and inanimate objects, followed this with sixty on plants, sponges and feathers, and ended his book with twenty-eight pages on the atmosphere, the stars and the Moon. Hooke's plan was to trace nature's path from simple objects to more complex organisms. So he began with the point of a needle, the edge of a razor blade and a printed full stop, demonstrating the roughness of even these apparently regular objects and reflecting on the impossibility of making a hard surface smooth by grinding or polishing. On almost every page he came up with an original idea, or a suggestion for future scientists or craftsmen to work on. His comparison of flax and silk led him to suggest that imitation silk could be made from some artificial glutinous substance, and his study of gravel in urine concluded with the idea that physicians should look for a harmless solvent that might be injected into the bladders of those suffering from the stone. If this had been possible at the time it would have delighted one of his first readers, Samuel Pepys, who had suffered painful and dangerous bladder surgery in 1658. A description of small glass drops led him into a discussion of the reasons for the expansion of heated material, and an account of his new method for standardizing the calibration of thermometers so that different instruments would give comparable readings.

One of his longest observations, 'Of the Colours observable in Muscovy Glass, and other thin Bodies', was of special significance for Hooke's career and reputation.[7] Muscovy glass was Muscovite or mica, a form of silicate of aluminium which was composed of many very thin and transparent layers, and which could easily be split into fine flakes which were themselves composed of several thinner layers. Looking at these flakes through the microscope, Hooke saw rings of colour in a succession of rainbow sequences around a white central spot. He found that by pressing the flakes he could move the coloured rings, and that by splitting them he could produce plates of a single colour. If Hooke had been a simple Baconian observer, the Royal Society ideal, he might have recorded these phenomena and moved on to his next observation. Being Hooke, he decided 'to examine the causes and reasons of them, and to consider, whether from these causes . . . may not be deduced the true causes of the production of all kinds of Colours'. It was clear to him that the colours he could see were not in the mica itself (which was colourless in thicker pieces) but produced by the juxtaposition of thin transparent slivers of a certain thickness. He observed the same effect with two pieces of glass pressed close together, with oyster shells, and with films of glutinous liquid on a polished surface. 'In general, wheresoever you meet with a transparent body thin enough, that is terminated by reflecting bodies of differing refractions from it, there will be a production of these pleasing and lovely colours.'

To explain the rings of colour in thin plates, Hooke presented his own explanation of the nature and behaviour of light. He put forward several important propositions, which amounted to a new wave theory of light, different in many respects from the abstract speculations of Descartes. Hooke's first proposition, that all luminous bodies had their particles in motion, was shared with Descartes. But the fact that a rubbed diamond shone indicated to him that this motion was not circular (as Descartes had said) but a short and rapid vibration. The rubbed diamond also suggested to him that luminosity did not involve the transmission of particles, since the diamond was not worn away. This example was important in his later dispute with Newton over the nature of light. Vibrations or pulses of light travelled through any transparent medium with unimaginable speed, but there was no reason to accept Descartes's supposition that their journey was instantaneous, 'for I know not any one Experiment or observation that does prove it'. In a homogeneous medium the pulses or vibrations of the luminous body would travel at equal speed and in straight lines in all directions, like the ripples created when a stone lands in water. As in water, the waves would spread in bigger and bigger circles, with the wave fronts at right angles to the direction in which the light was travelling. The water analogy implied that light waves came at regular intervals, but Hooke did not explicitly say so.

He then moved on to the question of colour, and the problem of the thin plates. His explanation was quite different from that put forward by Newton a few years later, but it was nevertheless plausible, ingenious and original. Refraction, in Hooke's theory, was the key to understanding colour. Undisturbed or unrefracted light was white, but when a ray of light of a certain width hit a refracting surface obliquely it was distorted or confused in various ways, and these confusions gave the impression to our eyes of different colours. When a wave of light hit a refracting surface at an oblique angle its leading edge was dulled or weakened by the resistance of the new medium, and its trailing edge entered more easily. As a result of this the two edges of the ray followed diverging paths through the medium, and travelled with different degrees of strength. Because the speed of the leading edge of the wave was affected before the trailing edge, the wave fronts were no longer at right angles to the path of the ray of light, and the light was confused and (to our eyes) coloured. The light was not broken into different pre-existing components (as Newton was later to demonstrate); rather, rays that had experienced varying degrees of resistance and distortion created different sensations on the retina. Summarizing a long and complex argument, Hooke concluded that 'Blue is an impression on the Retina of an oblique and confus'd pulse of light, whose weakest part precedes, and whose strongest follows. And, that Red is an impression on the retina of an oblique and confus'd pulse of light, whose strongest part precedes, and whose weakest follows.' In the middle of this spectrum there were paler blues and yellower reds, and shades of green when these two met.

To test this hypothesis Hooke applied it to the colours he had seen in thin plates of mica. This (he used Bacon's words) was the *experimentum crucis*, the crucial experiment, by which his new theory of colours would stand or fall. His argument was that the colours seen by the observer were a combination of rays reflected by the top surface of the plates, and rays which were reflected from the lower surface of the plates, having been refracted and weakened in the process. When the plate was thin the two rays travelled almost together to the retina, but with the stronger unrefracted ray just ahead of the weaker refracted one. The effect of this combination (a strong pulse followed by a weak one) was yellow. When the thickness of the plates was greater the distance between the strong and weak pulses was altered, and the effect on the retina changed. A bigger distance between the leading strong pulse and the following weaker one produced red, and when the weaker pulse was so delayed that it was halfway between two stronger ones the effect was purple. A thicker plate delayed the weaker refracted pulse so much that it reached the retina just ahead of a stronger one, producing blue or (if the gap between the two pulses was very short) green. In a wedge-shaped piece of mica the sequence was repeated, producing the rings of colour Hooke had described in the first place. All that was needed to complete his hypothesis, he said, was an accurate measurement of the thickness of the plates needed to produce different colours, but this was something which his imperfect microscopes did not allow him to do. This all sounds strange now, as most discarded scientific theories do. But although Hooke was mistaken in his assumption that refraction created colours rather than separating them he came close to identifying the fact that the difference between colours is their frequency or wavelength or (as he would have put it) the distance between their pulses. This idea was soon overwhelmed by Newton's erroneous but persuasive corpuscular theory (in which light was a flow of tiny particles), but when Thomas Young and Augustin Fresnel revived the wave theory and demonstrated its truth in the early nineteenth century they acknowledged their debt to *Micrographia*.

Hooke was bursting with speculations and ideas for further enquiry on almost every subject. His observations on charcoal drew him into an explanation of reflection as the cause of the colour of objects (charcoal looked black simply because it reflected no light, and marble was white because it reflected so much), and gave him a chance to expound his new ideas on the nature of fire. In his discussion of 'petrify'd shells' or serpentine stones found near Bristol he dismissed the accepted idea that these were nature's tricks, perfect copies of organic forms but without any organic origins of their own. It was plain to him that they were the remains of 'the Shells of certain Shell-fishes, which, either by some Deluge, Inundation, Earthquake, or some such other means, came to be thrown to that place, and there to be fill'd with some kind of Mudd or Clay'. How did the remains of sea shells find their way to the tops of mountains, and why did some stones bear the imprint of species which no longer existed? These were new questions, and Hooke returned to them repeatedly for the rest of his life.[8]

Observing the 'very pretty candied substance' he found when he broke open a flint, and the 'Cornish diamants' that were found in hollow rocks, Hooke was drawn into a discussion of the geometrical shapes of crystals. He reasoned that the external shapes of crystals reflected the inner arrangement of their particles, and tested and illustrated the proposition by making various structures from piles of spherical bullets. The way in which the bullets were arranged, in cubes, triangles, hexagons, and so on, would determine the final shape of the solid: 'the coagulating particles must necessarily compose a body of such a determinate regular Figure, and no other'. Only in the early twentieth century was his proposition confirmed, using X-rays, and the modern definition of a crystal ('a solid substance in which the component atoms are arranged in a distinct pattern and whose surface regularity reflects its internal symmetry') confirms his almost forgotten intuition.[9] Hooke did not have the time or assistance, he said, to pursue a systematic study of crystals, but (typically) he drew up a ten-point plan for others to follow. Four years later the Danish scientist Steno's book on crystals was published, entitling him to share with Hooke the honour of being the founder of the science of crystallography.[10]

In his observations on the texture of cork, Hooke stumbled on a secret whose significance he did not understand. Trying something new, he cut a very fine slice of cork with a razor-sharp pen-knife, and saw 'the first microscopical pores I ever saw, and perhaps, that were ever seen'. There were over a billion of these tiny boxes in a cubic inch of cork, and because they looked like the walled compartments of a honeycomb Hooke called them cells, the name they still bear. He had no way of knowing that he had seen the basic elements in organic structure, and his tendency to call cells pores (open structures through which fluids pass) shows that he did not grasp their function. The true scientific study of cells, cytology, did not develop until the 1830s. Hooke was chiefly interested in the mechanical function of the cell in giving cork its lightness and springiness, and thus its value as a float or stopper.[11]

For Hooke the microscope was a window through which he could peep into the workshop of a master mechanic. The plants he had studied – mould, moss, seaweed, nettles, wild oats – were all beautiful machines, as mechanical in their operation as a pocket watch. He was constantly on the alert for the lessons which he could learn from the divine mechanics of nature. He observed one impressive mechanism, the sting of the common nettle, using a single-lens microscope fixed into a little spectacle frame. As the nettle sting pierced his skin he watched the little bladders at the base of the sting pumping their stock of poisonous liquid into his body, and thought about ways in which the principle of the syringe or poison arrow might be used against gout, dropsy and other distempers. For 'good men might make as good a use of it, as evil men have made a perverse and

Diabolical'.[12] The beard of a wild oat, which bent when the air was damp because one part of its stem was porous and expanded when wet while the other did not, could teach him how to make 'contrivances that should thus wreath and unwreath themselves, either by heat or cold, or by driness and moisture, or by any greater or less force'. It might also help to explain the working of muscles, a subject of particular interest to him. If physicians studied nature instead of books of herbal remedies they might find there 'the most natural, usefull, and most effectual and specifick Medicines'.

When he moved on from the vegetable to the animal world, the beauty and complexity of God's design became more impressive and much harder to imitate. The sting of a bee (drawn and described in minute detail) and the structure of birds' feathers – strong, light, easy to repair and resistant to the air – both deserved the most careful study by those who wanted to understand practical mechanics and 'the nature of bodies and motions'. Among insects, he was mightily impressed by the wonderful mechanisms with which flies had been furnished by the Creator. For who could be 'so sottish as to think all those things the productions of chance?' A fly could walk on a ceiling or window, he said, thanks to the tiny talons on its feet, not (as was generally believed) because of 'an imaginary gluten' on them. The drone fly's huge eye, which was the subject of one of his finest drawings, seemed to be made up of 14,000 tiny hemispheres, which made the fly (Hooke joked) 'really circumspect', and compensated for the immobility of its head. He studied the intricate design of a fly's wings, covered in innumerable tiny bristles, and watched them beating when the creature was glued to the end of a feather. Hooke, who was very interested in the question of vibration, thought it likely that 'the quickest vibrating spontaneous motion is to be found in the wing of some creature', and believed it would be possible to count the number of vibrations of a bee's or fly's wing by comparing the sound they made with that of a vibrating musical string. When he met Samuel Pepys and Richard Reeve near the Temple in August 1666 he assured them that he could tell the speed of a fly's wingbeats from the musical hum they made. Pepys was interested but unconvinced: 'That, I suppose, is a little too much raffined; but his discourse in general of sound was mighty fine'.[13]

Although Hooke's main interest was in the outward appearance and mechanical structure of his insect subjects, it also delighted him to be able to study the inner workings of their transparent bodies without subjecting them to the tortures suffered by the Royal Society's unfortunate dogs. 'When we endeavour to pry into [Nature's] secrets by breaking open the doors upon her, and dissecting and mangling creatures whil'st there is life yet within them, we find her indeed at work, but put into such disorder by the violence offer'd, as it may easily be imagin'd, how differing a thing we should find, if we could, as we can with a Microscope, in these smaller creatures, quietly peep in at the windows'. All his life, Hooke was an advocate of the microscope, in preference to the knife or the pestle and mortar, as an instrument through which organisms and the structure of matter could be understood. He allowed a louse, a constant companion to most seventeenth-century Londoners, to feed on his hand, then watched as his blood made its way through its digestive system. 'And the Creature was so greedy, that though it could not contain more, yet it continued sucking as fast as ever, and as fast emptying itself behind: the digestion of this Creature must needs be very quick, for though I perceiv'd the blood thicker and blacker when suck'd, yet, when in the guts, it was of a very lovely ruby colour, and that part of it, which was digested into the veins, seem'd white'.[14]

Hooke was bound by the knowledge and assumptions of his day, but where it was possible he tried to approach his subject with an open mind. Why should we cling to an old and false explanation, he asked, 'where our senses were able to furnish us with an intelligible, rationall and true one'? He was unsure, for instance, about the prevailing theory of spontaneous generation, which claimed that some life forms developed from putrefying matter. He regarded this as a possible explanation for the growth of mushrooms and moss, but he believed that if microscopists looked hard enough they would discover the parents of some creatures whose birth had until then been regarded as spontaneous. He had been interested in tiny mites since 1661, when he had watched them wandering across the window of his room in Oxford, and he wondered whether he had found in them 'the vagabond parents of those Mites we find in Cheeses, Meal, Corn, Seeds, musty Barrels, musty Leather, &c'. Perhaps these creatures were attracted by the smell of rotting matter and decided to spend their final days there 'in very plentiful and riotous living', leaving their offspring behind them when they died. Hooke had not seen this happen, but he hoped that others, following his conjecture, would discover 'that many animate beings, that seem also to be the mere product of putrefaction, may be innobled with a Pedigree as ancient as the first creation'.[15]

The strangest transformations could take place, he discovered, by entirely natural and rational processes, if only we could see what was happening. He watched a collection of minute creatures in a jar of rainwater over several weeks and saw their bodies changing shape as they were transformed into gnats, a process which had never been described before. This led him to wonder 'whether all those things that we suppose to be bred from corruption and putrifaction, may not be rationally suppos'd to have their origination as natural as these Gnats, who, 'tis very probable, were first dropt into this Water, in the form of Eggs' which are too small for us to see. Gnats, flies and other such creatures had been endowed by their Maker with 'a faculty of knowing what place is convenient for the hatching, nutrition, and preservation of their Eggs and of-springs', and were driven by an impulse implanted in their mechanical structures to act in a way that would propagate their species.[16]

Micrographia was a work of such originality, scope, audacity and beauty that its impact and influence was almost bound to be great. It was reissued in 1667, using plates and printed pages which had been saved from the destruction of the printer James Allestry's premises in the Great Fire of September 1666, and excerpts were used in works on natural history or microscopes for the next 150 years. *Micrographia* promoted interest in the microscope as a scientific toy and taught its readers how to use it to good effect. No doubt most of these enthusiasts soon lost interest in their hobby, but *Micrographia* almost certainly influenced the greatest of the early microscopists, Anton van Leeuwenhoek, when he visited England in 1666 or 1667. It demonstrated as conclusively as possible that the scholars of the Royal Society had been involved in something more than weighing air and suffocating chickens over the past four years. The work was reviewed in the leading scientific journals in London and Paris, and a copy was sent to the great Huygens, who had read it, and written to the Society in praise of it, by March 1665. Huygens's opinion of Hooke's work was expressed much more frankly in a letter to his father, who had sent him some quotations from *Micrographia*: 'Thanks for the extracts from Hooke. I know him very well. He understands no geometry at all. He makes himself ridiculous by his boasting. I know his machine [for grinding lenses] well – it is quite inept. And a bad example of his mechanic algebra.' As Hooke's career shows only too well, there was a rough world of jealousy and ambition beneath the superficial politeness of seventeenth-century philosophical correspondence.[17]

One of Hooke's most critical and attentive readers was a young Cambridge graduate. Isaac Newton's seven pages of densely packed notes on *Micrographia* still exist among his papers in the Cambridge University Library. They show that he was interested in Hooke's theory of combustion, his ideas on the congruity of matter, and his description of the nettle's sting, and had thought of several ways of pursuing the problem of atmospheric refraction. The section of *Micrographia* which stimulated him the most, though, was Hooke's work on light and colours. Newton was not convinced by Hooke's theory that light was composed of waves, like sound, and he could not see why Hooke's 'weaker' waves should travel as fast as 'stronger' ones. 'Why then may not light deflect from streight lines as well as sounds &c? How doth the formost weake pulse keepe pace with the following stronger & can it bee then sufficiently weaker'? Still, he accepted that the rings of colour produced by thin plates (which we now call Newton's rings) might provide clues to the nature of light and colour, and set to work on them. The outcome of his investigations, when they were made public in 1672, brought about a disastrous collision between the two scientists, and cast a shadow over the rest of Hooke's career.[18]

7. Falling Bodies

(1665–1666)

SINCE 1348 LONDONERS had lived in fear of the bubonic plague. There were deaths from plague almost every year, and in the previous hundred years there had been eight serious outbreaks in London. The worst of these, in 1563, 1603 and 1625, had each killed between a fifth and a quarter of London's population. The last epidemic had been in 1635, and since 1650 there had been few plague deaths in the city, but the return of the disease was always feared and anticipated. During the winter of 1663–4 there was a severe outbreak in Amsterdam, and the arrival of the disease in London seemed imminent. Towards the end of April 1665 several cases of the plague were reported in London, and on 7 June Samuel Pepys saw two or three houses in Drury Lane marked with red crosses as a sign of infection, and read for the first time the dreadful words 'Lord have mercy upon us'. By the middle of that month recorded plague deaths were running at more than a hundred a week, and fear was spreading through the town.

In common with most men of wealth and good sense, the fellows of the Royal Society decided to evacuate the city until the epidemic had passed, leaving poorer citizens to take their chances in London. The Society held its final meeting on 28 June, and shortly afterwards its members dispersed to the safety of the countryside or to plague-free towns. By the time that they reassembled in March 1666 at least 100,000 Londoners – a quarter of the population – had died of bubonic plague, and probably the same number had suffered from the disease and survived. A large contingent of fellows joined Robert Boyle in Oxford, and carried on meeting there. The Royal Society's dutiful secretary, Henry Oldenburg, remained in London throughout the epidemic, looking after the Society's business from his house in

Pall Mall, where the plague was raging. It was less rampant in the City, and Hooke remained in Gresham College in early July, packing his instruments so that he could continue his experiments in the country. He was interested, of course, in discovering the causes of the disease, but by mid-July Londoners were dying at a rate of 2,000 a week and fear triumphed over curiosity. He wrote to Boyle on 8 July:

> I cannot, from any information I can learn of it, judge what its cause should be, but it seems to proceed only from infection or contagion, and that not catched but by some near approach to some infected person, or stuff: nor can I at all imagine it to be in the air, though yet there is one thing, which is very differing from what is usual in other hot summers, and that is a very great scarcity of flies and insects.

The idea that the flea, whose beautifully designed body had been one of the greatest wonders displayed in *Micrographia*, played any part in transmitting the disease never crossed Hooke's mind, or occurred to any of his colleagues.

It is possible that Hooke had visited the Isle of Wight for his mother's funeral in June 1665, but there was no question of returning there in July, since it had closed its ports after an outbreak of the plague in Southampton. His visit was delayed until October 1665, when restrictions had been lifted. In July he had intended to go to Nonsuch, a neglected royal palace between Ewell and Cheam, but at the last minute he changed his plans and went with William Petty and his old patron Dr Wilkins, along with the Royal Society's Operator, Richard Shortgrave, to Durdans near Epsom in Surrey, a house belonging to Lord Berkeley, a gentleman Fellow of the Royal Society. Writing to Boyle about these arrangements in July, Hooke still adopted the tone of a paid assistant to the great chemist:

> I hope, we shall be able to prosecute experiments there as well almost as at *London*; and if there be any thing, that you should desire to be tried concerning the resistance of fluid mediums, or any kind of experiments about weight or vegetation, or fire, or any other experiments, that we can meet with conveniences for trial of them there; if you would be pleased to send a catalogue of

them, I shall endeavour to see them very punctually done, and to give you a faithful account of them.[1]

The diarist and amateur scientist John Evelyn called on Hooke, Petty and Wilkins at Durdans at the beginning of August 1665 and found them 'contriving chariots, new rigging for ships, a wheel for one to run races in, and other mechanical inventions; perhaps three such persons together were not to be found elsewhere in Europe, for parts and ingenuity'.

Petty soon went off to Oxford, leaving Hooke and Wilkins to pursue their many interests freed from the pressure of the Royal Society's weekly demands. Wilkins finished off his book on a new language or 'Universal Character' (the whole manuscript was later destroyed in the Great Fire) and Hooke collected natural rarities (including 'shining animals whose blood, or juices, did shine more bright than the tail of a glow-worm') for his repository. Many years later, he recalled damming the River Mole near the point at which it started flowing underground, and putting his ear to the ground to hear the noise of the water 'as falling down a Precipice of a good height'.[2] His experimental programme was, as usual, ambitious and extensive. The neighbourhood offered 'such an opportunity, as is scarce to be met with in any other place I know', he wrote to Boyle in September.

I have in my catalogue already thought on divers experiments of heat and cold, of gravity and levity, of condensation and rarefaction of pressure, of pendulous motions and motions of descent; of sound, of respiration, of fire, and burning, of the rise of smoke, of the nature and constitution of the damp, both as to heat and cold, driness and moisture, density and rarity, and the like. And I doubt not but some few trials will suggest multitudes of others, which I have not yet thought of.[3]

Hooke discovered that on Banstead Downs, about two miles from Durdans, there were wells deeper than the height of old St Paul's. Hooke and Wilkins were at that time experimenting with a new type of carriage or chariot, in which a single horse could carry two men, one in a seat and the other riding above the horse on a springy saddle. It is pleasant to imagine the stooping scientist and the ageing clergyman (Wilkins was Dean of Ripon) riding from Epsom to Banstead on this bouncy chariot, but they probably transported their cumbersome equipment there in a more conventional carriage. In the Banstead wells Hooke and Wilkins tested the quality of the air at a depth of 315 feet by lowering candles, hygroscopes, thermometers and barometers on lengths of packthread. They dropped lead and wooden bullets to confirm Galileo's law of acceleration, tested the qualities of echoes, and weighed objects on the surface and at great depths to discover whether there was a measurable difference between weights at different levels. Their tests failed to confirm the recent findings of Henry Power (the microscopist) that objects weighed less at great depth than on the Earth's surface. Hooke still believed, as Francis Bacon had, that it was possible that gravity was weaker at depth (because part of the Earth and its attracting power was above the weighed object) but he thought that the difference would be so small that a more delicate measuring device, a sealed pendulum clock or a fine spring balance, would be needed to detect it. He designed a spring balance for these tests, which he described to the Royal Society on 21 March 1666, and he planned to use it on a trip to the Welsh mountains the following summer. It is certain that with the materials and techniques available at the time Hooke's balance would not have worked, but his principles anticipated those used in twentieth-century gravimeters.[4]

Hooke's ideas on gravity and its bearing on the great question of planetary movement seem to have developed rapidly between 1664 and 1666. If the gravitational attraction of the Earth was magnetic, as the English scientist William Gilbert had suggested in his book *De Magnete* in 1600, or a different force which was similar to magnetism in its qualities, as Wilkins had argued in his *Discovery of a New World in the Moon* (1640), then it would not be surprising to find its power reducing with distance above or below the Earth's surface. This was one of the propositions that Hooke, Wilkins and the others had been testing when they climbed to the top of St Paul's in August 1664, and tested again on Banstead Downs. Hooke's important paper of 14 December 1664, in which he questioned the search for a standard one-second pendulum by pointing out that gravity probably varied with height and latitude (things might weigh less near the equator, as well as at the top of cathedrals), marked a significant advance in his understanding of gravity.

The development of Hooke's ideas on gravity had been stimulated by the appearance of two bright comets in December 1664 and March 1665. Hooke and Wren were asked to collect and study the many observations that were sent in to the Royal Society, and to present their thoughts on the paths that comets took. At first both were inclined to the common view that comets moved in straight lines, but by the spring of 1665 their observations and discussions had led them to wonder whether they were travelling in some sort of circular or elliptical orbit, as the English astronomer Jeremiah Horrocks had suggested in the 1630s. Samuel Pepys heard Hooke lecture to this effect on 1 March, when he advanced the 'very new opinion' that the December comet had been seen before in 1618, and would return in forty-six years' time. 'I have I thinke lighted upon the trew Hypothesis wch. when it is riper & confirmed by your observations I shall send you', Wren wrote to Hooke on 20 April. His reply to Wren two weeks later implied that both men now thought comets moved in great circles or curves, returning to be observed again over a period of years. 'If I can be so lucky to meet with it again, I hope to trace it to its second appearing'.[5]

Why were the paths of comets curved? When Hooke returned from his six months with Wilkins at Durdans his views on this question had moved on a little further. On 21 March 1666, the second Royal Society meeting after the plague, Hooke spoke at length about his experiments in St Paul's and on Banstead Downs, and explained that they were a part of his investigation of gravity, 'one of the most universal active principles in the world'. His intention was to discover 'whether this gravitating or attractive power be inherent in the parts of the earth; . . . whether it be magnetical, electrical, or of some other nature distant from either' and 'to what distance the gravitating power of the earth acts'. To clarify his thoughts on gravity he began a series of experiments on magnetic attraction. Although he did not assume that gravitation and magnetism were identical, he saw that there was at least a useful analogy between the two forces. The Royal Society's Treasurer, William Balle (or Ball), the 'virtuoso' who had been the curator in charge of magnetical experiments in the early 1660s, had retired to his Devon estate during the plague, and did not return to London until 1667. Taking advantage of his absence Hooke took over the Society's work on magnets and turned it into a pioneering study of attraction at a distance. A letter was sent to Balle asking him to return the Society's experimental equipment, and Sir Robert Moray (one of the Society's leading courtier fellows) was asked to borrow the King's lodestone for Hooke's use. On 28 March 1666 Hooke showed the Society his scales for measuring the effect of distance on the power of magnetic attraction, and within a week he had produced a set of results for distances up to six inches. On 18 April he produced a watch powered by a lodestone, and in late April and early May he used iron filings and a spherical lodestone known as a 'terrella' (little Earth) to discover the shape of the magnetic forces emanating from the hanging sphere, and their effects on iron dust when it was sieved over it. These innovative experiments were far from conclusive, but they provided Hooke with enough ideas and information to enable him to clarify his account of gravity and celestial motion.

On 16 May, the Royal Society heard a paper from John Wallis, Professor of Astronomy at Oxford, on the relationship between tides and the common centre of gravity of the Earth and the Moon. When some fellows wondered how two separate bodies could have a common centre of gravity, Hooke offered to use pendulums to explain the motion of celestial bodies to the Society. In the next meeting, on 23 May 1666, he seized the opportunity to put on record his latest thoughts on gravity and celestial motion. 'I have often wondered', he began, 'why the planets should move about the sun according to Copernicus's supposition, being not included in any solid orbs . . . nor tied to it, as their centre, by any visible strings.' Why did they not move in straight lines, 'as all bodies, that have but one single impulse, ought to do'? What was the second impulse acting upon them? There were two possible explanations. Perhaps the fluid or medium (the aether) through which the planets travelled was of unequal density, preventing the body from moving away from the Sun and thus holding it in orbit around it. This might be the case if the aether further from the Sun was colder and therefore denser than that nearer to it. (Newton advanced a similar explanation of gravity in 1717.) The alternative explanation, favoured by Hooke, was that the Sun exerted some sort of constant attractive force which deflected the straight motion of the planet into a curve. 'For if such a principle be supposed, all the phenomena of the planets seem possible to be explained by the common principle of mechanic motions; . . . the phenomena of the comets as well as of the planets may be solved'. Centripetal force, rather than the centrifugal force proposed in Descartes's widely accepted 'second law of nature' ('things which move in a circle always tend to recede from the centre of the circle that they describe'), would explain the orbital motions of the planets and their satellites. This was a simple but brilliant and original insight, drawn from Hooke's study of comets and his discussions with Wren in 1665, from his work with Wilkins on Banstead Downs during the plague, from his reading of the writings of earlier scientists like William Gilbert and Jeremiah Horrocks, from his own work on magnets and from his intuitive grasp of mechanics.

Hooke was not capable of giving a satisfactory mathematical explanation of his proposed planetary system, but instead he demonstrated it with a practical analogy. He showed the meeting a large pendulum swinging from the ceiling in a circle or ellipse (the oval shape of a cross-section cut obliquely through a cone), to demonstrate that the motion of the bob was a compound of two forces: the tendency of the weight to move in a straight line or tangent, and the pull towards the centre exerted by the wire, representing the gravitational attraction of the Sun. Hooke demonstrated how pushing the bob with more or less strength produced elliptical orbits of an elongated or flattened shape, and showed that a circular motion was produced when the direct impulse and the inward pull of the wire were equal. Then he added a smaller pendulum to the long wire to create a satellite or moon for the larger weight, to show that the circle or ellipse was now traced by the centre of gravity of the two bodies.[6] Hooke was proud of his conical pendulum demonstration and used it often, although he knew it did not really represent all the forces involved in orbital motion: for instance, the pull of the wire did not diminish with distance, as gravitational attraction did.

Hooke's paper on 'The Inflection of a Direct Motion into a Curve by a Supervening Attractive Principle', read before the Royal Society on 23 May 1666, was the first correct account of the role of gravity in creating planetary motion. The paper was 'registered' by the Society but not widely read, and its full significance was not understood, perhaps not even by its author. Huygens and Newton were far more able mathematicians than Hooke and eventually explained what he could not, but for the next ten years or more they laboured along the false Cartesian track of centrifugal force and did not grasp the importance of Hooke's insight. Huygens, who invented the term 'centrifugal force' in 1659 and published it in 1673, still regarded it as the key to understanding planetary motion in the 1670s, and Newton had apparently not considered the possibility that planetary orbits were a compound of a tendency to direct motion and the centripetal force of the Sun until Hooke introduced him to the idea in a letter of 1679.[7] Hooke has often been criticized for his tiresome habit of greeting the inventions or discoveries of other scientists by saying 'I said it first', but in this case, at least, it was true.

By 1670, when he gave a Cutler Lecture at Gresham College on 'An Attempt to Prove the Motion of the Earth by Observations', Hooke had developed his ideas of 1666 into something more ambitious. His 'System of the World differing in many particulars from any yet known' was based on three suppositions. First, he proposed a theory of universal (or almost universal) gravitation. It was not only the Sun that exerted attraction over bodies in orbit around it, but

> all Cœlestial bodies whatsoever, have an attraction or gravitating power towards their own Centers, whereby they attract not only their own parts, and keep them from flying from them, as we may observe the Earth to do, but that they do also attract all the other Cœlestial Bodies that are within the sphere of their activity: and consequently that not only the Sun and the Moon have an influence upon the body and motion of the Earth, and the Earth upon them, but that Mercury also Venus, Mars, Jupiter and Saturn by their attractive powers, have a considerable influence upon its motion, as in the same manner the corresponding attractive power of the Earth hath a considerable influence upon every one of their motions also.

Hooke's second supposition was that all bodies 'that are put into a direct and simple motion' would continue to move in a straight line unless some other power deflected them into a circle or another curved path. His third was that these attractive powers were stronger near the centre of the attracting body, and diminished with distance from it. He was sure that there was a formula for calculating the relationship between gravitational attraction and distance, but he did not yet know what it was or how to discover it. 'Now, what these several degrees are, I have not yet experimentally verified'. Typically, his inclination was to discover the answer by experiment, rather than calculate it mathematically. Whoever discovered this vital principle would, he believed, have the key to understanding 'all the great Motions of the World', and achieving 'the true perfection of Astronomy'. He invited others less preoccupied than himself to undertake the great task. 'This I only hint at present to such as have ability and opportunity of prosecuting this Inquiry, and are not wanting of Industry for observing and calculating, wishing heartily such may be found, having myself many other things in hand which I would first compleat, and therefore cannot so well attend it.'[8] Hooke printed this lecture in 1674 and reprinted it in his collected Cutler Lectures in 1679, having made little further progress towards giving mathematical or experimental substance to his three suppositions. But when another scientist eventually took up the challenge, found the rule and used it to unlock the secrets of celestial motion Hooke was rather less delighted than he had promised to be.

By January 1666 the plague had subsided, and it was safe to return to London. As soon as he had arrived back in Gresham College Hooke was busy with plans for an improved repository and a new laboratory, to revitalize the Society after its eight months of inactivity. 'I am now making a collection of natural rarities, and hope, within a short time, to get as good as any have been yet made in the world', he wrote to Boyle on 3 February. 'I design, God willing, very speedily to make me an operatory, which I design to furnish with instruments and engines of all kinds, for making examinations of the nature of bodies, optical, chemical, mechanical, &c. and therein to proceed by such a method, as may, I hope, save me much labour, charge, and study'. A few weeks later, using a £100 gift from its Treasurer, Daniel Colwall, the Royal Society bought Robert Hubert's collection of natural curiosities, one of the best in England, and added it to the repository. Hooke wrote proudly to Boyle on 21 March: 'Our collection of rarities at *Gresham* college is now very well worth your perusal, and I hope to increase it every day'. Hubert's large accumulation of exotic animals and intriguing objects, which included an armadillo, a crocodile, a giant's thighbone and a stone in the shape of 'the secret parts of a woman', was stored in Hooke's Gresham College rooms for the next ten years. Partly because of the disruption caused by the Great Fire of 1666, and partly because Hooke could not find the time to organize the collection, it was not put on display in a separate gallery in the north wing of Gresham College until March 1676.[9]

In spring 1666, Hooke was occupied with his familiar clutter of Royal Society duties. In late February and early March he observed spots on the face of Mars and calculated from their movement that the planet rotated on its axis roughly every twenty-four hours, just as the Earth did. The figure accepted today is twenty-four hours thirty-seven minutes. With Wilkins, he perfected his new chariot and its springy saddle in preparation for a contest against Colonel Blount's rival vehicle on St George's Fields in Lambeth, on 26 May, three days after Hooke had delivered his important paper on planetary motion. In June, using Boyle's sixty-foot telescope, Hooke found a second spot on Jupiter to add to the one he had seen in 1664, and published his results in *Philosophical Transactions*.

He was fascinated by the time-keeping qualities of the circular or conical pendulum he had used to demonstrate planetary motion, and spent much of June and July developing a clock based upon it. Lord Brouncker, the President of the Royal Society, had the clock for four days and found it kept almost perfect time. By the end of August 1666 Hooke had a watch based on the circular pendulum which would, he told the Royal Society, keep accurate time on land and sea. This, the meeting was asked to believe, was the longitude timekeeper of his dreams.

At the same time he had been working on a small quadrant, in which a screw-controlled adjustment and a reflective device would give levels of accuracy previously seen only in much larger and more cumbersome instruments. Since improvements in the quadrant played a significant part in his career, it is important to know a little about the nature and purpose of the instrument. The quadrant was one of a family of instruments used by astronomers, navigators and surveyors for measuring angles between celestial bodies or other distant objects. Astronomy was a form of mathematics, and measuring angles accurately was of the utmost importance in establishing the geometry of the heavens. It was also one of the best ways for sailors in the open ocean to discover their latitude (by measuring the altitude of the Sun at noon, for instance) and eventually, given the right astronomical information, their longitude. A segment of a circle (quadrants were a quarter, sextants a sixth, octants an eighth) made in wood or brass was marked around its arc in degrees, and an arm pivoting from the centre of the circle could be moved in a sweep around the arc. This arm, known as the alidade, had sights at both ends so that the observer could point it accurately to any distant object. By placing his eye at the centre of the circle and pointing the fixed lower side of the quadrant or sextant at the horizon and the movable alidade at the Sun or Pole Star, then reading from the marked scale, the navigator could establish the angle of elevation of the Sun or star, and thus, with appropriate calculations, his own latitude. In astronomy, where greater accuracy was required and observations could be made in perfect conditions, the arc might be marked in finer subdivisions (minutes and seconds), and the quadrant could be of large proportions. The famous mural (wall-mounted) quadrant installed by Tycho Brahe in his Hven observatory in 1582 had a radius of more than six feet, enabling him to achieve wonderful accuracy without telescopic sights. These instruments could be used to measure the diameters of planets, map the stars, calculate precise celestial movements and so on.[10]

Obviously, instruments of such a size were not practical at sea, and Hooke introduced several improvements which would give sailors the accuracy they needed in a quadrant of convenient dimensions. In February 1665 he showed the Royal Society a quadrant of seventeen-inch radius with an arm or alidade that moved round the arc on a 'roller', which he claimed could be accurate to ten seconds, or 1⁄360 of a degree. At Durdans in August 1665, he had tried out a quadrant that was so accurate that he could 'tell the true distance between

Paul's and any other church or steeple in the city, that is here visible' to within twelve feet.[11] By March 1666 he had apparently replaced the roller with a micrometer screw adjustment, which not only enabled the observer to move the arm more delicately and precisely, but enabled him to calculate the exact position of the arm by the number of times the screw was turned and by calibrations on the large head of the screw. Hooke saw that the introduction of a thread onto the arc of the quadrant enormously increased its effective length and accuracy without increasing the size of the instrument, and this little quadrant, he said, could do the work of a much larger one, 'accurately dividing minutes into seconds'. This instrument probably had telescopic sights, but even with this advantage Hooke's claims were greatly exaggerated. Accuracy to one second of arc in a hand-held instrument was not achieved until the early 1920s, when the Swiss instrument designer Heinrich Wild produced the Wild Universal Theodolite.[12] In August and September 1666 Hooke introduced another important refinement, which enabled the observer to see both objects at the same time, instead of suffering the inconvenience and inaccuracy of making two observations. Hooke intended to present his 'new perspective for taking angles by reflection' to the Society on 12 September 1666, the Royal Society meeting disrupted by the Great Fire, and described it more fully in print eight years later. It seems to have involved the double-reflection principle adopted universally (and credited to others) in the mid-eighteenth century. A small mirror attached to the movable arm of the quadrant caught the image of the Sun or star and reflected it onto a second mirror which the observer viewed through a sight. The observer also saw beyond this second mirror to the horizon or other chosen object, and simply moved the arm, using Hooke's accurate micrometer screw, until the two images coincided. Perhaps because he failed to submit a written description at the time, the practical value of his reflecting quadrant was not recognized, and the instrument was forgotten. In 1699 Newton, innocently enough, described a very similar device to the Royal Society as his own invention. By this time Hooke was in poor health, but he came to the next meeting to assert his priority against his opponent's claim. Then it was forgotten again, despite its potential value to sailors in rough seas, until John Hadley demonstrated a similar instrument of his own to the Royal Society in May 1731.

The spring and summer of 1666 was a busy and productive period in Hooke's life. He had proposed an explanation for planetary orbits and all celestial motion which he hoped would make him one of the most honoured scientists of his or any age once its finer details had been worked out. Progress on a longitude watch based on the steady motion of the circular pendulum was encouraging, and it would surely not be long before its perfection brought the financial comfort and security that he desired. His other important work in the field of navigational and astronomical instrumentation, the reflecting quadrant with micrometer adjustment and telescopic sights, was nearing completion. His calculation of the rotation period of Mars, along with his discovery of the spots on Jupiter, had earned him a place alongside Hevelius, Huygens, Cassini and the other great astronomers of the day. In almost every investigation the Royal Society undertook in 1666, from tests on refrigerating salts to experiments on blood transfusions, he was a central or important figure.

In the mid-1660s Hooke's energy and inventiveness were more important than ever to the Royal Society's weekly meetings, but his versatility was not universally admired. His habit of keeping too many projects going at one time, though it was forced on him by his multiple obligations to the Royal Society, Robert Boyle, Sir John Cutler and Gresham College, frustrated even some of his admirers. Sir Robert Moray, one of the Society's leading lights, was irritated in November 1665 that Hooke would not provide the Society with observations which would settle a dispute between Auzout and Hevelius over the two comets of 1664–5, and wrote to Oldenburg about Hooke's 'excuses for his slackness'. 'I easily beleeve Hook was not idle, but I could wish hee had finisht the taskes lyet upon him, rather then to learn a dozen trades, though I do much approve of time spent that way too. But I will be glad to see a chariot such as wee would have'.[13] Like a donkey that is overloaded and then whipped for falling over, Hooke, the Royal Society's paid – and underpaid – servant, was given more tasks than one man could perform, and then blamed for leaving some of them unfinished.

In spite of these difficulties Hooke's progress seemed steady and predictable, in a career of intellectual adventure but not (it must be admitted) of great events. But at this point his life was given an unexpected twist by a careless baker's boy in Pudding Lane.

8. 'London Was, But It Is No More'

(1666–1667)

ROBERT HOOKE LIVED IN a medieval city. The City of London had been inhabited continuously since at least the late ninth century, when Alfred the Great captured it and refortified it against Danish invaders. Many of the City's streets and lanes, especially in the western half of the town and towards the riverside, were laid out by the late Saxon kings before 1066 and almost the entire street pattern was established and named by the thirteenth century. In a crowded city the roads were not allowed to take up more space than was necessary. It was not uncommon for medieval London lanes to be only five or six feet wide, hardly broad enough to roll a barrel. A walk around the little lanes next to St Mary-le-Bow, or between Cornhill and Lombard Street, or through the passageways around Cloth Fair, east of Smithfield, gives some sense of the dimensions of London's medieval lanes. Most of the houses that lined and overhung these narrow lanes were medieval or Tudor in style and origin, an exposed wooden framework filled in with laths and plaster. A few of these houses still survive in London, in a much reconstructed form: Staple Inn in Holborn, the gatehouse of St Bartholomew-the-Great in Smithfield, Prince Henry's Room in Fleet Street, or the plastered houses in Middle Temple Lane. Many more survive in York, a city which preserves the atmosphere and appearance (on a smaller scale) of the pre-Fire City of London. There were great stone buildings in the City, including the Guildhall, the Tower, Leadenhall, St Paul's Cathedral and over a hundred churches. Brick had been in fairly common use in important London buildings since the fifteenth century, but despite pressure from James I it was not used much for City houses, and certainly not for those on narrow lanes or alleys. Here the medieval timber and plaster style, with overhanging upper storeys almost touching above the roadway, was still the common rule, even in more recent houses.[1] We can see the Elizabethan and early Stuart City in all its wonderful density in London's earliest surviving maps and panoramic views. The sketch-books of Anthony van den Wyngaerde and the so-called Agas and Copperplate maps capture the crowded tiled rooftops of the City in the mid-sixteenth century, and a hundred years later the great Bohemian engraver Wenceslaus Hollar and the cartographers Richard Newcourt and William Faithorne show us a city which had changed little within its walls. Gabled wooden houses crowded together along narrow lanes, and the medieval skyline was broken by the towers and steeples of a hundred churches and by the vast bulk of the Norman-Gothic cathedral, spireless since the great storm of 1561.[2]

A city built in such a style, with few wide streets and hardly any open spaces, was bound to suffer serious fires from time to time. There were open flames in every house, for lighting, heating and cooking, and although the citizens feared and expected fires they regularly lit street bonfires during public celebrations. Most recently, the fall of the Protectorate and the return of King Charles had been welcomed with open fires on the City's tinderbox streets.

Pudding Lane, one of London's old Saxon streets, ran from Eastcheap down to Thames Street and the riverside wharves next to London Bridge. Its name had nothing to do with baking, but referred to the 'puddings' or entrails that Eastcheap butchers used to cart down the lane to their boats on the Thames. The fire that broke out in Thomas Faryner's bakehouse on Pudding Lane just after midnight on 2 September 1666 was not an unusual event, but circumstances combined to turn it into one. It started in a particularly crowded and combustible part of the town, where there were warehouses full of pitch, tar, oil, spirits, coal, timber and flax. The summer had been hot and dry, and a strong north-east wind fanned the flames and drove them down to the river and westwards along it. Perhaps the fact that the fire began early on a Sunday morning delayed Londoners' reactions to it and gave it time to gain a hold on Thames Street and along the riverside. By the end of its first day it had spread a quarter of a mile westwards through the wharves and alleys of the riverfront between the Bridge and Queenhithe, and it was starting to creep northwards through the lanes between Thames Street and Eastcheap. London's rudimentary water-pumping engines were too few and too feeble to stop the fire, and the great waterwheel on London Bridge, which pumped a supply into the centre of the City, was an early victim of the fire. The policy of pulling down or exploding houses to create a firebreak needed to be pursued with more organization and authority than the exhausted Lord Mayor, Thomas Bloodworth, was able to provide. On Monday 3 September the King's brother, the Duke of York, was trying to establish fire-fighting posts staffed by soldiers and parish officers. But now it was much harder to contain the fire because it had moved north from the river across Cannon Street and Eastcheap towards Lombard Street and Cornhill, in the very heart of the City. The fire engulfed Thomas Gresham's beautiful Royal Exchange, and reached the southern end of Broad Street, about 200 yards from Gresham College. Imagine the frenzy of activity in Hooke's rooms as he desperately tried to save his papers and instruments, his telescopes, microscopes, barometers, scales and quadrants, his air pump, weather clock and springy chariot as the fire advanced towards him. But Hooke and the Gresham professors were lucky. Although the worst day of the fire was yet to come it made very little progress along Broad Street or Bishopsgate after Monday afternoon. This was mainly thanks to a friendly wind, but there was also some determined fire-fighting in this part of the City. The massive stone walls of Leadenhall, the fifteenth-century market, resisted the fire more stoutly than almost any other building in the City, and a wealthy alderman whose name is now forgotten paid some citizens to stop the fire taking a hold of it. Leadenhall acted as a bulwark against the fire, and the north-eastern edge of the medieval city, from the Tower to Moorgate, was saved. It was not impossible for a determined group to stop the fire, at least where the wind was not too strong. East of London Bridge, where the fire spread more slowly, John Dolben, the energetic Dean of Westminster, led a party of scholars from Hooke's old school to save St Dunstan-in-the-East, and Samuel Pepys organized sailors to defend St Olave's Church and the Navy Office in Seething Lane.

On Tuesday the fire spread along the river to Blackfriars, then northwards through the little streets south of Ludgate Hill. At the same time it advanced from the south and east towards Cheapside, London's wide medieval market street. Once it was across Cheapside it spread north towards the Guildhall and Cripplegate, where it eventually stopped at the City wall. As the fire began to surround the Cathedral the printers and booksellers in the neighbouring streets abandoned their shops and piled their books and manuscripts in St Paul's Churchyard or in St Faith's Church, in the crypt of the Cathedral, where their safety seemed assured. But on Tuesday evening the dry roof timbers of the Cathedral were ignited, and flames spread up the wooden scaffolding which had been erected for renovation work. Lead from the roof ran through the streets like melted snow, and the roof collapsed into the Cathedral. The walls were too old and weak to withstand the enormous heat of the fire, and by Thursday the whole building was in ruins. Somehow the fire had spread into St Faith's, where a vast stock of books and papers caught fire and burned for a week. Among them were the unsold copies of *Micrographia* and the manuscript of poor Dr Wilkins' book on a universal language.

Neither the western City wall nor the River Fleet could stop the spread of the fire. On Tuesday it burst through Newgate and Ludgate and started to race through the tenements and alleys of Holborn and Whitefriars. Fleet Street and the Inner Temple were ablaze, and householders in Covent Garden, the Strand and even Whitehall started to load their goods onto carts. The threat to Westminster and Whitehall stirred the court into action, and fire-fighting parties were organized to hold the line from Smithfield and Fetter Lane to the Temple. A slackening wind on Tuesday night helped them to achieve this, and although there were fresh outbreaks on Wednesday it proved possible to contain them. The Tower, which was packed with explosives, was saved by the demolition of the houses that clustered around it. By Thursday morning the City was smouldering, not blazing. In places, small fires were rekindled weeks or months afterwards.

Those who tramped through the ruined City on Thursday or Friday saw a scene of almost utter destruction. Of the 350 acres within the walls, only the north-eastern quarter, about seventy-five acres, was saved. Another sixty-three acres of the western suburb, Holborn and Fleet Street, was also lost. Along with the Cathedral, eighty-seven parish churches and forty-four Livery Company halls were ruined, and the inhabitants of 13,200 lost houses were camped out in open space outside the City walls. As John Evelyn lamented on Monday night: 'London was, but it is no more.'

The Royal Society's meeting place, Gresham College, was in the unburnt north-eastern quarter, but the Society still had to find a new headquarters because the City corporation, burnt out of the Guildhall, took over its rooms. Luckily, Henry Howard, the future Duke of Norfolk, offered the Society the use of Arundel House on the Strand, between the Temple and Somerset House. For about five years Gresham College was transformed into a busy, noisy and smelly centre of administration and business, combining the functions of a city hall and a shopping centre. A hundred shops and stalls crowded into the quadrangle, which was lit with lanterns as the Royal Exchange used to be. For a while, until proper latrines were dug, a dunghill polluted the air of the College with its 'noysome noyse'. The Gresham professors were told that the Gresham committee, which had lost its main source of income with the destruction of the Royal Exchange, could only pay their salaries by renting out their rooms. Only Robert Hooke, with his useful City connections, avoided eviction. From January 1667 to November 1673, when the Society held its meetings in its old home again, he had to carry or cart his experimental equipment down Threadneedle Street and Cheapside, around the ruins of old St Paul's and along the full length of Fleet Street every Thursday afternoon.[3]

The rebuilding of the heart of London was an enormous task, and promised employment for carters, builders, labourers, carpenters, glaziers, brickmakers, surveyors and architects for years ahead. Among the many who seized this moment of opportunity was Robert Hooke. Immediately after the fire, the King and the City authorities agreed that the burnt area should be rebuilt to an entirely new plan, free of the inconveniences and manifest dangers of the old City. Within a few days of the fire three Fellows of the Royal Society had prepared and presented plans for the new city. Christopher Wren was first in the field on 11 September, and John Evelyn followed two days later. Both plans featured a series of piazzas linked by wide and straight main streets, with smaller side streets arranged in grids or radiating patterns, and both paid more attention to European models than to London's needs. Wren and Evelyn beat their competitors (to the chagrin of the Royal Society's secretary, Henry Oldenburg) by going straight to the King. Hooke, the loyal Society man, presented his plan to a Royal Society meeting on 19 September, though it had also been seen by City leaders. The plan is lost, but it seems to have been based on a grid pattern with 'all the churches, public buildings, market-places, and the like, in proper and convenient places'. The Society liked Hooke's plan, and so did the City's Lord Mayor and aldermen, who preferred it to the one prepared by the City Surveyor, Peter Mills.[4]

As it turned out, none of these hastily conceived plans was adopted. At least Hooke, Wren and Evelyn did not suffer the misfortune of Captain Valentine Knight, whose plan was so offensive that the King had him arrested. It soon became clear that the enormous cost of building a completely replanned City, and the urgent need to act quickly to restore its population and trade, made it essential to reconstruct the City on its old property lines. The King, acting more decisively now than he had in the first days of the fire, proclaimed the main outlines of the new city on 13 September. Building would be in brick or stone, narrow alleys would be kept to a minimum, a fine new Custom House and quayside would be constructed. The City authorities met the King's Council and agreed that a small commission of experts should be appointed to survey and map all City properties, establish a set of building rules that would create a safer, healthier, fairer and more convenient town, and oversee the whole process of rebuilding. The King's three nominees to this Rebuilding Commission were Christopher Wren, Hugh May and Roger Pratt. All three had substantial experience as architects, and all had been working on a survey of St Paul's before the fire. On 4 October the City chose two trusted and experienced employees, Peter Mills and Edward Jerman, and gave their third place to the scientist whose City plan had so impressed them, Robert Hooke. Hooke's close friendship with Wren, which would surely make cooperation between the Court and the City easier, might have influenced the City's choice.

This appointment transformed Hooke's life. His change of career was not as dramatic as Wren's, since he never gave up his scientific work and did not achieve immortality through his architecture. But the work he did took up a great deal of his time and made him a modest fortune. It also had a wider importance to London's history, since it was Hooke, as much as Wren, who was responsible for the almost miraculous regrowth of the City in the seven years after the fire. For most of the rest of the century Hooke was a familiar figure on the City streets, staking out roads, measuring building plots, writing certificates, settling disputes, hiring and supervising builders, watching over the progress of churches, wharves and public buildings, and collecting fees. By chance, hardly any of the buildings for which Hooke had sole responsibility still survive, and those which he worked on in partnership with Wren are, naturally enough, credited to the greater and more famous architect. Much of Hooke's work as a surveyor, though important, was humdrum in comparison to Wren's. Staking out new streets and writing foundation certificates hardly compares with designing St Paul's or St Mary-le-Bow.

Although Hooke had no experience in building or surveying he found the work as easy to master as everything else he had turned his hand to. Surveying and architecture were branches of mathematics, and Hooke was a professor of geometry. He understood the techniques of accurate measurement, and the astronomical and navigational instruments which he had played such an important part in improving, the sextant, quadrant and backstaff, were also the basic tools of the surveyor's craft. He was the inventor of various waywisers (devices for measuring distance), and in June 1671 he showed the Royal Society a new self-recording waywiser which would, when attached to a carriage, keep a paper record of all distances travelled and the angle and direction of all turns. He was also very fast to take up, refine and publicize the inventions of others, and to recognize their practical and scientific applications. He was an early user of that surveyors' standby, the spirit level or 'artificial horizon' (invented around 1661). Hooke and Wren produced variations and improvements on the original concept in the later 1660s, and in the 1670s Hooke added a spirit level to his quadrant. There is no evidence, though, that the methods Hooke used in his City surveying work were anything but traditional.[5] Furthermore, Hooke was methodical, energetic, fair-minded (except, perhaps, in some of his scientific priority disputes), honest, acutely intelligent and urgently in need of a second income. When his duties developed from surveying and dispute resolution to building and design, he was already equipped with the necessary skills. He was an accurate draughtsman, he had an almost unrivalled grasp of the tensions and forces that were involved in creating stable structures, he was the first man to explain the mathematics of arches, and he was an expert on the strength and elasticity of materials. In short, Hooke's qualifications for his new profession were the same as those of Christopher Wren.

A few of Hooke's Royal Society demonstrations reflected his new interest in building and surveying. As early as October 1666 he was testing different earths to see which would produce the most durable bricks, and the following April he described a method of making bricks at high speed. When his new brick engine was tested in May 1667 the Society decided that the method needed too much space for laying out the completed bricks, and he lost interest in the project. In November and December 1666 he produced a working version of Wren's 'artificial horizon', a new water-filled glass level of his own, and 'a new kind of back-staff for taking altitudes'. Later, in December 1670, he presented to the Royal Society's President, Lord Brouncker, a mechanical solution to the problem of designing arches to carry particular weights. He kept it secret from the rest of the Society, and at the end of his Cutler lecture on Helioscopes (published in 1676) he announced the solution to 'a Problem which no Architectonick Writer has ever yet attempted' in his usual anagrammatic code, in which all the letters in the Latin solution to the problem were arranged alphabetically: abcccdd, and so on. This was meant to establish authorship without revealing content. His solution, which is still regarded as correct, was that an arch should be shaped like an inverted version of a hanging chain. Hooke believed that Wren learned this law from him and used it in his church building, but it is equally possible that Wren – Hooke's equal in these matters – discovered it for himself.[6]

One of the first duties of the Commissioners was to devise a set of rules for leaseholders and freeholders to follow in rebuilding their houses. These rules, which were largely the work of Hooke and Mills, became part of the Rebuilding Act of February 1667. They established three sorts of houses, the 'first sort', with two storeys (plus cellar and garret) on lanes, the 'second sort', with an extra storey, on ordinary streets, and the 'third sort', with four storeys, on principal streets. Copying earlier rules which had never been rigorously enforced, the Commissioners insisted that all new houses should be of brick or stone, with no exterior woodwork, and that wall thicknesses, cellar and room heights and timber should all be of prescribed dimensions. Party walls, a source of conflict for centuries, were to be built equally on both properties and at fairly shared expense, and the inevitable disputes over light, access and drainage were to be settled by fair and efficient public arbitrators. The fast and cheap building which had helped create the sort of city that was destroyed in September 1666 was forbidden, though the new rules did not discourage speed. The legislation also created a Fire Court, in which some of the most distinguished judges of the day would provide fast, fair and authoritative settlements of the most difficult disputes arising from the rebuilding process.[7]

In March 1667 the Common Council of the City, assisted by Hooke and Peter Mills, reached agreement with the King on the extent to which each of the City's streets and lanes was to be widened, to make the City and its approaches safer from fire and more convenient for traffic. On 14 March Mills and Hooke were sworn in as City Surveyors, at a salary of £150 a year, backdated to September 1666. Their immediate task was to stake out the new wider streets, and their more extended one was to ensure that the houses built along them complied with the new safety and quality controls. John Oliver, a glass-painter, master mason and experienced surveyor, began helping Mills and Hooke in June 1667 and became the third surveyor six months later. Jerman died in November 1668 and Mills in 1670, so Hooke was the only City Surveyor to serve for the whole of the rebuilding period.[8]

Pepys gloomily predicted at the end of 1666: 'The City less and less likely to be built again, everybody settling elsewhere, and nobody encouraged to trade.' Householders, businessmen and the City authorities were desperate that the groundwork should be done quickly so that citizens could return and reopen their shops. If this did not happen then the City would lose business and population to Westminster and the other suburbs, or to a rival town. Staking out the streets began in late March 1667, and was mostly finished by the end of May. Mills and Hooke started with Fleet Street, the main approach road from Westminster, which was to be widened to about fifty feet. They took six carpenters to cut the stakes, and seven labourers to clear rubble from the land so that the street and plot boundaries could be identified, measured and staked. Pepys saw some of the streets staked out at the end of March, and his pessimism started to lift: 'if ever it be built in that form, with so fair streets, it will be a noble sight'. In all, and mostly by the end of 1671, the three surveyors measured and marked out 8,394 foundations, replacing the 13,200 lost in the Fire. Hooke's 3,000 foundations probably took him about an hour each, and at this rate he must have worked for an average of twenty hours a week, leaving aside his other duties as a City Surveyor, for the three busiest years.

When the staking of streets and measuring of plots was almost complete the surveyors' work entered its next phase. For rebuilding to take place quickly, the many disputes that arose between neighbours, over party walls, drainage, light and access, or between the City and housebuilders over compliance with the new building laws, had to be settled fast. Hooke, Mills and Oliver had to inspect or 'view' all disputed properties, and write reports for the City authorities or the Fire Court. From March 1668 to 1674, when the work diminished, Hooke was involved in hundreds (perhaps 500) of these views, and the job continued spasmodically until 1693, when Hooke, still working alongside his old colleague Oliver, did his last recorded view. The surveyors' reports, which were generally written by Hooke, show an admirable ability to get to the nub of intricate neighbourly squabbles, and to produce a crisp and judicious recommendation from a tangle of claims and counter-claims. Time and again, the work of Hooke, Mills and Oliver prevented disputes from turning into long-running conflicts, and kept the urgent process of rebuilding running fast and smooth.[9]

All this distracted Hooke from his work for Cutler and the Royal Society, but at least the City, despite its severe financial problems, paid well and on time. His salary as City Surveyor from 1666 to December 1673 amounted to £1,062, paid at the rate of £150 a year on every quarter day. After 1673 he was paid irregular gratuities by the City, amounting to around £150 over the next ten years. In addition he earned a total of around £1,600 between 1666 and 1674 in fees for views, and for the certificates he issued to tenants or property owners, which allowed them to begin rebuilding or to claim compensation for loss of land to road-widening or some other civic improvement. Sometimes these certificates were lost, and Hooke made another useful income by issuing replacements based on the records in his daily survey books, which he did not hand over to the City as he was meant to do. The survey books that he retained were later lost, with the result that most accounts of the rebuilding of London after the Fire emphasize the work of Mills and Oliver, whose books were kept in the City archives, and ignore that of Hooke.[10]

Hooke's work for the City also involved him in some major civic projects, including the rebuilding of some of the City gates, the canalization of the Fleet Ditch, the creation of a new Thames Quay and the erection of a monument to the Fire. Most of this took place after 1670, but some work on public buildings began in the 1660s. The rebuilding of the Royal Exchange, London's great trading centre, and the source of the income that financed Gresham College, was a matter of urgency, and in November 1666 the Joint Grand Gresham Committee (representing the Mercers' Company and the Common Council of the City) asked its three nominees on the Rebuilding Commission, Hooke, Mills and Jerman, to estimate for a new building. Hooke told the Committee that the Exchange could be rebuilt roughly to its old plan, reusing old materials where possible, for between £4,000 and £5,000. The Committee preferred Jerman's much more elaborate design, which ended up costing £58,000 and all but bankrupting the Mercers' Company. The reopening of the Exchange in 1671 and 1672 meant that the shops and stalls in Gresham College quadrangle were returned to their proper place, and College life could return to normal, though Royal Society meetings did not move back to Hooke's rooms until November 1673.[11]

One of the civic improvements included in the 1667 Rebuilding Act was the removal of markets from overcrowded streets to convenient off-street sites. In May 1668 Hooke and some members of the City Lands Committee, which had responsibility for public rebuilding works in the City, negotiated an extension of the land allocated to Billingsgate Dock to allow for storage, a market and street access. For the next five years Hooke, sometimes working with Mills or Oliver, was involved in measuring and preparing the ground for markets at Newgate (a live animal market south of Newgate Street), Woolchurch (the Stocks Market, where the Mansion House now stands), Honey Lane (a small meat market north of Cheapside) and Leadenhall, measuring, staking and preparing the ground, negotiating with local residents and the King's representative, Christopher Wren, and reporting on possible designs for the buildings.[12]

By the end of 1671 over 7,000 new houses had been built, and life in the City was beginning to return to normal. The re-emergence of London as a fine brick city within five years of its greatest disaster was one of the wonders of the age. A French visitor, Guy Miège, wrote: 'It is a matter of amazement to me to see how soon the English recovered themselves from so great desolation, and a loss not to be computed. At three years end near upon ten thousand houses were raised up again from their ashes, with great improvements . . . a full and glorious restauration of the City'.[13] The new houses were built to the best contemporary standards, roads were widened, gradients were reduced, inconvenient street markets were replaced by new market-places, and thousands of conflicts of interest were resolved. All this was done by the City itself, without undue intrusion into private property rights, without destroying the City's ancient street plan, and without financial help from the King or Parliament. In this process Hooke's work as a planner, surveyor, adviser and negotiator was of the utmost importance to the City authorities. The Royal Society, as its secretary Henry Oldenburg complained, was continually asked 'What have they done?' Here, from the Society's scholarly and perhaps comical Curator of Experiments, was a striking if unpublicized demonstration of the practical contribution that men of science could make to solving the problems of the wider world.

9. 'Noble Experiments'

(1666–1670)

Although Hooke was being drawn into the building world in the late 1660s, he continued to produce scientific work of great variety and outstanding quality. His research between the autumn of 1666 and the summer of 1668 covered almost the entire scientific spectrum, from anatomy to earthquakes. These were years of rapid advance in the understanding of the purpose of respiration and the composition of air, subjects in which he had been interested for years. Despite some misgivings, he was still involved in experiments on the blood and lungs of living animals, which were directed at the unknown relationship between respiration, heartbeat and the circulation of the blood. He was not alone in this field. The most successful anatomist and vivisectionist of the day was Dr Richard Lower, who had been a contemporary at Westminster and Christ Church. Lower had performed the first dog-to-dog blood transfusion in Oxford in February 1666, and the experiment was repeated by a Royal Society team led by Dr Edmund King on 14 November 1666. Hooke watched the experiment and told Pepys that he had great hopes of the benefits it would bring to mankind. The trial was successful (for the scientists and one of the dogs), and the team moved on to cross-species transfusions, starting with a dog and a sheep. Success in this led King and Lower to attempt a human transfusion, like the one which had recently been carried out in France. Hooke and the two doctors did not manage to persuade the physician at Bethlem, London's lunatic asylum, to supply a subject, but they found an impoverished and mentally unsound divinity graduate, Arthur Coga, who was happy to be given the blood of a lamb, with its reassuring New Testament connotations, for a fee of £1. He survived two transfusions, one on 23 November and another on 12 December 1667, but a man died during a transfusion in France two months later, and the experiments stopped.

Hooke left most of this work to Lower and King. He was more interested in respiration, and in finding out what part it played in sustaining life. Since Harvey had explained the circulation of the blood in 1628 doctors and natural philosophers had been puzzled by the relationship between circulation and respiration. Was it the supply of air or the pumping motion of the lungs that kept the heart beating? Did air mix with the blood, and if so, where, how and to what purpose? And was the function of respiration to cool the blood and the heart, or (remembering the part air played in producing fire) to warm it up? Hooke may not have been a great anatomist or an enthusiastic one, but he had an unrivalled eye for the crucial experiment, and knew better than anyone how to relate laboratory work to key scientific issues. After his horrible experiment with a dog in November 1664 and his work on the ingredients of gunpowder in January and February 1665 Hooke believed that there was a 'nitrous quality' in the air which sustained life, and 'which being spent or entangled, the air becomes unfit'. He was unwilling to repeat his canine experiment for the entertainment of fellows who had forgotten it, and he was keen that Dr Lower, whose dexterity would make vivisection a less painful experience, should be taken on as a second curator, in charge of anatomy.[1] But in August 1667 Hooke's nitre theory was challenged in a book by the anatomist Dr Walter Needham (another Westminster scholar), who argued that the function of respiration was mechanical, not chemical. The motion of the lungs stirred or agitated the blood, he said, but there was no evidence that air entered the bloodstream. Spurred by this challenge, Hooke and Lower joined forces, and on 10 October 1667, in front of the Royal Society, they repeated the November 1664 experiment. This time Hooke joined a second set of bellows to the first 'by a contrivance I had prepared', and pricked small punctures in the dog's pleural membrane with 'the slender point of a very sharp pen knife', so that the creature's lungs could be kept permanently inflated. 'This being continued for a pretty while, the Dog lay still as before, his eyes being all the time very quick, & his heart beating very regularly'. This clearly demonstrated that it was air itself that kept the dog alive, and that 'the motion of the lungs did not contribute to the circulation of the blood'. Furthermore, by cutting off a piece of lung, Hooke showed that blood circulated through the lungs, enabling it to mix with the air. This unpleasant but highly effective experiment established for the first time that the body needed air (or some component of air) rather than lung motion to survive, and that some sort of interaction between blood and air took place in the lungs.[2]

Pursuing the same idea, Hooke tried several times in November 1667 to arrange the circulation of a dog's blood so that it went from the right side of the heart to the left without going through the lungs, but he found the operation too difficult. When Needham insisted in April 1668 that the motion of the bellows rather than the air had kept the dog alive, Hooke and Dr King responded in May with a privately performed experiment in which a dog's lungs were repeatedly filled with stale air from a bladder and bellows connected by a brass tube to its windpipe. After it had been forced to rebreathe its own exhalations for about eight minutes, the dog's struggles and convulsions subsided, and it seemed near death. King and Hooke pumped fresh air into its lungs and sewed up its throat. 'Then we untied his mouth, and he presently fell to licking himself, as not much concerned: but we concluded, that if he had stayed but one minute more, before we let in fresh air, in all probability the dog's life would have been quite lost.' At the Society's insistence, the two repeated the experiment later in May, this time continuing until the dog was dead.[3] At about the same time Hooke devised an experiment to discover whether blood changed colour, from dark venous blood to bright red arterial blood, when it mixed with air in the lungs, rather than when (as some argued) it was 'heated' or 'fermented' in the heart. The Royal Society urged him to perform the experiment, which involved examining the blood as it passed from the lungs to the left auricle of the heart, but Dr Lower no longer attended Society meetings, and Hooke did not care to do the job without his help. Instead, Lower did the experiment himself, and in late 1669 brought out his important tract *De Corde* (*On the Heart*), which used this and other experiments, especially Hooke's of 10 October 1667, to show that the blood's colour was changed by air in the lungs not heat in the heart, and to explain and describe the muscular structure of the heart. This breakthrough was achieved by a combination of Lower's anatomical skill and Hooke's experimental brilliance, and Lower fully acknowledged his debt to his school and university friend, 'the very famous Master Robert Hooke', in *De Corde*.[4]

The Hooke–Lower vivisection of 10 October 1667 inspired another important tract, this time from John Mayow. Largely on the strength of Hooke's experiment, Mayow argued that the active element in the air, aerial nitre, carried in the blood, was essential in causing the muscles to contract, and thus in causing the heart to pump. Hooke, as far as we know, was quite satisfied with the acknowledgement and never complained that Lower or Mayow had cheated him of due recognition. Once again, just as in his collaboration with Wren, we find a more cooperative and sharing side to his character. No doubt he did enough in his career to earn his reputation as a competitive and jealous rival, but like most reputations it does not do justice to the whole man. And we should not forget that the work he had done, often rather unwillingly, between 1664 and 1668 had provided the experimental basis for some of the most important advances in the explanation of respiration and circulation, and opened the way to a new understanding of the lungs and heart.

After May 1668 Hooke abandoned vivisection and returned to experiments on animals in confined spaces. In July 1668 he established that the length of time birds could survive in airtight conditions was directly proportional to the volume of air in the vessel that held them. His next contribution to the problem involved placing a human subject – himself – in an airtight container. In June 1667 the Society had agreed to pay £5 to make an air pump connected to an airtight wooden box large enough to hold a man. Within a few weeks Hooke had managed to make the box airtight by filling in the pores in the wood with some paste or 'cement', and was waiting for cooler weather to test the effects of reduced air pressure on temperature and respiration. 'There will be, I doubt not, very many noble experiments discovered by it, of which I have conceived a good number', he wrote to Boyle. As usual, his declaration of success was premature, and the planned experiments were delayed for several years, no doubt because the box was not completely airtight. In January 1671 Hooke announced that he had a design for a new airbox, and by 9 February the machine was ready. This time there were two large casks, one to hold the man, and the other, surrounding it, to encase the inner cask in water. The inner cask had an air-pressure gauge and a cock to allow its occupant (for this was a man, not a mastiff) to let air in as he needed it, and was connected to a system of bellows and valves to extract the air from it. To avoid the risk of damaging the airbox on a bumpy journey to Arundel House the experiments were done before witnesses in Hooke's rooms in Gresham College. In the first test, in mid-February, Hooke crouched in a slightly evacuated cask for fifteen minutes without discomfort, and in the second, about a week later, the loss of a tenth of the air in the cask caused a little pain as his ears 'popped'. The final test came on 13 March 1671, when he stayed in the cask for over fifteen minutes while a quarter of its air was removed, suffering nothing more than painful ears and a brief loss of hearing. A candle that he had taken into the cask with him (the first time an animal and a flame had been tested together under reduced pressure) went out before his ears started to hurt. At this point he gave up such experiments for good, either because he thought he could discover nothing else from them (where were all the 'noble experiments' he had promised Boyle?) or because he did not like crouching in a cask with earache.[5]

Hooke's other scientific activities in the late 1660s took him into more comfortable and familiar territory. One of his most consistent interests was in timekeeping. He managed to maintain interest in his longitude timekeeper by developing new circular and inclining pendulums, which would keep time as accurately as longer perpendicular ones. He explained the principles behind his circular pendulum to the Society on 21 November 1666, using arguments probably drawn from Galileo's *Discourses on Two New Sciences*, which was published in Latin in 1638 and in an English translation in 1666. Two weeks later he produced an adjustable pendulum which could easily be set to beat a required time, and on 2 January 1667 he showed the Society a new clockwork, based on an adjustable circular pendulum, through which 'the clock-motion should be reduced to an exactness, which it had not had before'. Hooke's device for ensuring the regularity of the pendulum's swing was to confine it within the curved surface of a paraboloid bowl, an idea picked up and published (to Hooke's irritation) by Huygens in 1673. A little later, on 21 March 1667, Hooke produced a design for an instrument which made it possible to draw accurate dials for clocks, sundials or scientific instruments, using the motion of the Sun. This device, perhaps the earliest attempt to mechanize this complex and important task, incorporated a simple version of Hooke's universal joint.[6] At the same time, Hooke was working on improvements to the interlocking cogs and wheels that transmitted the movement of the pendulum to the hands of the clock. In September 1667, during the Society's summer recess, Hooke wrote optimistically to Boyle on his advances in clockwork:

> I have lately contrived a new way of wheel-work for clocks, watches, &c., which I think does much excel all the ways yet known; and indeed I think it the very perfection of wheel work, and capable of the highest perfection, that can be expected in that kind. there has been nothing like it yet practised. Many other things I long to be at, but I do extremely want time.[7]

What Hooke meant by this, he explained in a Cutler lecture published in 1674, was that he could produce a wheel with, say, a thousand teeth by screwing ten identical circular plates together, cutting a hundred teeth in this composite wheel, then unscrewing them and fixing them to their spindle or arbor in a staggered formation, 'that the Teeth may gradually follow each other'.[8] On 31 October 1667 he showed his 'perfect wheel-work' to the Society, claiming that it was 'so made as equally to communicate the strength of the first wheel to the last, the teeth of it being always taking, so that before one tooth had done taking, it was passed a good way into another'. About six years later Hooke made a wheel-cutting machine which would produce more accurate clock wheels than the usual hand-cut ones, reducing friction and irregularity even more. His diary for 18 March 1673 mentions that he 'began wheel cutting engine' that day, and he told the watchmaker Thomas Tompion about an 'engine for finishing wheels' on 2 May 1674. By 1675, according to John Smith's *Horological Dialogues* of that year, the tooth-cutting machine was well established: 'there is no man that can cut them down by hand so true and equal as an ingine doth'. The importance of Hooke's contribution to the success of this useful machine is uncertain.[9]

Early in 1668 Hooke showed the Royal Society a new device to keep a pendulum swinging regularly in unsteady conditions, and on 20 February a Florentine scientist, Lorenzo Magalotti, saw a demonstration of a watch incorporating this improvement. Luckily he kept a note of his visit, because the demonstration is missing from the Royal Society minutes.

> We also saw a pocket watch with a new pendulum invention. You might call it with a bridle, the time being regulated by a little spring of tempered wire which at one end is attached to the balance-wheel, and at the other to the body of the watch. This works in such a way that if the movements of the balance-wheel are unequal, and if some irregularity of the toothed movement tend to increase the inequality, the wire keeps it in check, obliging it always to make the same journey. They say that if you keep it hung up, the invention works well and that it corrects the errors of the movements as well as a pendulum, but that if you carry it in your pocket the temper of the wire changes in accordance with the temperature of the body, and getting softer, allows the balance wheel to turn with more freedom.[10]

It seems that Hooke was making progress on a longitude timekeeper that would correct for the movement of a ship, but not for temperature variations. Some years later, when he was in bitter dispute with Huygens and the Royal Society's secretary, Henry Oldenburg, over who had first invented a spring-driven 'longitude' watch, the absence of this demonstration from the Society's minutes (Oldenburg's responsibility) took on a rather sinister appearance. It is possible, though, that Magalotti was shown the watch outside the formal meeting, or that Oldenburg, who had spent two months locked in the Tower of London as a suspected enemy agent in the summer of 1667, had lost some of his customary efficiency.

Hooke was not sure that his improved spring-regulated watch would keep accurate time at sea, and in April 1669 he produced a watch controlled by a magnetic balance, perhaps a refinement of the one he had made three years earlier. This, he told the Society on 20 May, 'should move in all positions, with any kind of motion, without stopping, or being disturbed.' But when he showed it to the Society a week later it did not keep good time, and he took it away to make further improvements. It was not until November 1670 that he announced a new type of longitude watch to the Royal Society, again claiming that it 'would go equally in all positions and motions at sea'. Once more the performance of the new watch failed to impress, and he went off to work on it again. This is the last we hear of his longitude watch for several years, though he also developed various accurate and long-running pendulum clocks for astronomical use.

In the meantime Hooke's main rival in the race to find a longitude timekeeper, Christiaan Huygens, was making great strides towards their mutual goal. His pendulum clocks were given three sea trials in 1668–71, each trial leading to improvements in the pendulum design and in ways of suspending the clock to resist the motion of the ship. And in the early 1670s it began to dawn on Huygens, whose grasp of the mathematics of pendulums far outstripped Hooke's, that a balance spring might beat time as accurately as a pendulum. Hooke heard about these sea trials from Huygens's letters to Oldenburg, and it must have been clear to those who heard the letters read out, and saw Hooke's repeated timekeeper experiments, that the race was going to be close.[11]

Hooke was so active in so many fields that it must have been easy to overlook one of his many demonstrations. In February 1668, when Magalotti saw his watch, Hooke was also working on a cider-making engine and a wind-measuring machine, as well as continuing his work on the lungs and circulation of dogs, and trying to perfect his dividing engine for marking precise scales on measuring rods or quadrants. He had also begun a new project, aimed at discovering more about the nature of matter. On 13 February he reported on an experiment to discover whether gold and mercury, mixed together, might produce a substance heavier than gold alone. Would the mercury 'penetrate into the pores of the gold' and thus increase its density? Since solids as well as liquids and gases were composed, in Hooke's corpuscular theory of matter, of particles vibrating in space (or in aether), such interpenetration was feasible, so long as the two substances were 'congruous', or had a tendency to unite with one another. Congruity was an idea that Hooke had introduced in his first work on capillary action and developed in *Micrographia*, where he explained it as a harmony or sympathy between the vibrating motions of different bodies. On 20 February he tried mixing copper and tin, and in May he weighed a mixture of water and sulphuric acid, and another of nitric acid and mercury. By weighing the liquids separately and in a mixture in a hydrostatic balance (in a glass ball suspended in water) Hooke claimed that he had shown that water and sulphuric acid had lost volume and gained density when mixed, but that the reverse had happened with water and nitric acid, which expanded and became more rarefied on mixing. Such experiments, he said, might 'afford an excellent clue to lead one further into the recesses of nature, and to inform us of the internal texture and component parts of bodies'.[12] In short, they could reveal an inner world which even the best microscope could not show us. This was an important idea, and one which he returned to in the late 1670s. For the time being, though, he let it drop.

Hooke recognized that precise measurement of time, weight, air pressure, angles and distance was a key to advancement in the new science, and he devoted much of his energy to developing the accurate instruments he needed – spirit levels, clocks, pendulums, hydrostatic balances, dividing engines, telescopic sights, quadrants and barometers. In the late 1660s he had a particular interest in the measurement of planets. In January 1667 Oldenburg read a letter from Adrien Auzout, a French scientist who had already crossed swords with Hooke, announcing that he had found a way to measure the diameters of planets with great accuracy. Hooke and Wren, who had both used and improved upon the telescope eyepiece micrometer first invented by William Gascoigne around 1640, rightly replied that this was nothing new to the English. Typically, Hooke promised on the spot that he could use an eyepiece micrometer to prove the annual motion of the Earth, something that had never been done, and thus settle the dispute between the followers of Copernicus (who believed the Earth moved) and Tycho Brahe (who did not).[13] His idea was that the Earth's movement could be proved by the familiar phenomenon known as parallax – the apparent movement of a stationary object in relation to another when the position of the observer changes. Look at your finger, blinking first one eye then the other, for a simple demonstration of the effect. If the Earth moved, long and accurate observation of a fixed star would show tiny changes in its position, resulting from the Earth's orbit. He was right, but the changes are in fact so small that even with modern instruments it is hard to see them.

To measure changes of perhaps twenty seconds (1/180 of a degree), which would be invisible to a naked-eye astronomer, telescopes with micrometer eyepieces of the greatest possible accuracy would be needed. Richard Towneley, the Royal Society's resourceful Lancastrian correspondent, sent a description of his version of Gascoigne's micrometer to the Society. Hooke made a working version of the device and, typically, an instrument of his own which he claimed was even better. When he showed Towneley's micrometer to the Society on 25 July 1667 he compared it with his own version, 'consisting of two threads and a ruler, whereby an inch is diagonally divided into five thousand parts, and might be with the same ease divided into forty thousand or more at pleasure'. With this device, which was 'of more plain and easy use' than Towneley's, the exact position of any star could be found. The two micrometers were compared several times over the next year, with Hooke adding refinements on almost every occasion. By 1670 he had found, he said, at least thirty ways of making a micrometer, 'some by the help of screws, others by the help of wedges, some after the way of proportional Compasses, others by wheels, others by the way of the Leaver, others by the way of Pullies'.[14]

Through the letters of Henry Oldenburg, news of Hooke's achievements was transmitted to the Royal Society's overseas members. In June 1668 one of Europe's greatest astronomers, Johannes Hevelius of Danzig, asked Oldenburg to send him the best telescope he could find, at the lowest possible price. The whole business, Hevelius told Oldenburg, should be entrusted to Hooke, 'who is so very experienced in that art'. Oldenburg should 'ask him not only to buy some excellent lenses for me, but to arrange for the preparation of the complete tube, properly and correctly put together from planks, with all the necessary apparatus, just as he employs it himself in observing the heavens'. Hevelius's Danzig observatory, the finest in Europe, was equipped with many large sextants and quadrants, as well as telescopes of up to 150 feet long. But he did not use telescopic sights, relying instead on the size and accurate calibration of his instruments and on his excellent eyesight. Using these, he had studied sunspots, discovered comets, produced the first detailed map of the Moon, and established his reputation as the best practical astronomer of his time.

Hooke's approach was completely different. He had argued in *Micrographia* that scientific progress would depend on replacing or reinforcing weak human senses with mechanical or optical devices.

The accuracy of traditional quadrants could never be greater than that of the sharpest human eye, which was incapable, in his view, of distinguishing angular differences of under a minute (a sixtieth of a degree). To get beyond the naked-eye observations of Tycho Brahe, astronomers would have to use telescopic sights with micrometer eyepieces. Wren and Seth Ward, the Oxford astronomers who taught Hooke his craft, had used these in the 1650s, and Hooke may have written a lost discourse on the technique in 1661. Much of Wren and Hooke's astronomical work in the 1660s, including their observation of the path of a comet in 1665, was done with telescopic sights, or by using two telescopes hinged together at their tops, so that the angle between two objects viewed through the two telescopes could be precisely measured, as in a traditional quadrant.

Hooke was younger and less experienced than Hevelius, but he was keen to tell him about his advances in the design of astronomical instruments. Before Hevelius had sent his letter of June 1668 he received a paper from Hooke (transmitted through Oldenburg) giving him advice on his way of fitting six-foot telescopes to quadrants and sextants, in place of ordinary naked-eye sights. Hooke assured Hevelius that with telescopic sights 'an instrument of one span [about nine inches] radius can be made much more accurate than another of sixty-foot radius, however good, having common sights'. He told Hevelius of the advantages of using crossed threads to mark the centre of the lens, and carefully explained how the two telescopes should be set to the angle of a quadrant or sextant. Hevelius was not convinced of the true benefits of Hooke's proposal. Small instruments would be much more convenient, he conceded, but in practice the telescopes were bound to shift, making nonsense of Hooke's fine calibrations. 'For many things seem most certain in theory, which in practice often fall far enough from the truth.' Until Hooke could provide him with actual measurements of stellar distances (between Aldebaran and Pollux, perhaps, or Altair and Marcab), Hevelius would reserve judgement on telescopic sights. In December 1668, in a letter answering Hooke's questions on his observations of the shape of comets and thanking him for his new telescope, Hevelius renewed his challenge. Exaggerating as wildly as Hooke, he claimed that with his large naked-eye instruments he could measure angles 'to the fifth or tenth part of a minute, or sometimes to single seconds of an arc, as many thousands of observations made by myself clearly show'. The tone of these letters seems friendly enough, but this disagreement over telescopic sights led to a bitter and damaging dispute between the two men five years later.[15]

Ignoring Hevelius's challenge, Hooke planned to set up a large vertical telescope in Gresham College and make precise observations of Gamma Draconis, the bright star at the head of the constellation Draco, the Dragon, which passed directly above London. By observing a star that was directly overhead he avoided the problem of atmospheric refraction, which would have thrown out his fine calculations, and also simplified the task of erecting his telescope. Since he wanted to place his perpendicular or 'zenith' telescope inside his Gresham College rooms and leave it there for months, he decided that it would be more convenient to use a tubeless telescope. He made a foot-square hole in the roof of his lodgings and fixed a ten-foot tube into it, so that it protruded a little into his first-floor room. When it rained the top of the tube could be closed with a string-operated lid. Near the top of this tube he fitted a brass frame holding an object lens with a thirty-six-foot focal length. Two long plumb lines hung from the brass frame through a foot-square hole in the intermediate floor down to the ground floor, where Hooke positioned his eyepiece, using the plumb lines as a guide. His micrometer, a brass circle divided by four fine hairs, gave the telescope the accuracy he needed to measure minute variations in Draco's apparent position as the Earth moved around its orbit. Every time he made an observation, Hooke had the trouble of adjusting the telescope again, to take account of movements in the roof or warping in the tube. He looked back with longing to the deep dry wells on Banstead Downs, or wished that he lived near 'some great and solid Tower' in which a longer telescope could be left permanently erected, and where constant corrections would not be necessary.

Hooke's observations of Draco's brightest star began on 6 July 1669, and his fourth and last was on 21 October. On this occasion he saw Gamma Draconis in bright sunlight, the first time a star was seen in this way, and noticed that it looked smaller and more clearly defined than at night, 'the spurious rays that do beard it in the night being cleerly shaved away'. Then the weather began to make further observations difficult, and his health, which was always poor, became so bad that he was unable to continue. Worst of all, the object lens he had been using, the only one that suited his indoor telescope, was accidentally broken. So his announcement in a Cutler Lecture in 1670 that there was a difference of about twenty-two seconds between the July and October positions of Gamma, enough to confirm the movement of the Earth and the Copernican system, was uncharacteristically tentative. The title of the 1674 published version of the lecture also suggests a degree of uncertainty: *An Attempt to Prove the Motion of the Earth from Observations.* Later the first Astronomer Royal, Flamsteed, found similar variations in the Pole Star, but when Samuel Molyneux and James Bradley (a future Astronomer Royal) repeated Hooke's observations of Draco's star from a house on Kew Green in the winter of 1725–6 they found puzzlingly different movements, which Bradley eventually explained in 1727 as being the result of a phenomenon he called the aberration of light. Simply, this was an effect produced by the velocity of the Earth, which alters the apparent speed and direction of the light from an observed star (and thus the apparent position of the star), much as cycling through rain alters the apparent direction of rainfall. Bradley discovered a second contributory factor which had already occurred to Hooke: slight oscillation in the Earth's axis (known as nutation) affected the fall of the plumb lines, and thus distorted Hooke's delicate measurements. So Hooke *had* found evidence for the orbital motion of the Earth, but not the evidence he thought he had found.[16]

Far away in Derby young John Flamsteed, the future Astronomer Royal, was keen to imitate Hooke's techniques, to grind his own lenses, to make giant telescopes, and (most of all) to see stars in daylight. Between 1669 and 1672 he often wrote to Hooke, usually via Oldenburg, for advice and information, and we can see his admiration turning into irritation, and the seeds of their later enmity being sown, as Hooke repeatedly failed to tell him his secrets. Hooke's lens-grinding engine (described in *Micrographia*) was a wonderful thing, Flamsteed wrote to Lord Brouncker in November 1669, but what sands or powders did he use? 'He affirms to know several secrets for the meliorating and improving of optics, of which yet we have had no treatise . . . Why burns this lamp in secret?' In February 1671, Flamsteed wanted to know how Hooke had seen stars by day, so that he could copy his method:

> If Mr Hooke esteeme not his invention as a secret to be concealed, I would begg that he would bee pleased to gratifie me with a communication of his method of observing, and he shall find me alwaies readie to recompense . . . if hee can performe well in the day, his way will be of singular use and great encouragement to such Astronomers as love the science, yet like not the practise; because it requires them to break theire reste in makeing nocturnall observations: . . . pray therefore if Mr Hooke be not nice of his method desire him to impart it to me; if hee be, pray urge him not much; 'aut ab alio discam, aut inveniam' ['either I shall learn it from someone else or invent it'].[17]

The central idea of the mechanical philosophy, and the principle at the core of most of Hooke's work, whether on planets, clocks, combustion or the circulation of the blood, was matter in motion. Galileo had begun the study of the laws of motion, Descartes had proposed seven laws of motion which were almost entirely wrong, Huygens and the young German mathematician Leibniz were working on them and so, in his Cambridge isolation, was Newton. But the laws still needed experimental investigation, and Hooke, intermittently, tried to provide it. In October 1666 the Royal Society had been shown some simple experiments on impact and the propagation of motion involving two swinging wooden balls colliding with each other, but its Council was not sure whether its meetings should go over territory already covered by Huygens. When Hooke suggested a proper experimental investigation of the laws of motion in October 1668 Lord Brouncker, the Society's president, told him that Wren and Huygens had already arrived at a theory that would explain all the phenomena of motion. Hooke disregarded Brouncker's doubts, and at the next meeting, on 29 October, he proposed a series of experiments on springiness, and the tendency of hard bodies to rebound when dropped. Was this because they consisted of springy particles, or because they contained air? At the following meeting he repeated the October 1666 experiment with three swinging balls, a device we now call Newton's cradle. When one of the side balls was lifted and dropped against the other two the farther ball swung up to nearly the same height, and the middle one remained almost still. Their appetites for an investigation of the fundamentals of the mechanical philosophy whetted at last by this early prototype of an executive toy, the fellows pressed Hooke for more experiments to test the hypothesis 'that no motion dies, nor is any motion produced anew'. He responded with a series of experiments on the collision of springy bodies and the reflection of motion. In the first meetings of 1669 Hooke showed two experiments to demonstrate that the 'force' in a moving body was proportional to the square of its velocity. This principle was not original to Hooke but it was important to his later thinking, and he devised persuasive demonstrations of its truth. On 14 January he produced two experiments to show that to double the speed of a moving body the force causing the movement had to be quadrupled. One of these was a pendulum in which an eight-ounce weight produced twice as many vibrations in a set period as a two-ounce weight, and the other was a water vessel with a stopcock at its base. When the head of water in the vessel was quadrupled the speed of a measured flow from the stopcock doubled. Then the vessel sprang a leak and the experiment was aborted.

Testing these hypotheses often demanded great precision in measurement or observation, and Hooke usually failed to achieve anything like the levels of accuracy that he claimed were possible. When he tried in January 1671 to test the assertion that a falling and a projected ball would hit the ground simultaneously his witnesses, the Fellows of the Royal Society, could not decide after several attempts whether the two balls had landed together or not. Another interesting idea occurred to him on 9 March 1671, when he demonstrated for the benefit of two visiting Italian noblemen the effect of striking or vibrating a broad shallow glass dish containing some flour. When the vibration caused the flour to rise up the glass and run 'like a fluid' over its side, Hooke mentioned 'that it might . . . suggest an hypothesis for explaining the motion of gravity'. Two weeks later he was still absorbed by the possibilities offered by the experiment, 'upon which he thought considerable things in philosophy depended'. On 30 March he demonstrated that different strokes on a glass bell-jar made different sounds, and that each sound produced its own pattern of movement in the flour it held. This phenomenon, he told the Society, 'might much contribute to the explication of the nature of the internal motion in bodies'. Hooke, as we shall see later, had a strong interest in vibrations of all kinds, and wondered whether they might be the key that would unlock the secrets of gravity, springiness, the congruity of matter, and even the mystery of the human memory. His demonstrations with flour or sand on vibrating plates were repeated and developed in 1787 by E. F. F. Chladni, to show the patterns caused by different musical vibrations.

As usual, Hooke shared his time between these philosophical interests and more practical work in technology and scientific instrumentation. So we find him in March 1667 telling the Society about a new sort of hemispherical lamp in which the pressure of a second pivoting hemisphere would keep the wick constantly supplied with the right amount of oil. The device was described at length in a published Cutler lecture, *Lampas*, in 1677. In the same month, he described a method of making a short telescope containing two mirrors, which would have the power of a conventional telescope of three times its size. When he produced the new telescope on 16 May 1667 the poor quality of the mirrors spoiled its performance, and he was ordered to remedy its defects. Undaunted, he returned three weeks later with the suggestion that the loss of light could be turned to advantage, and would make the instrument ideal for observing the Sun. His new device, with four black glass mirrors to absorb all but a tiny fraction of the light, was revealed to the reading public in *A Description of Helioscopes and some other Instruments* that year.

Almost every month there was something new to amuse and impress Hooke's colleagues and employers in the Royal Society. This can be illustrated by a selection of his work in the first half of 1668, when his duties as a City Surveyor were very heavy. On 2 January he described a way of predicting the weather at sea with a new marine barometer, and on the 16th, as well as showing two new barometers, he produced one of several models of his cider-making and apple-cutting engine. The distinctive features of these barometers were not described in Oldenburg's minutes, but only in Hooke's handwritten note which was kept in the Society's archives. The first was a wheel barometer with a small stopcock at the bend in the pipe, in order to stop the mercury from rising and falling with the motion of the ship, 'the key of the stopcock [being] just so farr turn'd as to leave the least imaginable passage for the mercury between the two stemms'. This effective device was ignored and forgotten, and invented again in 1770 by Edward Nairne. The second barometer was intended to enable mariners to calculate the true atmospheric pressure by removing the inaccuracy caused by changes in temperature. It combined a sealed spirit thermometer based on the freezing point of water (Hooke's own invention) with an ordinary mercury barometer, and a set of calibrations which he had arrived at by standing them both in water of various temperatures on a day when the mercury stood at a normal 29.5 inches. This simple but clever instrument, the first to give fairly accurate readings in all temperatures, was forgotten until Hooke recalled it in a lecture in December 1690.[18]

In February 1668 Hooke began a series of experiments on the weight and interpenetration of alloys, devised a new wind detector, and gave the demonstration of his new pendulum watch described by Magalotti. In March he showed how his wind detector (a complex structure of cones, fans, valves and partitions) could be adapted to serve as an ear trumpet or 'otacousticon', and proposed a way of making glasses for seeing in the dark. The barometers and the wind gauge were components of the comprehensive 'weather clock' which Hooke was supposed (as the Royal Society sometimes reminded him) to be preparing. In April he devised the anatomical experiment, carried out by Lower, to show that a dog's blood was transformed from dark venous to florid arterial by mixing with air in the lungs, and in May he carried out two experiments to show that a dog could not live by rebreathing the same air, and began cataloguing the books in the Arundel House library. In June he brought in an account of his discovery and microscopic examination of moss seeds, which were so exceedingly tiny (many billions to the ounce) that they could be carried invisibly in the air or rain, which had created the popular but false impression that moss, mushrooms and mould were spontaneously generated in the soil. In the same month he began a series of studies of the system of pores and valves by which trees and other plants transmitted juices to their upper branches, drew up a design for the Royal Society's proposed college in Chelsea, and made a new mercury and water barometer. This was the first practical double-liquid barometer, in which the presence of the lighter liquid magnified the change in levels and made the barometer easier to read. This barometer became very popular despite its weaknesses, which included the tendency of the water to evaporate and react to changes in temperature.[19] Then, to make sure that his days were full, the Society ordered him to try growing moss on a dead man's skull. In the last weeks before the summer recess, Hooke tested the prolonged survival of a chick and a flame in compressed air, demonstrated an improved bubble-level using water in a curved tube, produced and discussed petrified shark teeth, developed his analysis of the circulatory system of large plants, published an essay on a device to project the image of any object onto the wall of a light room, and completed a series of lectures putting forward the entirely novel and potentially heretical idea (anticipated only in his own *Micrographia*) that earthquakes and volcanic eruptions were responsible for the uneven face of the Earth and for the presence of petrified marine species in all inland areas.[20]

Hooke needed to keep his employers happy with promises of accurate magnetic watches, perfect lens grinders, astronomical measurements of unprecedented precision and so on, and he was always willing to try something new. But he was an opponent of thoughtless dabbling, and one of the leading advocates in the Royal Society of a systematic programme of experimental investigation based on or in search of a sound theoretical understanding. His ambitious schemes for the investigation of air and the weather, though they were never completely carried out, show the orderly cast of his mind. If his philosophical programme was not to be repeatedly derailed by Royal Society fellows' demands for curiosities and trifles, it was important for him to establish a sensible research strategy and to implant it in the minds of his colleagues and masters. At some time between 1666 and 1668 he produced a long essay, *A General Scheme or Idea of the Present State of Natural Philosophy*, which set out some of the guiding principles of the new science. The essay was not printed in Hooke's lifetime and it is not dated, but his reference to his discovery of Jupiter's spot and therefore its rotation as having happened 'two Years since' probably places it in early 1666, before his discovery of a second spot on Jupiter in June 1666, and even before his discovery of the rotation of Mars in February and March. If the paper had been written later it is unlikely, knowing Hooke, that he would have forgotten to mention these achievements.[21]

In the essay, he urged natural philosophers to turn away from the practices of the ancients, who had clung to false and untested theories and distorted the few natural observations or experiments they made in order to force them into compliance with their own hypotheses. The ancients 'maintained their Opinions more because they had asserted them than because they were true, they studied more to gain Applause and make themselves admired ... than to discern the Secrets of Nature'. Logic, of which the ancients were masters, had its value in modern science, but a new method was needed if natural philosophers were at last to unlock the secrets of nature. This method, which Hooke said he had drawn from the works of Francis Bacon, was the Philosophical Algebra. The Philosophical Algebra was a set of rules and procedures for gathering, recording and processing scientific information, guiding the diligent scientist from the first observations to the final creation of general laws or axioms. The modern scientist had to recognize the weakness of his own intellect, especially his memory, and follow systematic procedures which would compensate for these infirmities. In the *General Scheme* Hooke dealt with the methodical collection of information and with the most useful way to record and arrange it, but not with the final stage of scientific enquiry, the derivation of general laws from the information collected.

One of the essential qualities of the natural philosopher, he wrote, was a mind cleared of prejudices and preconceptions. Although he should have a knowledge of the work of past philosophers, he should not be the servant of their theories, and should be able to distinguish their best experimental findings from those that were of little value. He should also 'remove those Puerile and Childish Fancies that we suck'd with our Milk, and learnt with our Language'. The ideal natural philosopher should be competent in mechanics and mathematics, so that he would be able to cope with the practical and theoretical problems of experimental work. He should be able to design and make experimental apparatus, and draw well enough to record and communicate his results. He should be able to accept help from other scientists, and to reveal his findings 'freely and impartially', in plain and accurate language. The natural philosopher had three faculties, sense, memory and reason, and all of these, Hooke said, could be improved. The senses, he explained yet again, could be assisted by new technology, as well as various processes designed to intensify the qualities being observed. What the microscope and telescope had done for the eyes other instruments, as yet uninvented, would do for the other senses. One day, Hooke predicted, doctors would be able to listen to 'the several Offices and Shops of a Man's body, and thereby discover what Instrument or Engine is out of order'. 'Methinks I can hardly forbear to blush, when I consider how the most part of Men will look upon this', he said, but 'tis common to hear the Motion of Wind to and fro in the Guts, and other small Vessels, the stopping of the Lungs is easily discover'd by the Wheezing, the Stopping of the Head, by the humming and whistling Noises, the sliping to and fro of the Joynts in many cases, by crackling, and the like'. It was one of his most striking characteristics that he would never dismiss any advance as impossible, 'though never so much derided by the Generality of Men, and never so seemingly mad, foolish and phantastick'. Hooke knew plenty of eminent doctors, but they preferred to make their diagnoses by asking questions and tasting urine than by listening to the workings of the body. The scientific revolution made extraordinarily slow progress in the medical profession, and doctors took another century to grasp the possibilities of listening as a diagnostic device. It was not until 1816 that a French doctor, Laennec, invented the stethoscope – simply a short wooden tube – and revolutionized medical diagnosis.

The way to achieve scientific advance, Hooke argued, was through a systematic programme of enquiry. One of the fundamentals of his Philosophical Algebra (along with the recording of results in symbols) was a minute subdivision of the natural world into a hierarchy of components, and the preparation of exhaustive lists of questions to guide the investigation into the 'history' of each of them. Hooke listed twenty-nine distinct natural 'histories', from comets and stars to mosses and earthworms, and eleven qualities (light, transparency, colours, sounds, tastes, smells, heat, gravity, density, flexibility and brittleness), and followed this with three large pages listing every known occupation and craft, from chemists to wrestlers, all of which also needed to be studied. Each topic suggested dozens of distinct questions: on the atmosphere and the weather, he listed ninety-three. It was an agenda for many lifetimes, not one, but Hooke made a study of almost every one of the twenty-nine natural histories and eleven qualities, and dozens of the crafts, in his career.

Hooke's advice for the investigator was to learn nature's secrets by observation where you could, but to interrogate her by experiment where it was necessary. It was important to measure accurately where measurement was possible, and to use agreed standards where they existed. He suggested that 'the Degrees of Light may be determin'd by Comparison to the Light of a Candle, of a determinate Bigness', anticipating the creation of the science of photometry by J. H. Lambert a hundred years later. Experiments should be repeated to eliminate the chance effects of particular circumstances, and familiar things should be observed as though they were being seen for the first time. Hooke urged the true natural philosopher not to take anything for granted, not to dismiss seemingly trivial things as unimportant, and not to prefer the world's judgement to his own. 'Observations, Experiments, or Circumstances are not to be esteem'd according to the common Opinion of the World, nor are they to be look'd upon as they are curious or not, or pleasant, or strange, or gainful, or sumptuous, or esteem'd by the great'. 'Those things which others count Childish and Foolish, he may find reasons to think them worthy his most attentive, grave, and serious Thoughts; and those things which some are pleased to call Swingswangs to please Children, have been found to discover Irregularities even in the Motion of the Sun it self.' Later in his career, when chemists and others outshone him with their dramatic demonstrations, Hooke repeatedly assured his audience that simple experiments could tell them more than 'pompous' ones, and that simple inventions, though they might not impress the world, could change it.

When the great storehouse of information had been gathered together it was the philosopher's task to make sense of it, to use it to build general truths and universal axioms. To help this enormously complex process it was important that information should be recorded briefly and plainly, without 'Oratorical Garnishes, and all sorts of Periphrases and Circumlocutions'. Shorthand or abbreviation – perhaps in a new universal language – might be used to compress the material into the smallest space, and to give the final and most difficult stages of the enquiry the simplicity of geometrical algebra, in which complex qualities were expressed in 'a few obvious and plain symbols'. This process of simplification and compression was the beginning of the final stage of the Philosophical Algebra, which Hooke intended to describe in the second part of his unfinished

General Scheme. By this means all the information relevant to a particular problem could be stuck onto two facing pages of a very large book (as Hooke had done with his weather records), so that the philosopher could see at one view everything he needed to know. This juxtaposition would assist the feeble memory and stimulate the sluggish intellect, allowing the philosopher to discover comparisons, contrasts and patterns that he might otherwise have overlooked, and thereby to arrive at general laws.[22]

Hooke could see that the ideal programme of research he had described was impossibly vast, a task for thousands of workers toiling for generations, not a small society in seventeenth-century London. His own instincts, in any case, led him in an entirely different direction, guided by intuition and inspired guesswork rather than enormous accumulations of information. In a lecture on earthquakes in 1667 or 1668 he proposed a way of reducing the work to more realistic dimensions. If the Society would retreat a little from its Baconian stance, and accept that it might be permissible to gather information with a specific aim in mind, rather than collecting material on every conceivable subject, the work might be completed more quickly. 'I mention this, to hint only by the by, that there may be use of Method in the collecting of Materials, as well as in the use of them, and to shew that . . . there ought to be some End and Aim, some pre-designed Module and Theory, some Purpose in our Experiments'. This was certainly the principle that guided Hooke in most of his work. He did not have the temperament to wait until all possible evidence had been accumulated before he ventured to speculate about general causes, and usually designed experiments to test or illustrate an intuition that had already occurred to him. His long lists of causes or explanations were often more for show than anything else, like a conjuror's pack of marked cards. He knew which answer he would choose, but wanted his audience to believe that his method was more systematic than it really was.

Hooke's love of building great speculations from a fairly small collection of evidence is well illustrated by the series of lectures on earthquakes he presented to the Royal Society in 1667 and 1668. He had written briefly about the organic origins of petrified shells in *Micrographia*, and since then he had continued his study of stones in the shape of shells, fish, wood and other natural substances. The previous summer (probably 1667) he had revisited his family home on the Isle of Wight, and while he was there he studied the cliffs near the Needles, a mile or two from Freshwater. High above sea level, and running for several miles through the cliff, there was a layer of rock containing shells. He dug into it, and found petrified cockles, periwinkles, and many other shellfish. Rocks from the cliff were constantly falling onto the beach, and he broke some of them open to find the remains of shells inside them. The impressions left by the shells on the surrounding rock were exactly like the shapes that earlier authors had called tricks of nature. He was in no doubt that these stones, like the many others that had been sent to the Royal Society bearing impressions of giant spiral shells, teeth, crabs, fish and wood, were 'nothing but the Impressions made by some real Shell in a Matter that was at first yielding enough, but which is grown harder with time'. The accepted idea that they were 'figured stones' created by some 'plastic virtue' of the Earth in imitation of natural shapes was plainly nonsense, one of those 'Puerile and Childish Fancies' which scientists had to sweep away. Why should we doubt that these were organic remains, and therefore evidence for the natural history of the world, any more than we doubt that discoveries of coins and urns are evidence of its ancient civilizations? For these records were 'written in a more legible Character than the Hieroglyphicks of the ancient Egyptians, and on more lasting Monuments than those of their vast Pyramids and Obelisks.' Breaking ranks with the Christian orthodoxy that had prevented such speculations during and since the Middle Ages, Hooke was sure that natural philosophers could discover the history of the Earth not by reading the Bible, but by studying the Earth itself.

Watching the cliffs of West Wight crumbling and tumbling into the sea led Hooke to an idea of fundamental importance, 'that a great Part of the Surface of the Earth has been since the Creation transform'd, and made of another Nature: that is, many Parts which have been Sea are now Land, and others that have been Land are now Sea; many of the Mountains have been Vales, and the Vales Mountains'. How, otherwise, could we explain the presence of oyster shells on the top of Banstead Downs? He had several suggestions as to how this great transformation might have taken place. Earthquakes and volcanic eruptions might in the past have raised or lowered large areas of land, produced new mountains, deposited thick layers of earth, created new types of stone and brought ores and minerals from the depths of the planet to its surface. Islands had been created and disappeared. He quoted the accounts of ancient writers, especially Pliny, and modern travellers to show that such things had happened, and made it clear to his listeners that he rejected the concept of a planet that was essentially as Adam and Eve had known it. Earthquakes had 'turn'd Plains into Mountains and Mountains into Plains; Seas into Land and Land into Seas; . . . overturn'd, tumbl'd and thrown from place to place Cities, Woods, Hills . . . and have been the great Instruments or Causes of placing Shells, Bones, Plants, Fishes, and the like, in those places, where, with much astonishment, we find them.' It was likely that the British Isles themselves were raised out of the sea by an earthquake or volcano, just as some of the Canaries and Azores had been. More gradual processes were also at work. Tides, rivers and winds were constantly moving earth from one place to another, smoothing out the jagged edges produced by earthquakes. As to the conventional Christian explanation of these great changes, Noah's Flood, Hooke was not convinced that an event which had lasted only a few months could have brought about the great changes to which the fossil record bore witness. He was more attracted to the dramatic possibility that over time the Earth's centre of gravity had changed, causing great shifts in the waters on the Earth's surface. The known wandering of the magnetic poles made this a plausible idea, though it seemed an extraordinary one. Perhaps, he said, the Earth itself was slowing down, like a top that was set spinning ages ago. This was a simple intuition based on straightforward mechanical analogy, but modern research has shown that he was right.[23] He was not sure how long these changes had taken, and in general he felt obliged to stick to the accepted biblical timescale, which dated the Creation at 4004 BC. But he warned his listeners not to take this for granted. 'Who told them where England was before the Flood; nay even where it was before the Roman Conquest, for about four or five thousand Years, and perhaps much longer'? Egyptian and Chinese histories told us 'of many thousand Years more than ever we in Europe heard of', though Hooke questioned these, too.

Moving boldly onto more difficult territory in his Christian age,

Hooke suggested that the great changes in the Earth's surface must have destroyed some species of land and sea animals, accounting for the petrified remains of creatures that were now unknown. Furthermore, it was likely that new varieties of these extinct species had developed, taking advantage of changes in soil and climate. 'And this I imagine to be the reason of that great variety of Creatures that do properly belong to one Species; as for instance, Dogs, Sheep, Goats, Deer, Hawks, Pigeons &c. for since it is found that they generate upon each other, and that variety of Climate and Nourishment doth vary several accidents in their shape.'[24] Hooke was not a seventeenth-century Darwin, but his suggestion that species developed and disappeared in response to changing climatic and environmental conditions has a very modern ring to it, and showed that he was prepared to suggest possibilities that very few of his contemporaries would consider. His agile and unorthodox mind led him over and over again into speculations which would one day become important areas of scientific study, and into others, admittedly, which led nowhere at all.

Hooke's ideas on earthquakes and fossils were not unknown in the ancient world, but in his own day he was a pioneer. In the words of Robert Westfall, a modern expert who is by no means an unconditional admirer, as a geologist Hooke 'yielded pre-eminence to no man in the 17th century'. His closest rival was the Florentine Nicolaus Steno, whose *Prodromus* was published in 1669, shortly after Hooke's *Micrographia* and lectures on earthquakes. Steno shared Hooke's views on fossils and geological strata, but he was less inclined to question the significance of the Flood and the accuracy of biblical chronology, and he had been able to read *Micrographia*. The immediate impact of Hooke's ideas in England was small. The Royal Society's two leading experts in fossils and animal life in the 1670s, Martin Lister and Robert Plot, both believed that the petrified remains of unknown creatures, such as giant ammonites, were tricks of nature without organic origins, and some of the most respected historians of the Earth at the end of the century, Thomas Burnet, William Whiston and John Woodward, were still happy with their stories of a great flood and of figured stones that mimicked natural shapes. It was not until the middle of the eighteenth century that most geologists accepted the ideas that Hooke had proposed in the

1660s. And by this time, of course, they had forgotten all about Hooke.[25]

*

THE VAST PROGRAMME Hooke had in mind obviously had to be the work of many hands, but in practice the Royal Society's experimental programme in the late 1660s rested largely on Hooke's efforts, and when he failed to deliver the promised experiments fellows were usually reduced to listening to Oldenburg's correspondence and admiring collections of natural curiosities. Schemes intended to encourage other fellows to shoulder some of the burden, like Brouncker's 'medals for experiments' idea in April 1668 and Oldenburg's plan in February 1669 to set up two experimental committees (one in Westminster, one in the City), had little impact. Hooke often had the assistance of the Society's Operator, Richard Shortgrave, but proposals to appoint a second Curator of Experiments were dropped. The two men most likely to have been offered the job, the physicians Richard Lower and Walter Needham, both made too much money in medical practice to be tempted by the Royal Society's meagre stipend. Driven by the Society's needs and by his own reluctance to recognize another man's superiority in any branch of science or technology, Hooke tried to cover the whole spectrum on his own. The effect of his versatility on Society meetings was not always appreciated. In December 1666 the Society's Council had discussed the problems of an experimental programme that seemed to flit from one problem to another in a 'promiscuous way', instead of following up selected issues more systematically. But in practice whenever Hooke tried to abandon a proposed investigation the Society chased him for results. In April 1667 he claimed to be able to measure the circumference of the Earth with a telescope and three stakes in St James's Park, and four years later he was still being exhorted (for the umpteenth time) to do what he had promised.

Productive as it was, Hooke's habit of running a dozen projects at the same time, and finishing only half of them, irritated the leaders of the Royal Society. A few unkept promises seemed to overshadow (but not outnumber) his many real achievements, and the sheer multitude of his contributions had the effect of diminishing their individual value. These discontents came to a head in November

1670, when the Society's Council resolved that at the next meeting of the Royal Society its Curator of Experiments would be censured for neglecting his office. But the next meeting took place, and nothing happened. Evidently, Hooke was not alone in leaving jobs unfinished. As someone gloomily observed at the meeting of 9 February 1671, 'very many things were begun at the society, but very few of them prosecuted'.[26]

10. A New Career

(1668–1675)

It is not really surprising that the Royal Society complained about Hooke's waning dedication to his experimental work. By 1670 his science had to compete in his increasingly busy life with his duties as a City Surveyor, and with a new career as an architect. The Royal Society was itself partly responsible for launching this third career. For some years the leaders of the Society had hoped to build their own college, and the loss of the Gresham meeting room turned this hope into a definite plan. Henry Howard, one of the Society's very few generous aristocratic patrons, offered to donate a site in the grounds of Arundel House, the Strand mansion in which the Society held its meetings from 1667 to 1673. In January 1668 the Council asked Hooke to draw up a plan, and in May both Hooke and Wren were invited to submit costed designs. Wren's ambitious plan would have cost well over £2,000 (twice the amount the Society had been able to collect from its fellows and patrons), and on 22 June the Council accepted Hooke's more modest proposal. Over the following weeks Hooke was asked to make a model of his building, to provide a drawing of its riverside elevation, and to start looking into the cost and availability of materials, workmen and a foreman. If anyone could find men and materials at this time of exceptionally high building activity Hooke could, but demand must have affected prices. It was probably the cost that led the Council to defer the project on 10 August and then to abandon it altogether, but it is also possible that they dropped the idea because it was causing angry divisions within the Society, as noted by Pepys on 2 April 1668.[1]

The draft of a new college for the Royal Society was probably Hooke's first commission for a building outside the City, but many years later, in June 1689, he spoke to the Royal Society about subsoils that were incapable of supporting heavy buildings, and told them of his own experience with Berkeley House in Piccadilly. Its walls were built on a bed of gravel one foot thick, he said, but the builders did not trust this to support the chimney stacks, so dug down to the stiff clay underneath. As a result the chimney stacks sank, and had to be demolished and rebuilt on piles. Berkeley House was built in the mid-1660s, on a 200-foot frontage between Stratton and Berkeley Streets. Its architect was Hugh May, Charles II's Paymaster of the Works and one of his three Rebuilding Commissioners, but Hooke's recollections in 1689 strongly suggest that he played a part in building the house, which burnt down in 1733.[2]

In 1669 or 1670 Hooke built some stables for Somerset House, the residence of Charles II's long-suffering queen, Catherine of Braganza. His first important private commission, as far as we know, was offered to him in December 1670, when he was appointed to manage the building of a new Royal College of Physicians in Warwick Lane, to replace the building lost in the Great Fire. The work began in 1671 and continued until June 1679, with Hooke planning and supervising every stage of the process. It is likely that he also provided the designs for the College, and it is certain that he designed its most striking and prestigious building, the anatomy theatre. The College was a fine three- and four-storeyed building round a large quadrangle, with houses for the College officers on the north and south sides (designed and built by Hooke), a great hall and dining room on the west, and in the east, on Warwick Lane, Hooke's distinctive and much-admired octagonal domed entrance and porch, with a light, spacious and acoustically efficient anatomy theatre on its first floor. The building was used by the physicians till 1825, and then spent its last forty years as a meat market, brass foundry and pawnbroker's shop before being destroyed by demolition and fire in the 1860s and 1870s. Hooke's style, in this and most of his other buildings, was influenced by Renaissance and seventeenth-century French and Dutch designs he knew from the books and prints in his collection. It is possible that he visited Holland as a young man, perhaps in the early 1660s, when he was at sea in a ship called the *Katherine*. Victorian writers attributed the college to Wren (who had hardly anything to do with it), and routinely praised its combination of ostentation and utility, its lofty hall, its magnificent staircase, its famous stuccoed dining hall, its unusual porch, 'a *tour de force* of the ingenious architect'.[3] It is unlikely that Walter Thornbury, the author of *Old and New London*, would have praised the building in these fulsome words had he known that its architect was not Wren but a forgotten scientist whose name is not mentioned once in his six heavy volumes.

What the physicians got when they hired Robert Hooke was an honest, hard-working and highly competent surveyor and designer with a good knowledge of the latest European styles, and extensive experience in dealing with craftsmen, suppliers of materials and building regulations. They also got a natural philosopher who was keen to use his scientific and mechanical knowledge to improve building techniques at a time of enormous activity and competition in the building world. The development of sash windows, which slid up and down with ease and stayed open without pegs or locks because they were balanced by ropes, pulleys and counterweights hidden in their frames, took place in the 1670s. Early examples, in which two narrow windows were separated by a central mullion (a vertical bar) containing the weights, were installed in Whitehall Palace in 1672 (possibly a year or two earlier), and in Ham House in 1672 or 1673. Hooke seized on the idea and improved on it for the windows of the Physicians' College, probably in 1673 or 1674. He instructed the joiner Roger Davies to do without the central mullion and install six large sash windows (about three feet six inches wide and ten feet high) with their counterweights in the side frames in the central block of the College, thereby establishing a strong claim (alongside Wren and the Master Joiner Thomas Kinward) to be regarded as the inventor of the modern sash window. Hooke spread the new idea by using it in his other town and country houses in the 1670s and 1680s, and by teaching the technique to the joiners he employed. His Montagu House was probably fitted with many sash windows of the original mullioned type in the later 1670s, and when it burned down in 1686 some of these windows were salvaged and re-used at Ralph Montagu's Boughton House, in Northamptonshire, where Roger Davies was the principal joiner. In the 1680s, when Hooke became one of England's leading country house builders, he and his joiners took the new idea with them and passed it on to local craftsmen. Ragley Hall, Ramsbury Manor and Escot House were all fitted with the modern sash windows Hooke and Davies had pioneered in the Royal College of Physicians.[4]

In the public sphere, Hooke's building work in the early 1670s ranged from the squalid to the sublime. His work for the City Lands Committee involved him in the rebuilding of two City gates, Newgate and Moorgate, and gave him responsibility for finding locations for laystalls (rubbish dumps), public latrines, slaughterhouses and water cisterns.[5] At the other extreme, he was involved in the design and construction of fifty-one City parish churches to replace the eighty-seven destroyed in the Fire. In 1670, when Wren was appointed by the three Commissioners for Churches (the Archbishop of Canterbury, the Bishop of London and the Lord Mayor of London) to supervise the design and building of the new City churches, Hooke and the surveyor and draughtsman Edward Woodroffe were named as his assistants. Woodroffe, a man even more forgotten than Hooke, concentrated on his duties as Wren's assistant at St Paul's until his death in 1675, but was also involved in other design and drawing work for Wren. His replacement in 1676, John Oliver, was a practical craftsman and City Surveyor, so Hooke remained Wren's chief creative associate until Nicholas Hawksmoor joined the team as a junior, and eventually as Hooke's replacement in the mid-1690s.

The nature of Hooke's partnership with Wren has often been misunderstood. The cast of mind that has assigned to Hooke the scientist the role of Newton's foot-rest has also assumed that Hooke the builder was little more than Wren's clerk or foreman. It was not anything said by Wren, but Wren's son's account of his father's work in *Parentalia* (published in 1750), that fixed this impression in the public record. In this account it is asserted that Wren reserved 'all the publick works to his own peculiar care and direction', and even Hooke's independent responsibilities as City Surveyor ('measuring, adjusting, and setting out the Ground of the private street-houses') were presented as if they were undertaken as Wren's assistant. These sweeping claims have had an influence which is far beyond their real worth as objective evidence of Wren's authorship of the fifty-one parish churches.[6]

The true division of the work on the City churches is obscure, but it is generally understood that it was impossible for Wren to have designed and built all of the churches that were rebuilt between 1670 and 1695, when he was also working on St Paul's Cathedral, St James's Piccadilly, Chelsea Hospital, Hampton Court, Greenwich Royal Naval College and several other commissions. Many of the smaller City churches, on their cramped and obscure sites, were exactly the projects that a brilliant and ambitious man like Wren would delegate to a hard-working associate, or to a trusted master craftsman. The vestry minutes and churchwardens' accounts published by the Wren Society do not offer much help in identifying Hooke's role. Eleven City churches paid Hooke a few shillings for his services, usually as City Surveyor, and large payments were rare. But it is plain from Hooke's diary and several surviving drawings that the two friends worked in close partnership on many churches, and that Hooke's role was not always limited to dealing with the practicalities of measuring sites, choosing craftsmen and supervising construction. It is now thought likely that some 'Wren' churches, especially St Benet Paul's Wharf (1678–84), St Edmund the King (1670–74), St Martin within Ludgate (1677–86), and perhaps St Michael Crooked Lane (1684–9, demolished in 1831), were predominantly Hooke's creations, and that some others, including many that are now demolished, emerged from a creative collaboration between Wren and his assistant.[7]

Perhaps the best indication of the extent of Hooke's work on the City parish churches is the salary he received for it, since unlike many of his high-born contemporaries Hooke was never paid for work he did not do. In 1671, 1672 and 1683 Hooke was paid £250 by Wren's office, and in every other year up to 1688 he earned from £100 to £150. From 1691 to 1693, when his activity was less, he was given £50 a year, and in August 1696 he received his last payment for work on churches. His total salary for his part in building City churches was at least £2,820, more than all the money he ever received from the Royal Society and the Cutler lectureship, a little more than the amount he made in salaries, gratuities and fees as City Surveyor between 1666 and 1677, and about the same as the total value, in money and lodgings, of his Gresham professorship for nearly forty years. Hooke's relative earnings in his various professions are not a true reflection of the amount of work he did in each, since some employers were more generous than others, but it is fairly clear from these figures that his career as a City church builder was as important as his career as a scientist, a lecturer or a City Surveyor. The money he was paid came from a tax on the City's coal imports, which funded the whole churchbuilding programme.[8]

The lack of formal documentation on many of the City churches makes it difficult to separate Wren's contribution to their design from that of Hooke, Woodroffe and the craftsmen they employed, and the fact that for economy's sake many of the towers and steeples of the churches were built a decade or so after the rest of the building adds to the difficulty of attributing authorship to a single architect. Paul Jeffery, a leading expert on Wren's churches, argued in 1996 that although Wren was plainly the creative and organizational force behind the rebuilding programme and the architect of the finest churches, there was no reason to regard him as the designer of most of the individual buildings:

> Wren's definite contribution to parish church design, as indicated by the documentary evidence (parish records, payments to him, patronage and so on) amounted to no more than about half a dozen churches. For most of the others no clear evidence of authorship has been found . . . Evidence of some kind, indicative but far from convincing, exists to suggest that a few more may be by Wren. The rest may be by Hooke, but there also remains the possibility, or even probability, that some of these designs are the result of a collaboration between the two men. Evidence for Hooke's involvement in the church-building programme beginning in 1670 is overwhelming.[9]

Considering the sixteen new churches in which building started in 1670 or soon after, Paul Jeffery suggested that Wren and Hooke had divided the burnt area between them, with Wren taking the southern and western district, and Hooke his familar parishes in the north and east of the City, the area he was responsible for as City Surveyor. Jeffery was also inclined to attribute to Hooke the smaller, more humdrum and irregular City churches, especially those which were decorated externally with devices of a Dutch origin, such as carved keystones, swags and festoons (chains of flowers, leaves or ribbons), and urns and pineapples, leaving Wren with the prestigious showpiece churches on larger sites, like St Mary-le-Bow and St Bride Fleet Street. These arguments are speculative, and even a biographer wishing to magnify the achievements of his subject would be reluctant to claim half of Wren's City churches on the strength of a few pineapples. Hooke's strongest claim to the authorship of one of the early churches is to St Edmund the King (1670–4), for which his drawing of the west front, signed and approved by Wren on the King's behalf, still exists.[10]

About forty City churches were built between about 1672 and 1683, the period covered by Hooke's first diary. Hooke recorded visits to about thirty churches, and these visits, especially when they were numerous (St Martin within Ludgate, St Michael Bassishaw, St Lawrence Jewry, St Magnus-the-Martyr, St Mary-le-Bow), strongly suggest that he played an important part in creating these churches, even if he was not their 'architect'. His claim to St Martin within Ludgate, which he visited thirty-one times, is probably the strongest, but surviving drawings also indicate his authorship of St Benet Paul's Wharf, on Upper Thames Street, which is not mentioned in his diary at all. By the time his diary reappeared in 1688–93 Hooke's role in church-building had plainly diminished, but it was not until 1695 that the rising star, Nicholas Hawksmoor, started drawing the annual salary of £60 or so that Hooke had previously been paid from the coal dues. Jeffery wonders whether some of the steeples that had been delayed until after 1687 (when the coal tax was renewed) were built to Hooke's designs, especially St Mary Abchurch and St Margaret Lothbury. Lack of evidence makes most of these attributions doubtful, but they serve the purpose of highlighting Hooke's enormous contribution, as surveyor, negotiator, supervisor and designer, to an enterprise which has for so long been credited entirely to Sir Christopher Wren.

Strangely, Hooke, who was notorious for his priority disputes in the scientific field, never insisted on his right to be recognized as the creator of particular churches or buildings which were later credited solely to Wren. He probably did not see his building career as being as important to his personal reputation as his scientific work was, but it was also true that he was always able to work with Wren without animosity. Their close collaboration in scientific and architectural matters provided very many opportunities for disputes over priority or authorship, but none ever took place. Hooke, who was so angry when Newton's scientific achievements eclipsed his own, apparently never resented Wren's growing reputation as an architect, even when his own accomplishments were overshadowed by it. Wren, for his part, freely provided Hooke with proposals for new scientific and mechanical projects which he lacked the time or inclination to pursue for himself, and helped him to develop some of his most important ideas. Like Hooke, Wren did not claim the authorship of any of the City churches. This lifelong cooperation should make us wonder whether Hooke really possessed the difficult, egocentric and argumentative character that thumbnail sketches routinely attribute to him.[11]

Wren and Hooke also worked together on two ambitious but not entirely successful civic projects, the canalization of the Fleet River and the construction of a Thames Quay along the City's river frontage. Both the King and the City had an interest in these two schemes, and Wren and Hooke worked on them not because they were friends, nor because Hooke was Wren's 'assistant', but because they were respectively Surveyor-General of the King's Works and City Surveyor. A second Rebuilding Act (drafted in 1668 but not finally enacted until 1671) proposed that the Fleet River, which had become a stinking and almost stagnant ditch, choked by sewage, offal, mud and street refuse, should be cleaned, dredged, and turned into a navigable canal forty feet wide. For almost half a mile, from the Thames to Holborn Bridge, the new canal's banks were to be lined with thirty-foot wharves and underground storerooms. On 22 March 1671 Wren and Hooke provided the City Lands Committee with a technical report on the canal project, allowing the negotiation of contracts to begin. In fact the Fleet Canal was a very complicated project, and difficulties arose which could not be anticipated in paper calculations or scale models. In October and November 1672, Hooke and Wren persuaded the City Lands Committee to recognize that the task was more difficult than they had expected, and to increase their payments to their main contractor, Thomas Fitch.

One of the main problems was that the Fleet Canal was only the final section of a longer river. When sections of the wharf were constructed to test the Hooke–Wren plan, the structure proved too weak to cope with the mud and water flowing into the Fleet from the surrounding land and the upstream waters of Turnmill Brook, which was beyond Holborn Bridge and outside the City's direct control. The canal banks and wharves had to be made much stronger than its designers had anticipated, and at their lowest point the final canal walls were almost ten feet thick. To remedy the flow of mud and rubbish from the north, part of Turnmill Brook was covered in brick arches, with a strong metal grating at its southern end to hold back solid matter. The work was poorly done (under the general supervision of Hooke, Wren and Oliver), and the blocked grating became a dam, holding back the winter rainwater until it burst in December 1672, taking much of the brick arching with it into the lower Fleet. The work was redone at great expense, but in heavy rains on 1 January 1674 water rushed over the top of the blocked grating, damaging the brick arches and depositing three feet of mud and silt along the whole length of the River Fleet.

Thomas Fitch, an honest and competent builder who took on the whole project in May 1672 and completed it, using 200 men, by October 1674, saved the situation for Hooke and Wren, and earned himself a fortune and a knighthood. The final cost of the canal, over £50,000, was never justified by the tolls paid by the shipping that used it, and the flat wharves were more popular as a route for carts and carriages than for unloading cargo. So in 1733 the northern part of the Fleet was bricked over for use as a market, and in 1766 the whole canal was covered to provide an approach road for the new Blackfriars Bridge.[12]

The attempt to create a wide quayside along the City bank of the Thames from London Bridge to the Temple was equally unsuccessful. To achieve this by taking over privately owned river frontage was too costly for the City's meagre resources, but in 1670 a bolder scheme, involving the reclamation from the river of a stretch of land between twenty and eighty feet wide, was adopted. Hooke provided a map of the new waterline, handled the City's negotiations with Wren, and helped to win the King's approval for the scheme in December 1671. The usual legal and financial problems delayed the work, and in 1673 the Lord Chancellor, Lord Shaftesbury, intervened to get things moving. Hooke's diary for May 1673 shows him coming under pressure from an impatient Shaftesbury to produce a map of the quay and speed up the clearance of the riverside land. The fundamental problem, the refusal of the existing users of the waterfront, the wharfingers, to give up their plots without compensation, and the City's inability to pay them off, was beyond Hooke's control.[13] The Lord

Chancellor's anger was more than offset by the favour shown him by the King: '(in the councell chamber King smild on me) Lord Chancellor rattled for nothing. fetchd map. met him on water. To Whitehall with Chancellor . . . walkd in park with Dr Wren. King seeing me cald me told me he was glad to see me recoverd'. In August and September Hooke bore the brunt of Shaftesbury's anger at the lack of progress in the clearance of the quayside. He was criticized in his absence at a Council meeting in early August, and on 3 September he was called to attend the Council in person. 'The Lord Chancellor accused me with great bitternesse and craven but grew milder'. Hooke saw Shaftesbury in the Council several times in September, and did his best to speed up the clearance and paving of the quays. On 7 November 1673 Shaftesbury was stripped of his office, a fact which Hooke recorded, perhaps with relief, two days later. After that, the grand scheme was quietly dropped, though parts of the quay were built, and Hooke's duties as City Surveyor continued to include dealing with docks and riverside stairs and keeping the riverfront clear of new buildings.

The diary that Hooke kept from March 1672 shows how important his work as a surveyor and architect had become. His first brief diary entries in March and April show that he was already working on some of the City churches: 'Mr Watts gave me the Griffins Claw . . . With Dr Wren viewing the churches. Dined at Vulture with St Michaels'. This was probably the vestry of St Michael Cornhill, a church which had not been completely destroyed, and was being repaired and partly rebuilt by 'skilful workmen' in 1669–72. Hooke was responsible for repairing the bell, which he fitted with a new clapper on 16 August, and it is likely that he did other work on the church. In August and September 1672 the diary's description of his daily appointments gets fuller, and we can see that his life was one of almost incessant activity. This was his day on 10 August 1672: 'Controuler, Warwick Lane. Knight, Newgate. Sir W. Jones, Dr Tillotson, Coxes, Guys, Mr Boyles, Storys, Dr Wrens, St Pauls, Bow, Home. Marks brought Mr Balls quadrant. Slept well after milk.' He often visited St Paul's with Wren, who had been testing the subsoil, surveying and clearing the site, and making plans for a new cathedral. He made several trips to St Magnus the Martyr in Lower Thames Street (near the Monument and London Bridge), where

Wren's Commission had just taken over existing building work, and went to St Mary-le-Bow in early October with the Danish sculptor Caius Cibber. A week later Hooke and Cibber were at the Guildhall discussing the repair or replacement of its giants, Gog and Magog. Hooke later employed Cibber on the Monument, which has Cibber's bas-relief carvings round its base.

In addition to his work on the Fleet Canal, the Thames quays, the City streets, markets, sewers, gates and churches, Hooke supervised the construction of the tallest stone monument in the world. Like virtually everything else built in the City after the Fire, the Monument is usually credited to Wren, and there is no doubt that he bore the ultimate responsibility for it. It is impossible to disentangle the collaboration between Hooke and Wren, but Hooke's drawings of possible designs for the column, including one with brass flames springing from its sides, still exist, with Wren's signature indicating his official approval rather than authorship. The aldermen instructed the City Lands Committee to build the pillar 'with all expedition', and in October 1672 Hooke and Oliver fenced off the piece of land that would be needed for the work on the base of the column to begin. In October 1673 Hooke completed a model of the column, the next stage in his preparation for its construction. Incidental references in a lecture he gave many years later to the Royal Society show that he was responsible for, or involved in, the preparation of the Monument's foundations, which rested on a bed of gravel six feet thick.[14] His diary does not show that he took daily charge of the building of the column, but he visited it often, and when it was nearing completion in late 1674 he was responsible for designing and making the railings and gilded bronze urn that topped the enormous structure off.[15]

Hooke's architectural work in the early 1670s involved designing and supervising the construction of several huge and important London buildings. He did some work on the Bridewell workhouse, probably helping to repair the damage it sustained in the Great Fire, and helped several City Livery Companies with their rebuilding work. To judge from the visits he made and the company officials he met in the early 1670s, he was involved in the Barber-Surgeons' hall and theatre, and to a smaller extent the Mercers' and Grocers' halls. In 1673–4 he rebuilt the Merchant Taylors' Company school, designed the company's gardens, and designed a wooden screen for their restored medieval hall. The screen was destroyed in 1940. His largest and most impressive project was the design and construction of the Bethlem Hospital on Moorfields (just across the City Wall from Gresham College), a vast building which was compared in scale and appearance with the Tuileries in Paris. Between 1674 and 1680 he designed and built Montagu House in Bloomsbury, on the future site of the British Museum. He was involved in some way in rebuilding the Navy Office in 1674, and he played a part in repairing and rebuilding Christ's Hospital School, Newgate, which had been largely destroyed in the Great Fire.[16]

The beautiful new City which Hooke had played such a great part in creating announced its rebirth to the world by commissioning a new map. Early in 1672 John Ogilby, a publisher who had lost everything in the Great Fire, persuaded the City authorities to support and finance his production of a large-scale map of the City, based on an accurate survey of the rebuilt houses and streets. A experienced surveyor, William Leybourn, was put in charge of the map, and a City committee, of which Hooke was a member, was put in charge of Leybourn. Leybourn's professional skills, along with Hooke's active interest in the project, ensured that the final map was based on sound modern surveying techniques, in contrast to the bird's-eye-view pictorial maps that had always been produced before. On 14 August 1673 Hooke 'designd sheets for London' with Ogilby in Garraway's, in one of about sixty meetings the two men had while the map was being prepared. Two days later Hooke met the experienced Bohemian engraver and map-maker Wenceslas Hollar, and they seem to have decided the map's 100 feet to the inch scale together. Hooke had some new ideas about ways of shading maps and of using movable type as an economical alternative to engraving text directly onto each map, and on 14 October 1673 he showed Ogilby 'the way of Letters for marking his map and also the way of shadowing'. Ogilby was also working on a road map of Britain, and it seems that Hooke tried to persuade Ogilby to try out a carriage-mounted waywiser for measuring turns and distances, the very latest thing in fast and comfortable surveying, as demonstrated before the Royal Society in June 1671. Ogilby was unconvinced, and when the first volume of *Britannia* appeared in 1675 it was illustrated with a traditional pedestrian waywiser, which the preface declared was preferable to 'any Coach or Chariot Mensurator'.

Sitting in Garraway's or the Spanish coffee house, Hooke checked the first sheets of the map as they were completed, and helped Ogilby draft reports and letters to the Court of Aldermen and London Livery Companies to persuade them to give the project adequate financial support. 1675 was a busy year for him, but he maintained his interest in the new map. On 29 January 1675 he checked the carving of the 'marginall metal for Ogylbys Letters and borders', and eleven days later he met Ogilby and Gregory King, one of his surveyors, and 'shewd designe of Epitomising [condensing] the sheets to serve for universal and particular mapps which I directed him how to doe in plates and to fold'. What Hooke had in mind was a smaller travellers' edition of the map, an early version of the pocket *A–Z*. Ogilby died in September 1676, when the City map was almost finished. His step-grandson, William Morgan, prepared its twenty large sheets for publication, and it went on sale in selected London coffee houses and taverns in January 1677. The map had its faults, and some unfinished features, including the Fleet Canal, the Thames Quay and St Paul's, were shown in an idealized form. But in general it was a fine piece of work, and its unprecedented scale and careful surveying methods made it the best and most accurate map of the City that had ever been produced. Among other things, it was a huge and beautiful picture of Hooke's all-but-forgotten work as a builder and City Surveyor in the ten years after the Fire, a picture that survives when most of the work itself is gone.[17]

11. Physicians, Lovers and Friends

HOOKE BEGAN HIS DIARY on 10 March 1672, mainly as a record of weather conditions and barometric readings. At first its foolscap pages were divided into two columns, with brief descriptions of the weather ('clear', 'fair', 'a cold cloudy dark day') and barometric readings on the left and cryptic notes on his appointments, health and expenditure on the right. Either to save paper or in pursuit of his belief that as much information as possible should be displayed on a single page, he crammed as much as he could onto each sheet of paper. Gradually his weather records diminished, and by the autumn of 1672 the diary had become a record of his daily meetings, his work as a surveyor, architect and scientist, his book-buying, his food, his frequent illnesses, and his prodigious and varied intake of remedies for his chronic digestive and sinus problems.

Hooke's diary was not a work of literature or introspection, nothing to compare with those of his two Royal Society contemporaries, Samuel Pepys and John Evelyn. He was not a natural diarist, and does not seem to have written for pleasure. The diary was written hurriedly in moments snatched from a busy and uncomfortable life, often several days after the events it described. It was apparently not intended to be read by others in his lifetime but was literally a collection of memoranda, a record of things that he did not want to forget. Hooke often mentioned the weakness of the human memory, including his own, and did not discount the possibility that chemical cures for the problem might exist. Taking a small pinch of fine silver filings every now and then sounded like a good idea, when it was suggested to him in September 1677. But he believed that the best way to train his mind and compensate for its shortcomings would be to record daily events in a book. The diary served an immediate and a long-term purpose. It reminded him of books or money he had lent, of the effects of various foods and medicines on his body, of bills paid and payments received, and of work he had done and needed to be paid for. It also provided a record from which an autobiography might one day be compiled. Hooke was concerned about his reputation as a scientist, and he wanted to be sure that his inventions and discoveries would be given proper credit when he was no longer alive to speak for himself. He was keenly aware of the danger that his inventions would be claimed by others, and that the Royal Society's minutes were an unreliable record of what he had achieved. A note of the exact date on which he had first propounded a particular idea or made an instrument might be crucial to establishing his priority.[1]

Hooke's is the only known diary of a seventeenth-century English scientist, and in some respects he wrote it as a natural philosopher, recording his own life and movements as he might have recorded those of a gnat under a microscope or a finch in a glass. Daily entries are brief, objective and unelaborated, saying little about his attitudes or state of mind, except when he is angry with someone: 'Audibrasse a Buffoon', 'Titus a Dog', 'Mullet a cheat'. He lists his visits and meetings, but rarely describes his conversations or activities in any detail. No doubt there were lighter moments in his social life, but he did not often describe them. At the back of the book he jotted down three stories he had heard in coffee houses, perhaps intending to repeat them. Two of them, not surprisingly in a society which had been introduced to the pleasures of sugar but was still waiting for improvements in dentistry, involved men who were fooled into taking comical cures for toothache. The more Chaucerian of the two had a painter persuading a City alderman to drop his breeches and stand with his back to the fire with his shirt tail nailed to the mantelpiece and his mouth full of water, which the painter told him would soon heat up and cure his tooth. While he waited for this to happen, first his maid and then his wife came into the room, forcing him to stamp, gesticulate, and finally to eject the water. In the end he 'curses his wife for making him spit out his water before it boyles and soe could not be cured of the tooth ake'.[2]

Hooke kept a careful record of the working of his own body, and gives us a disturbing insight into the laughable incompetence of seventeenth-century medical practice. He was plagued by a number of minor ailments which were uncomfortable rather than dangerous. He suffered from vertigo (or 'giddiness'), indigestion, flatulence, blockages in his nose and ears, occasional loss of the sense of taste and smell, headaches, heart palpitations, sore and watery eyes, noises in the head, fevers, chills and insomnia. He did not endure his discomforts passively, but assaulted them with every orthodox or unorthodox remedy he could discover. No doubt some of his problems made him less than perfect company. On 2 February 1672, the day on which he completed his 'arithmeticall engine in the contrivance for a product of 20 places', he tried to cure his blocked sinuses by 'spitting much and also voyding by the nose but the following night slept very little and disturbed, but I sung much which made me not hear the noyse in my head'. (This is the only reference in Hooke's diary to any music-making by him.) His diary is a meticulous record of the potions he took whether he was ill or not, and of the effects they had on his unfortunate body. He took advice from physicians, apothecaries, and men and women of his acquaintance. Mrs Tillotson, the wife of his friend, the great preacher and future Dean and Archbishop of Canterbury John Tillotson, was a frequent and trusted adviser. And Hooke, who had tried and tested almost every cure available in Restoration London, in turn gave advice to others.

His greatest problem was his digestive system, and he tried an enormous range of diets, drinks, purgatives, laxatives and emetics to keep it running smoothly. If the medicine had the expected effect and 'wrought well', causing him to vomit, urinate or defecate freely, he was satisfied. If it did not, he complained to his diary that he had been 'cheated of a shitt', and looked elsewhere for future relief. He tried mercury, flowers of sulphur, 'Aldersgate', Andrew's Cordial, senna, stewed prunes, steel and mineral water. 'Steel', which he quite often took, could be steel filings, water or wine in which red-hot steel had been quenched and left to stand, or perhaps iron chloride, which was known as 'tincture of steel'. Herbal remedies were in fashion (Chelsea Physic Garden was founded in 1673), but he was happy to try chemical potions, too. Fear of poisoning or other unwelcome side-effects rarely entered his head, though he always noted whether the medicine had caused him suffering or relief. At times, like a modern sufferer from bulimia (though his aim was comfort, not weight loss), he used a feather or whalebone to make himself vomit. Sometimes he thought he had discovered the perfect remedy for his condition, but nothing satisfied him for long. On 30 July 1675 he took spirit of sal ammoniac (ammonium chloride) before bedtime on the advice of Robert Boyle, and woke 'strangely refresht'. The next day he was still delighted with its effects: 'In a new world with new medicine'. And the next: 'I purged 5 or 6 times very easily upon Sunday morning. This is certainly a great Discovery in Physick. I hope that this will dissolve the viscous slime that hath soe much tormented me in my stomack and gutts. *Deus Prosperat.*' Hooke was generally in good health during the summer and autumn of 1675, but when his problems returned in December and January he took senna, chocolate and 'sturgeon vinegar and sugar', not Boyle's wonder cure. He was addicted to self-medication, but not to any particular drug. From time to time (especially in spring 1672) he used laudanum or 'syrup of poppies' to counteract the effects of strong Turkish coffee, but he would then move on to a less familiar remedy, perhaps nutmeg, ginger or tobacco in the nostrils or a glass of wine mixed with spirit of amber.

Hooke was not ill all the time, but enjoyed periods of reasonable health interrupted by longer periods of sickness. His health was particularly poor in the summer of 1672. On 27 July his friend Sir John Hoskins, a future President of the Royal Society, wrote to John Aubrey of 'the indisposition of Mr Hooke', 'the most dire news I ever heard of the RS', and offered him a cure for consumption of the lungs, or tuberculosis, a condition from which he was not suffering.[3]

If we follow Hooke's diary for a few months that summer we will get a picture of his typical medical regime. For the first four days of August he took a daily dose of iron and mercury, and then he moved on to a diet of boiled milk, which seemed to agree with him. Later in the month he was 'much distemperd with noyse and Rheume', and on 25 August he tried to clear his system with two quarts of Dulwich water (from Dulwich Wells in South London), which 'wrought well'. On 31 August a tart disagreed with him and he 'rectifyed' his stomach with distilled white aniseed. The next day drinking 'steel' (in wine) 'benummd' his head, and Dr Goddard gave him tincture of wormwood. On 3 September he took an infusion of crocus metal (probably oxysulphide of antimony), which made him vomit repeatedly the next day, when he felt 'disorderd somewhat by physick'. He reverted to milk and Dulwich water, which apparently restored his health for the next two weeks. On 22 September and again a week later he took syrup of poppies (opium) to help him sleep, but had 'frightfull dreams'. And on 29 September a congenial evening of medical gossip at the Tillotsons' yielded the disturbing news that 'many had died of Fluxes and some of the Spotted feaver upon drinking Dulwich water', which he nevertheless continued to take in large quantities. Mrs Tillotson recommended sal ammoniac instead, but he did not recognize it as a miracle cure at this stage. Instead he drank stale Chester beer, and took spirit of urine and laudanum with milk for three nights in a row. The first three weeks of October passed quietly, but on the 23rd he took 'conserves and flowers of sulphur', which seemed to agree with him. He took it again two days later, but this time it gave him bloody dysentery so badly that he 'swouned and was violently griped'. Dr Goddard's syrup of poppies settled him down. On 4 November he had a painful nosebleed: 'Smelling quite lost, noyse in my head'. A week later he took flowers of sulphur and a 'pippin posset', and then came down with a cold which had been brought on, he thought, by wearing a quilted hat over his head, which had been shaved for a periwig. Noise and numbness in his head bothered him again at the end of November, and he was therefore interested in his friend Haak's account of Colonel Blunt's remedy for the problem, 'honey put warme into the ear and stopt with wool and wrigled with the finger'. On 8 December he was 'taken at Garways [coffee house] with vertigo and vomiting' in the company of three eminent doctors, who advised their favourite remedies, 'clyster and blister'. Thanks to these two, an enema and something (maybe Spanish fly) to raise skin blisters, Hooke 'wrought all night'.

For much of December 1672 he slept poorly, and suffered from persistent giddiness or vertigo, which often kept him at home. He thought it might have been caused by 'drinking milk and posset drink at night and partly from coldness of perruke [wig]', and tried to relieve it with tincture of amber, conserves of rosemary flowers and 'two aloes Rosata pills'. He was also experiencing trouble with his eyes (which got worse over the years) and Dr Goddard fitted him with spectacles on 20 December. In the last days of the year his giddiness got worse, and he drew on a wider range of medical advice.

His friend Jonas Moore told him of a woman who had been cured by 'stone horse [stallion's] dung', but instead of trying this he went to the surgeon Mr Gidley, who drew off seven ounces of blood for half a crown. Hooke, who was a traditional believer in the influence of the four humours (blood, phlegm, choler and melancholy) over the body, thought his blood looked 'windy and melancholly'. Sniffing ginger cleared his nose of 'a lump of thick gelly', but the next day he was still giddy, and Dr Goddard prescribed 'amber and ale with sage and rosemary, bubbels, caraways and nutmeg steepd and scurvy grasse'. The following day, Christmas Eve 1672, he was worse than ever. 'The worst night I ever yet had, mellancholy and giddy, shooting in left side of my head above ear'. On Boxing Day Dr Goddard's pills 'wrought 14 times', but drinking ale left him 'strangely inlivernd and refresht' and he slept well and 'dreamt of riding and eating cream'. The vertigo persisted, and although he was bled twice from his shoulders by Mr Gidley and put bitter almond oil in his right ear he went into 1673 with vertigo, a noise in his right ear, a blockage in his right nostril, and a distorted eye. His problems seemed to ease during January and February, but they soon came back again in one form or another, and bothered him intermittently for the rest of his life.[4]

Hooke could get medical advice from several physicians who were also fellows of the Royal Society. These included Dr Daniel Whistler (an expert on rickets), Dr John Mapletoft, Sir Charles Scarborough (the King's physician) and the distinguished Dr George Ent, President of the Royal College of Physicians. The physicians he relied on most were Dr Thomas Coxe and Dr Jonathan Goddard FRS, Gresham Professor of Physic and the inventor of Goddard's drops, who had once been Protector Cromwell's doctor. Whether their advice was worth taking was another matter. Soon after the diary began Hooke's old friend John Wilkins, the Bishop of Chester, who had taken Hooke under his wing at Oxford, kept him company during the plague, and shared his interest in flying, life on the Moon and a universal language, fell desperately ill with kidney stones and 'stoppage of urine'. Four red-hot oyster shells in a quart of cider and blistering with cantharides (Spanish fly) were strangely ineffective, and Wilkins died, much lamented, on 19 November 1672. A post-mortem showed that the original diagnosis had been wrong, and that Wilkins, who was only fifty-eight, had probably been killed by bad doctoring. As Hooke noted, 'twas believd his opiates and some other medicines killd him, there being noe visible cause of his death'. Another founding father of the Royal Society and friend of Hooke's, Sir Robert Moray, died the following July. Moray choked to death after having dinner with the Lord Chancellor, and a post-mortem found his body 'very intire and sound and nothing amiss'. Hooke, who shared the general affection for Moray, thought that 'he dyed poor and seem to have been much afflicted with somewhat that troubled his mind'.[5]

Even in his new periwig, Robert Hooke was apparently not very attractive to women. He never married, and in the period covered by the diary he never had a sexual relationship with a woman who was not dependent upon him for shelter or income. He does not seem to have sought the company of prostitutes, who were plentiful in Restoration London, though it is possible that the diary entry on 2 July 1673, which mentions two orgasms and 'first pierce', is a reference to sex with a virgin. Hooke recorded his orgasms (denoted by the zodiac symbol for Pisces) as carefully as he recorded his medicines, and it is not clear whether all those that appear without a woman's name alongside them were solitary experiences or involved prostitutes. His sexual companion in the early 1670s was his servant Nell. Nell slept with him on 12 September 1672, and again on 8 October. Their third recorded encounter on 28 October was vigorous: 'Played with Nell – ♓. Hurt small of Back', and their fifth, on 2 December, shows that even scientists like variety: 'Slept ill with Nell ♓ supra [on top]'. Nell seems to have been an independent spirit, and she could imagine something better than serving Hooke's needs for £4 a year. To his displeasure she had other male friends, and on 14 July 1673 she 'lay out all night', probably with a barber named Young, the man she was about to marry. Her daughter was born nine months later. On 13 August 1673 Hooke wrote: 'Nell went out yesterday stayd abroad all this Day. I suppose today marryed'. She came back the next day and Hooke noted in his diary that they made love early the following morning. They slept together again a week later, but she left Gresham College for good at the end of August: 'Nells husband returned, shee lay with him that night and ever since.' She set up house as a needlewoman near Fleet Ditch, where Hooke visited her often and apparently innocently, sometimes to get clothes made or repaired, and sometimes to talk or eat with her as a friend. Nell also visited him in Gresham College, bringing news from the town and even from the Isle of Wight, and the two were still good friends in the 1690s.

Hooke had difficulty in finding a replacement. Nell's cousin Bridget arrived from the country at the end of August 1673 and left (perhaps after an affair with Hooke's lodger, Richard Blackburne) two months later. In October he hired Doll Lord, who soon took on all Nell's duties. They had some sort of sexual encounter on 27 December 1673, and Hooke 'first felt Doll – ♓' on 7 January. Within six weeks Hooke had decided to get rid of Doll, 'she being intollerable', and in March 1674 she went off to her uncle's. Her replacement, Bette Orchard, turned out to be clumsy and 'intollerably carelesse', and Hooke apparently had a brief sexual relationship with her, too. What was happening when he 'wrastled' with her in June is unclear, but his diary entry for 26 July 1674 – 'First saw Betty ♓' is less ambiguous. She was replaced on 30 September 1674 by Mary Robinson, who stayed on as his servant (and nothing more) for the rest of the decade.

As well as his servants, Hooke's household included various lodgers, relatives and scientific assistants. From the start of the diary he was responsible for the care of Grace Hooke, the daughter of his brother John, who had a grocery business in Newport, on the Isle of Wight. Grace was twelve on 2 May 1672, and went to a school in the City. Hooke received money from his brother for her board, and seems to have spent far more than this on clothes and gifts for her. He took her to Bartholomew Fair, London's greatest and oldest fair, in August 1672, and spent 2s 6d on the shows there. His gifts to her included a looking glass, a muff, books, speckled silk stockings and a pair of red pendants. Grace was under some sort of contract to Sir Thomas Bloodworth, who had been Lord Mayor of London during the Great Fire, perhaps an agreement to marry his son. Hooke's diary mentions the steps he had to take to release her from this obligation (he called it her 'divorce') in 1673–5. Hooke's relationship with Grace became closer, and in the end indecently close, as she grew up. On 14 June 1676, when his niece was just over sixteen and he was almost forty-one, Hooke 'Slept with Grace'. His words on 16 October and 11 November were more explicit – '♓. Grace in Bed'. In December 1676 Hooke decided to send Grace back to her father, but by March

1677 everything between them was wonderful, at least as far as he was concerned: 'Grace *perfecte intime omne.* ♓. Slept well.' No doubt Grace did not see her uncle as the perfect match, and Hooke had to struggle to stop her going the way of Nell. On 30 June 1677 he found her in the cellar with 'Pettis', and in October he was disturbed to hear from Nell that Grace, who was staying on the Isle of Wight, was being courted by Sir Robert Holmes, a fifty-five-year-old naval hero and Governor of the island. These relationships did not last, and Grace lived with Hooke as his mistress and housekeeper until she died in 1687. Despite the irregularity of their relationship Grace was the love of Hooke's life, and her death was a terrible disaster for him. In the 1650s Hooke's incestuous relationship with his niece would have been a capital felony, but Puritan moral legislation had been repealed after 1660, and he would only have been answerable to an ecclesiastical court if his offence had been discovered. The age difference between him and Grace was commonplace, and would not have upset his contemporaries as it does us.[6]

In December 1672 Hooke took a lodger, Richard Blackburne, a young Cambridge graduate with ambitions to become a physician. Hooke started to take lessons in High and Low Dutch from him, perhaps because he wanted to read Dutch architectural books or the letters and writings of the great microscopist Leeuwenhoek, who did not know Latin. Hooke's lessons seem to have petered out after a few weeks, but he bought Dutch dictionaries and grammars, and probably carried on studying without a teacher. In his papers in the British Museum there are notes in his handwriting taken from Dutch newspapers, and in 1682 his Dutch was good enough for him to translate a letter sent to the Royal Society by Leeuwenhoek. Blackburne left in November 1673, and went to Leyden to study medicine. After qualifying as a doctor he returned to London in 1676, and resumed his friendship with Hooke.

A much more important lodger, who arrived shortly after Blackburne, was Harry Hunt, who became Hooke's apprentice, assistant, colleague and friend for the rest of his life. Hooke already knew 'Harry', perhaps as a relative, when he moved in in January 1673, and he immediately started training him as his assistant. He showed him how to look after the Royal Society's collection of curiosities (its 'repository'), trusted him with his furnace and speculum (the mirror for his reflecting telescope), taught him to prepare experiments and make instruments and clocks, observed Jupiter and Venus with him, employed him in surveying and drawing work, and sent him on errands to craftsmen, colleagues and clients. Harry Hunt was an excellent and versatile assistant, and his artistic skills gained him an offer of employment and a visit to Italy from Hooke's wealthy client Ralph Montagu in 1675. Hooke thought Hunt could easily earn £150 or £200 a year through painting, but Hunt refused this and other offers, and instead took a £40 a year position as the Royal Society's Operator when Richard Shortgrave died in 1676. The following year Hooke took in another assistant, Thomas Crawley, but Hunt stayed on too. Hooke valued and admired Hunt's growing abilities, and as Secretary of the Royal Society he made sure that Hunt's work was properly credited in the Society's minutes.[7] The fact that Harry Hunt, whose many talents could easily have earned him a good independent livelihood, chose instead to spend so long as Hooke's friend and assistant suggests that Hooke had charms which his diary, with its catalogue of illnesses, orgasms and arguments, does not often convey.

In spite of its brevity, Hooke's diary allows us to see a real man with a life beyond the laboratory and the Royal Society, and with emotions as well as ideas. Certainly, he spent many hours in his rooms conducting experiments and preparing architectural drafts and models (or 'modules'), and his cultural life beyond science and building was sparse. But his work and his many friendships took him all over the city, and he was an early enthusiast for the new coffee-house culture that had developed so rapidly in London since the 1650s. London's first coffee house, Pasqua Rosee's Head, opened in St Michael's Alley, off Cornhill, in 1652, on a site later occupied by the Jamaica Wine House. Coffee houses sold exotic and increasingly popular commodities, including Arabian coffee, West Indies sugar, Virginian tobacco, Chinese tea and South American cocoa, and they offered comfortable, lively and relaxed meeting places which suited to perfection the needs of middling Londoners (rich enough to spend a few pence a night, but too poor to entertain lavishly at home). When licensing was introduced in 1663 there were eighty-two in London, and by the time of Hooke's death in 1703 there were at least 500, perhaps many more.[8]

Between 1672 and 1680 Hooke visited about sixty coffee houses and nearly eighty taverns, mostly in the City and Fleet Street, but also in Westminster and Southwark. He used coffee houses for conversations with friends and meetings with Royal Society colleagues, for negotiations with builders, lens grinders and watchmakers, and discussions with clients of his architectural practice. Occasionally he even conducted scientific experiments in coffee houses, where he could be sure of plenty of witnesses. He sometimes went to Man's coffee house in Chancery Lane, to the Guildhall coffee house, where books were auctioned, and to the Grecian coffee house in Devereux Court (near Temple Bar), a Royal Society favourite. But his 'local' until 1677, and the place where he expected to meet his friends and pick up or pass on the latest gossip, was Garraway's, one of the greatest and longest-lived of all London coffee houses. It was established in 1669 by Thomas Garraway in Exchange Alley, in the little network of back streets between Cornhill and Lombard Street which also housed Pasqua Rosee's Head, Jonathan's, the Jerusalem and the Rainbow. The alleys are still there today, with plaques to long-gone coffee houses set into their walls, and with the George and Vulture tavern, where Hooke supped with the Merchant Taylors on 15 October 1673, to remind us of how things used to be. The alleys were only about five minutes' walk from Gresham College. Garraway's was the first place in London to sell tea (in leaf and liquid form), and it was becoming a well-known centre of commerce. But Hooke could be almost sure of meeting several of his cronies when he went there, as he did most evenings. His diary often said simply 'At Garways', with nothing about who he met or what was said, but sometimes he was more expansive. On 24 February 1674 he was

> At Garways in the Little Room. Left with Cap. Gorge, Hoskins, Wild, Lodowick, Godfry, Diodati, Spencer. Mr Hoskins told Kings bitt by mad doggs cured by a felon in the neck. That severall tradesmen in the Indies use their feet as well as their hands for holding things. Gorge told of the cabbidge [palm] tree its prodigious height, that its out side is exceeding hard as iron its middle all pith that it hath many white circles whence the leaves fall off. That the neagers will clime to the top by the help of withes . . . Of the extreme deliciousness of the queen pine apple.

On 9 July 1677 a new coffee house, Jonathan's, opened round the corner from Garraway's, and Hooke took to it at once. Within a few weeks 'At Jonathans' or 'at Jon' replaced 'At Garways', and Hooke's friends moved there too. Within a few decades Jonathan's was famous for its stock dealers and share speculators, but not in the 1670s or even in the early 1690s, when Hooke was still picking up the hot news there from his familiar friends. Monday 8 May 1693: 'At Jon. Pif, Wal, Blackw, Lod, Gof, Hally, Hain: the Grand Vizier strangled'.

Hooke is remembered as a difficult and argumentative character, but he had many close and enduring friendships. Two of the companions who were with him at Jonathan's in 1693, Godfrey (Gof) and Lodwick (Lod), had been with him in Garraway's twenty years earlier when the talk had roamed from mad dogs to pineapples. Mr Godfrey was Hooke's regular companion on his walks across Moorfields or through the city, and one of his closest friends. Hooke never uses Mr Godfrey's first name, and he is not an easily identifiable scientist or Royal Society fellow, as most of Hooke's friends were. Nevertheless, many references in the diaries suggest that he was a member or servant of the Mercers' Company, and involved in the running of Gresham College. He ordered repair work on Hooke's rooms in December 1688, restored the College's water supply (from the New River) in April 1689, and pulled down the coach houses in the courtyard, following the Gresham committee's orders, in August 1681. There is no real doubt, then, that he was John Godfrey, clerk to the Mercers' Company from 1656 to 1697, a useful and natural friend for Hooke to have. John Godfrey had sons (mentioned in the diary on 23 September 1689), but his family commitments did not prevent him from sharing tea, dinner and coffee with Hooke on countless occasions, or keeping him company on hundreds of walks.[9]

Francis Lodwick (or Lodowick), another of Hooke's walking and coffee-drinking companions, was a merchant from a Dutch family, part of the community that worshipped in the Dutch Church of Austin Friars. In 1647 he had written a book on *A Common Language* which had influenced Wilkins, and he became a Fellow of the Royal Society in 1681, when he had just turned sixty. He shared Hooke's interest in a universal language and the replanning of London, and helped him translate from Dutch. Two others who were at Garraway's in February 1673, Sir John Hoskins and Edmund Wild, were also close friends of Hooke's. Hoskins, one of the Royal Society's enthusiastic amateur scientists, had been at Westminster School with Hooke, and often exchanged scientific ideas and information with him in coffee-house conversations. Hooke recommended him as Wren's assistant surveyor in 1675, when Woodroffe died. In September 1676 Hooke rode with Hoskins to Banstead, where he and Wilkins had conducted experiments during the plague. This time he was struck by 'the cloud of smoke over London ½ mile high and above twenty miles long', and by the jolting of his saddle, which bruised his testicles. Later in his career Hooke thought that Hoskins, a leading figure in the Royal Society, did not give him enough support in his various priority disputes, especially with Newton, but they seemed to remain friendly. Wild (or Wyld) was a wealthy amateur inventor and horticultural innovator, a Fellow of the Royal Society, with a fine house in the new suburb of Bloomsbury. Wild and Hooke had a mutual friend in the antiquarian John Aubrey, an amiable spendthrift who had wasted his fortune and tried to keep out of debtors' prison by borrowing money from his friends. Despite Hooke's contributions (all carefully tallied, but without expectation of repayment), Aubrey was briefly arrested for debt in March 1673. He continued to borrow Hooke's money and use his lodgings, and in return gave Hooke loyal support in all his scientific controversies and included a generous account of his career in his famous *Brief Lives*. Another everyday friend was Abraham Hill, a well-off merchant and book collector, a Gresham College resident and founding fellow of the Royal Society. And for quiet evenings at home there was always Theodore Haak, a German-born scientist who had been part of London's scientific world since the 1630s, and made a living by translating from Dutch and German. Haak was thirty years older than Hooke, but the two remained close friends and regular chess companions until Haak's death in 1690.

Hooke's network of friends, colleagues, clients and employees was large, and like Pepys's it stretched from the Court to the brickyard. He was on familiar terms (or so the diary suggests) with Lord Mayors and Aldermen of London, and even the King took an interest in his watches and experiments, chatting with him in St James's Park and Whitehall. Hooke's career occupied the middle ground between natural philosophy and mechanics, and he mixed freely and comfortably with all the London craftsmen whose skills he needed: bricklayers, engravers, carpenters, joiners, plumbers, lens grinders, watchmakers, apothecaries, plasterers, masons, smiths and so on. He worked very closely with some of these craftsmen, and sometimes their shared time and interests led to friendship. Some would say that his paid position in the Royal Society, and his attitudes towards profit, competition, secrecy and trust, gave him more in common with London tradesmen and craftsmen than he had with 'gentlemen philosophers' like Moray, Brouncker and Boyle.[10]

Hooke's long relationship with Thomas Tompion was based on respect and mutual dependence. Tompion's mechanical ingenuity helped to turn Hooke's paper projects into fine working machines, and Hooke's torrent of ideas and devices helped to put Tompion ahead of his rivals and to earn him fame as the 'father of English watchmaking'. Hooke's conversation with Tompion was not restricted to clocks and watches. In October 1674 he told him 'about plug for wind pump, and about the fabric of muscles. About Cork bladders leather &c. About fire engine the way of making it.' They met in coffee houses and taverns, as well as in workshops. Sometimes Hooke lost patience with Tompion's slowness, and no doubt Tompion had his criticisms of Hooke, but the two worked together for over twenty years, often spending long days in each other's company. In 1675, as we shall see, they formed a highly effective alliance in defence of Hooke's claim to have invented the spring-driven 'longitude' watch ahead of Huygens. Hooke's relationship with his favourite lens grinder, Christopher Cock, was almost as close, and involved a similar exchange of talents. On 18 January 1676 Hooke went to Cock's and 'saw him polish an excellent 12 foot glasse by changing place of the tool. Smokd with him three pipes . . . Promised to grind me a glasse of any shape if I would shew him my new way . . .'

Naturally, Hooke's best friends were members of the Royal Society. In the mid-1670s, when Hooke's relations with the leaders of the Society were bad, he formed small scientific clubs to discuss philosophical matters in the relative privacy of a tavern or coffee house. The men he invited to join these clubs – Boyle, Wren, Sir Jonas Moore, Dr Daniel Cox, Nchemiah Grew, Theodore Haak, Edmund Wild, Francis Smethwick, John Aubrey, John Hoskins, William Holder – were the Royal Society fellows he most trusted and liked.

Boyle had now moved from Oxford to his sister Lady Ranelagh's mansion in Pall Mall, and Hooke was a frequent dinner guest there. Now and again things went badly. After a visit in January 1674 he wrote: 'Dind at Lady Ranalaughs. Never more.' And on 20 June 1678: 'I will never goe neer her againe nor Boyle'. These were only hiccups in a long and close friendship. Hooke's relationship with Wren, his companion in both science and building, was even closer. Since their Oxford days the two had shared almost all their interests, and much of Hooke's work, including *Micrographia*, the laws of planetary motion and the weather clock, involved completing projects initiated by Wren. They met almost daily at Wren's home, in his office in Scotland Yard, in coffee houses, or at one of the many buildings they worked on together. Hooke went with Wren to St Paul's and to the new City churches very often in the 1670s, and was paid a salary of at least £100 a year from Wren's office 'on the City Churches account' between 1671 and 1688. Their partnership was so close that Hooke's diary, which mentions more than 1,000 meetings between them, is a key source for the study of Wren's career. Hooke was ready to gather scientific ideas from anyone, and to offer appropriate information in exchange, but his fullest and most valued discussions were with Wren, who was his equal in scientific insight and ideas, though not in publication or completed experimental projects.

Hooke was always working. Whenever he met Royal Society colleagues or skilled craftsmen his ears and eyes were open for information and ideas that would help him in his scientific and architectural work. He was still watching, listening and questioning, just as he did when he learned to make clocks and mix paints as a child, or when he worked in Peter Lely's studio and Boyle's laboratory as a young man. His range was so wide that almost every craftsman's trick or scientist's observation could be turned to account, and information drawn from one area of his work could easily be applied in another. Advice on making tin plates or polishing iron picked up in conversation with Wren, Hill and Hoskins in April 1676 might be useful in producing decorative metalwork for the roof of the Royal College of Physicians, or for making metal mirrors for reflecting telescopes. Techniques of printing, engraving and staining which he learned from Wren or from the engraver William Faithorne might be passed on to printers and map-makers he was working with, to painters decorating his buildings, or to the weekly meetings of the Royal Society. Information might be traded in a coffee house or workshop to induce a reluctant scientist or craftsman to part with some useful advice of his own. Hooke was an expert in the art of coaxing ideas out of unwilling informants. 'Hee makes questions to those he knows are Skilfull in them, & theire answers serve him for assertions on the next occasion', the astronomer John Flamsteed complained.[11]

Hooke enjoyed the friendship of eminent men, but he knew that skilled craftsmen had almost as much to teach him as scholars had, and that time spent in a London workshop might be as fruitful as an evening in Garraway's or an afternoon at the Royal Society. He had a particular interest in glass, both as a builder and a scientist. In July 1673 he went with Wren to the new Savoy glassworks of George Ravenscroft, one of the 'inventors' of flint or crystal glass in England, and saw 'calcind flints as white as flower, Borax, Niter and tarter with which he made his glasse' and 'pretty representastions of Agates by glasse', and asked him whether arsenic was used in the process. In February 1674 he picked up tips on making glass plates from Mr Reine's glassworks: 'saw his joyning glasse plates but grinding them together either by a square joynt or by an oblique joynt thus or by an undulated joynt thus. They did all very well and were very strong and but little to see. I suppose the whole secret consists in the make and heating of the furnace and cooling it which is neer a week in doing.' Three years later (on 30 March 1677) he went to Reine's again, and came away with a very detailed sketch of the furnace he used for cementing glass plates together. In September 1677 he watched the chemist Krafft fire gunpowder by warming it over coals and touching it with 'a 10th part of a graine of his white powder, which he putt into the pith of a quill and it immediately took fire.' The white powder was phosphorus, making one of its first appearances in England, and Hooke found these experiments 'exceeding strange and much more than I had ever seen'. He gained practical information on printing and dyeing cloth (another subject he had spoken of in Royal Society meetings) from visits to Philip Barrett's workshop on Moorfields. On 11 March 1674 he saw a 'tryall of Golding flowerd Shift which succeeded', and on 29 December he was impressed by the quality of Barrett's varnish. Four years later (on 4 January 1679)

Hooke and Barrett met at Garraway's and discussed ways of staining satin with lead moulds and copper plates, and a few weeks later he 'saw his way of Dying flowers'. Hooke's best informant on pottery and stoneware (important in building and in many of his experiments) was an old Christ Church friend, John Dwight, who had a patent for the manufacture of 'Chinese porcelain' in England. Hooke was interested in discovering the secret of Dwight's transparent china, and often discussed it in coffee houses in the early 1670s. When he visited Dwight's factory in Fulham on 16 May 1674 he claimed to be unimpressed by his new pottery – 'nothing but Tobacco pipe clay' – and guessed that Dwight's process involved 'burnt allum chalk or lime'. But nearly four years later he was still trying to find out the secret of Dwight's clay pipes: 'I suppose his way is mixing the powder of tobacko pipe clay once burnt and heat with other washt tobacco pipe clay, that salt helps the glazing to run, that the greatnesse of the fire is the secret, and the way of making the fornace.'[12]

In a society in which new ideas and techniques were being developed at an unprecedented rate, Hooke was at once a producer, a collector, a manipulator, a purveyor, a communicator and a user of new knowledge. His diary is like a collector's album of scientific, medical and manufacturing information. For example: 'Also an easy way of getting a mans children taught divers languages by exchanging the children of other countryes' (27 January 1673); 'Man told me that Quick Lime white of eggs and blood or slime of snails makes a cement for water pipes as hard as stone' (26 February 1675); 'T. Hewk told me that Pumice Stone melted with fritt [the materials which make glass] would make a ruby colourd past of Glasse' (26 February 1675); 'Cut off the top of my thumb, but cured it in 4 days by *Balsamum Peruvianum* from Shortgrave' (30 December 1675). Once collected or discovered, information could be displayed before the Royal Society, passed on to a more exclusive group, used, exchanged, sold or stored away. Hooke liked to boast and to show off his knowledge, but there was often an element of secrecy in his communication. Sometimes information was released in code, only revealed in part, or given to trusted friends. Even close friends like Wren and Abraham Hill might be kept in the dark. In January 1676, talking about ways of improving musical instruments, Hooke 'told them I could make a string vibrate without giving any sound but told them not how'. Coffee houses had not yet been turned into members-only clubs, and Hooke often worried that things being told to selected friends might be picked up by inquisitive outsiders. When he was telling Edmund Wild about his way of flying at the end of 1675 in Joe's coffee house (later the Mitre) he broke off when 'a company of 3 strangers interposed'. Secrecy was not unusual among seventeenth-century scientists, but sometimes Hooke's secrecy could be infuriating, and created the impression that he was a fraud, unable to deliver what he promised.

12. Heat and Light

(1671–1673)

BETWEEN 1666 AND 1669, while Robert Hooke was busy with a dozen scientific and building projects, Isaac Newton devoted much of his time and intelligence to one great question. Newton and Huygens had both read Hooke's *Micrographia* with care and interest. Both had been struck by Hooke's work on the coloured rings seen when thin plates were pressed together, and both had quickly worked out a way of measuring the thickness of the film of air between the two plates, something that Hooke had been unable to do. By placing a lens with a specific curvature on a piece of flat glass, Newton was able to use simple mathematics to calculate the distance between the two surfaces. This allowed him to study the circles of coloured light that Hooke had noticed and relate them to the new ideas about the nature of light and colour that he had been pursuing since 1666.

As we saw in chapter 6, Newton was not convinced by Hooke's assertion that light was composed of waves (a development of Descartes's idea that light was instantly transmitted pressure), and preferred Gassendi's theory that light was a stream of particles or 'corpuscles'. It never occurred to him – or to Hooke – that light could be composed of particles *and* waves. He also disliked Hooke's argument that colours were produced by the confusion of white light by refraction and reflection. Newton's hypothesis was that white light was not 'pure' or homogeneous, as Hooke and his predecessors assumed, but heterogeneous, a mixture of many colours. Refraction, instead of distorting or confusing the purity of white light, broke it down into the colours from which it was actually composed. It revealed colours, rather than creating them. Between 1666 and about 1669, when he first gave his Cambridge lectures on light, Newton devised wonderful experiments to test and demonstrate the truth of his hypothesis. He used a prism to refract a narrow beam of sunlight passing through a small hole in a window shutter, roughly as Descartes and Hooke had done, but rather than looking at the refracted beam within a few inches or feet of the prism, as they did, Newton used a wall twenty-two feet away. This allowed him to see the full rainbow spectrum, which was not circular, as the original beam had been, but an oblong thirteen inches long, which suggested to him that each colour refracted differently. Using a second board with a hole in it and a second prism, he could reverse the effect of the first prism (producing a circle of white light), or refract each coloured beam in turn, showing that they each refracted to a different degree but were not otherwise altered by a second refraction. This last test was Newton's *experimentum crucis*, and seemed to lead to one clear and unambiguous conclusion, that white light was not pure or homogeneous, but 'a heterogeneous mixture of differently refrangible rays', each of which gave the appearance of a different colour. There were five primary colours – red, yellow, green, blue and violet-purple – and an infinite number of intermediate colours made by the mixing of these five.

The realization that light broke into its component colours on refraction led Newton to abandon his search for a perfect refracting telescope, and to make a small reflecting telescope instead, using an arrangement of mirrors and lenses rather different from the one suggested by James Gregory in 1663. In the Gregorian telescope, the primary mirror, a large concave which collected the light, had a hole in the middle of it containing the convex lens eyepiece, which was thus located in its conventional position at the end of the tube. The light was reflected back to a small concave mirror in the middle of the telescope tube, which focused and reflected the light back to the eyepiece. Spherical aberration was eliminated by this arrangement of concave mirrors, and chromatic aberration was slight. In Newton's telescope the primary concave mirror was set, unperforated, at the end of the tube, and reflected the light back to a small flat mirror about halfway down the tube. This flat mirror was set at forty-five degrees, and thus reflected this light up to an eyepiece set into the side of the tube. At the end of 1671 Newton sent his new reflecting telescope to London in the care of Isaac Barrow, along with a letter in Latin explaining how it worked. The Royal Society gave the

1. 'A haven of a pretty breadth'. The River Yar and All Saints' Church, Freshwater, Isle of Wight, where Robert Hooke spent his childhood, pictured in 1834.

2. Dr Richard Busby (1606–95), headmaster of Westminster School for almost sixty years and prebendary of Westminster; Hooke's teacher and lifelong friend.

3. Dr John Wilkins (1614–72), Warden of Wadham, Bishop of Chester and a founder of the Royal Society. Hooke's patron and friend in the 1650s and 1660s, sharing his interest in flight and a universal language. Engraving from his *Mathematical Magick* (1680).

4. Robert Boyle (1627–91), Hooke's employer and patron, with his first air pump, which was constructed and operated by Hooke. William Faithorne's engraving (1664) used Hooke's sketch of the pump.

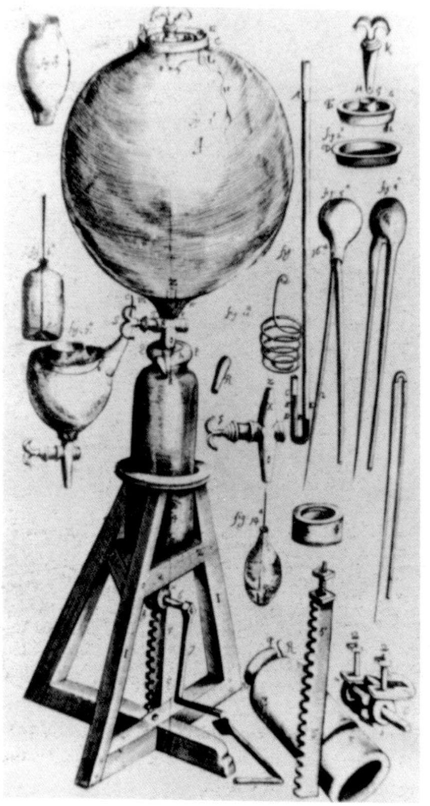

5. Boyle's first air pump, 'contrived and perfected' by Hooke and pictured in Boyle's *New Experiments Physico-mechanical* (1660). See page 20.

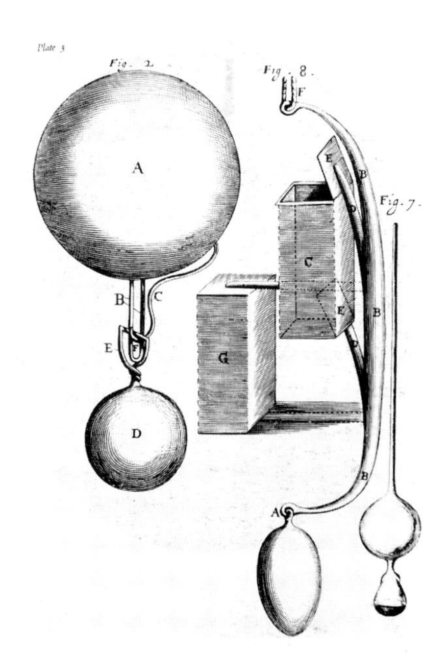

6. Hooke's oceanographic instruments, first proposed in 1663. In the depth sounder the weight (D) uncoupled when it hit the ocean floor, allowing the hollow ball (A) to float to the surface. In the water sampler the bucket was opened by the flow of water as it sank, but closed when it ascended, trapping water from the ocean floor.

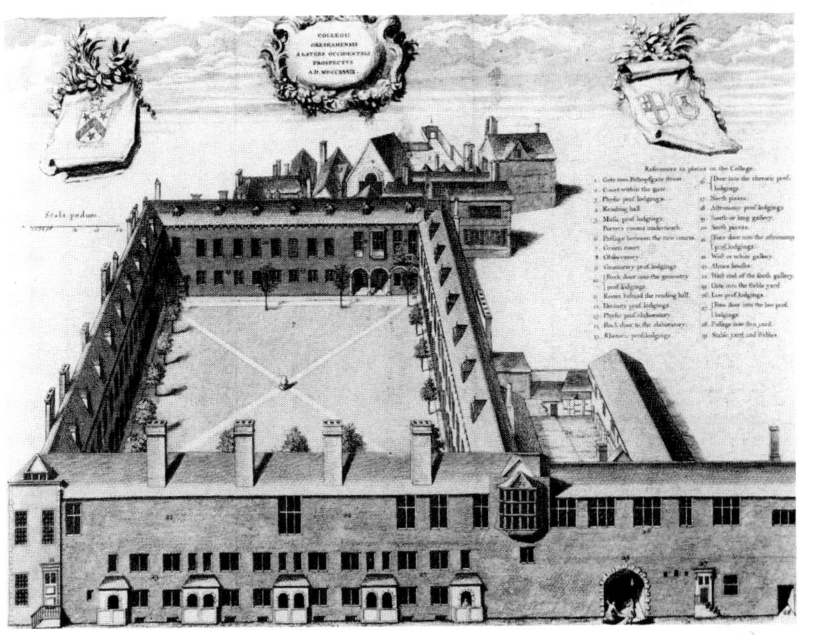

7. Gresham College, the Royal Society's usual meeting place, seen from Broad Street. Hooke's rooms were in the far right-hand (south-east) corner of the quadrangle. The lecture hall is the high-roofed building behind the quadrangle.

8. The plague subsided in December 1665 and those who had fled London started to return. John Dunstall's engraving shows people crossing the open fields north of the City.

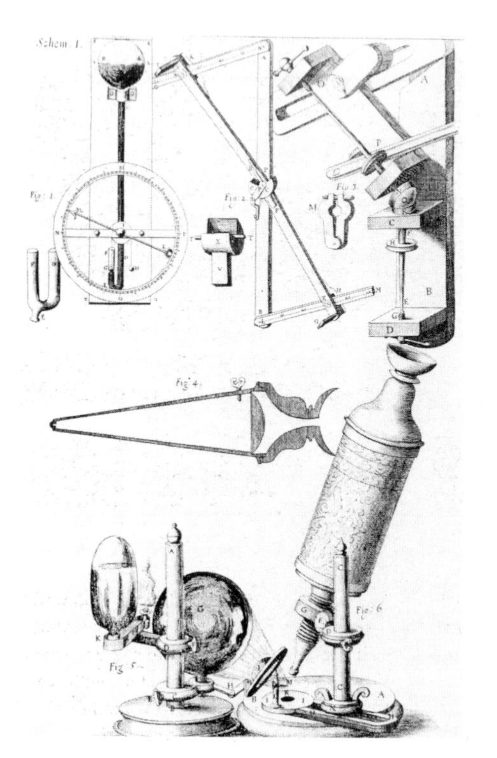

9. Some of Hooke's inventions from the early 1660s, pictured in *Micrographia*. Figures 1–6 represent a wheel barometer, a refractometer, a lens-grinding engine, a water-filled two-lens microscope, a light condenser and Hooke's main compound microscope on an adjustable stand.

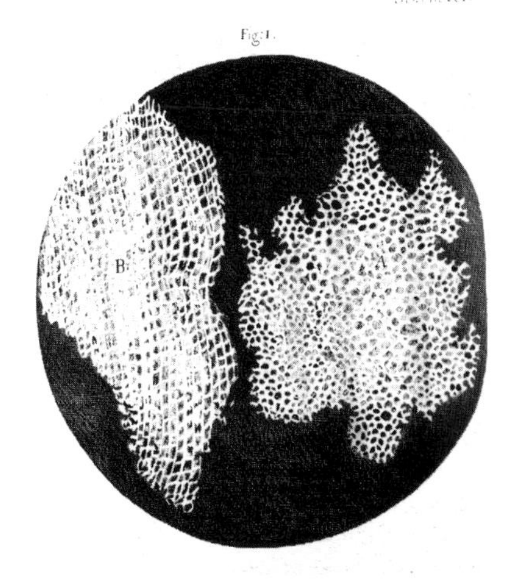

10. 'The first microscopical pores . . . that were ever seen'. Hooke called these tiny compartments in a sliver of cork 'cells', because each seemed enclosed and separate. *Micrographia.*

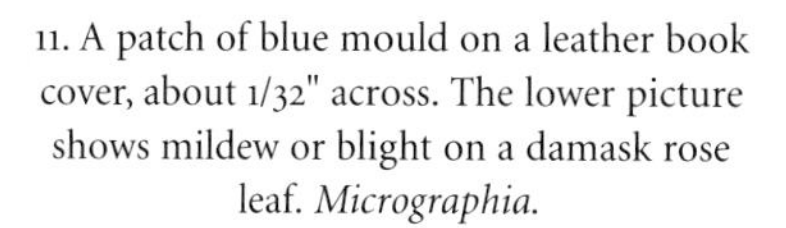

11. A patch of blue mould on a leather book cover, about 1/32" across. The lower picture shows mildew or blight on a damask rose leaf. *Micrographia.*

12. A nettle sting viewed through a powerful single-lens microscope. Below, a beard of wild oat, shown straight and bending when wet, and in cross-section, along with the hygrometer (for measuring humidity) Hooke made. *Micrographia.*

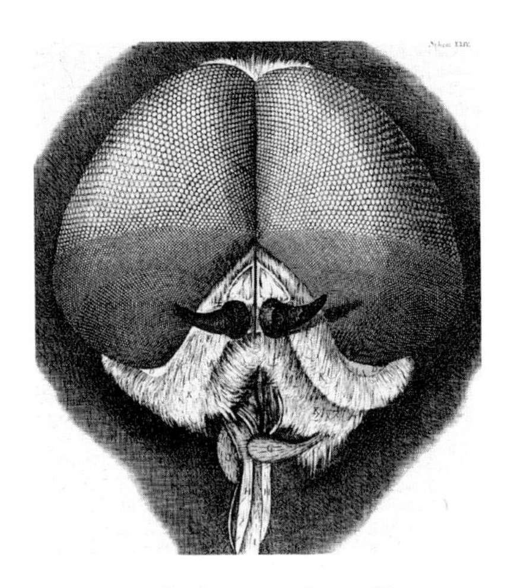

13. 'I took a large grey Drone-Fly . . . and cutting off its head, I fix'd it with the fore-part or face upwards upon my Object Plate . . . Then . . . by varying degrees of light, and altering its position to each kinde of light, I drew that representation of it which is delineated.' *Micrographia.*

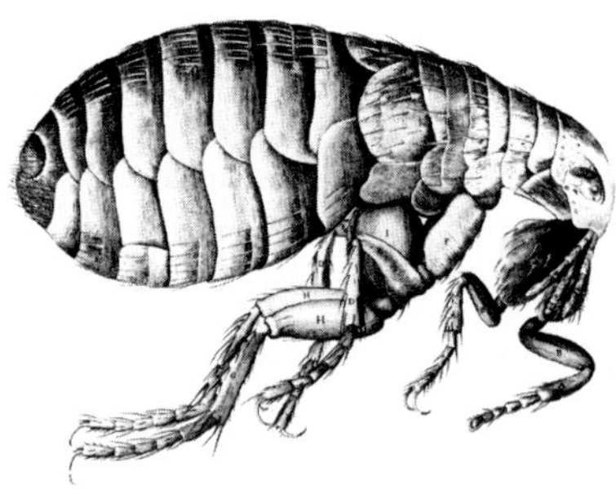

14. Hooke's drawing of a flea. 'This busy little creature' was busier than Hooke or his colleagues knew in 1665, the year of the Great Plague. *Micrographia.*

15. A London coffee house, around 1705. Anonymous drawing from E. F. Robinson, *The Early History of Coffee Houses in England* (1893).

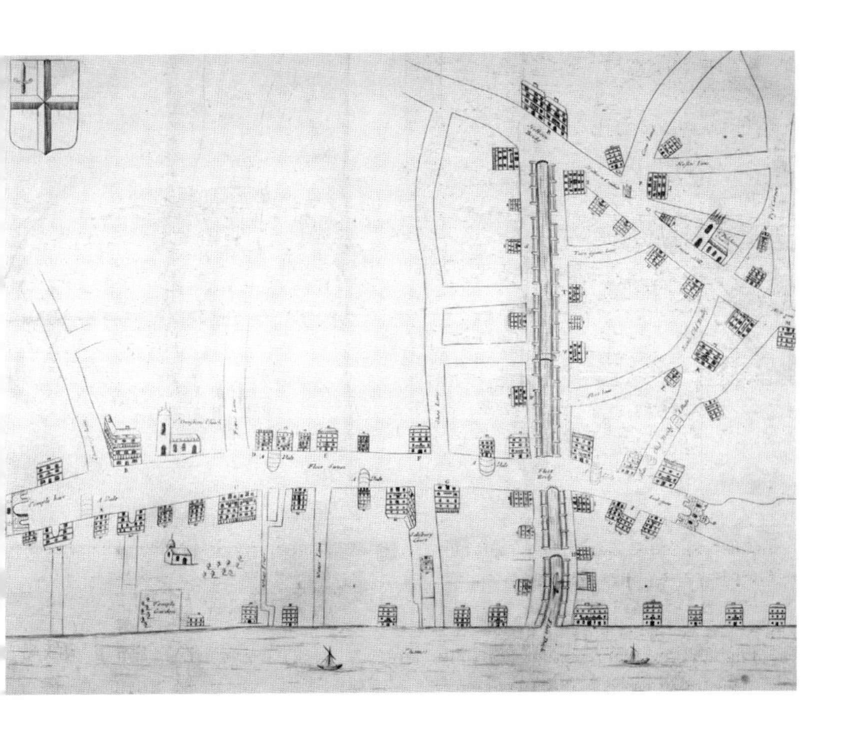

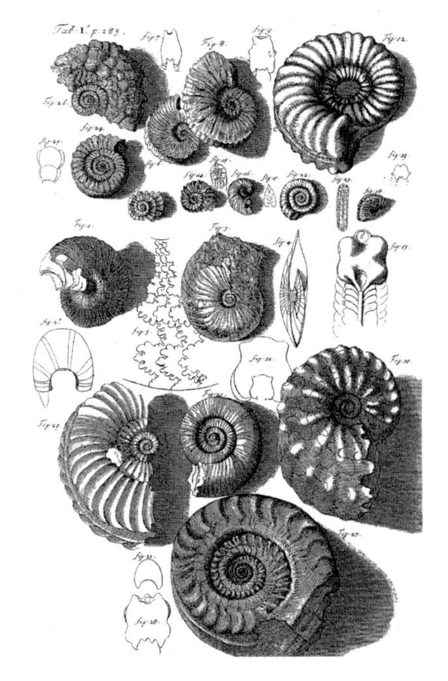

16. A map of Farringdon Ward Without, showing the new Fleet Canal and its wharves, built by Hooke, Wren and Thomas Fitch, and preparations against a possible Papist attack on London. Engraved by Andrew Yarranton, 1681.

17. Fossilized shells and ammonites found and drawn by Hooke, and used by Richard Waller to illustrate Hooke's 'Discourse of Earthquakes' (1668). Hooke, *Posthumous Works* (1705).

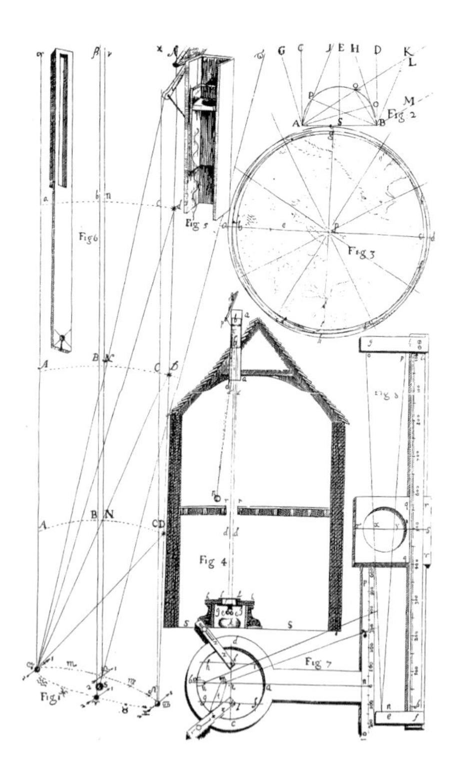

18. Hooke's 36-foot vertical tubeless telescope, erected in his lodgings in 1669 to measure tiny variations in the path of the constellation Draco. The rope (n) closed the lid when it rained and the plumblines (d), resting in water, enabled Hooke to position the eyepiece with speed and accuracy. From *An Attempt to Prove the Motion of the Earth* (1674)

19. Hooke's first major private commission, the College of Physicians in Warwick Lane, built 1671–9, destroyed 1860s–70s. Engraved by William Stukeley in 1723.

telescope to a committee of Wren, Brouncker, Moray, Sir Paul Neile and Hooke, for testing. The committee found that Newton's six-inch reflecting telescope magnified thirty-eight times, producing a larger image than a refracting telescope four times its size. Oldenburg at once sent a description of it to Huygens, to secure Newton's recognition in Europe as its true inventor, and on 11 January 1672, the day the telescope was presented to the Royal Society, Newton was elected a fellow.

Some years earlier, Hooke had also taken up James Gregory's idea of using metal mirrors to produce powerful telescopes of a more convenient size than the forty- and sixty-foot monsters he had installed at Gresham College. He made 'severall tryalls' of the unsatisfactory reflecting telescope made for Gregory by Richard Reeve in 1664. In the preface to *Micrographia* he talked briefly of having used reflecting microscopes, but he had put aside his work on reflecting telescopes around 1665, he explained in his first letter to Newton in 1672, because he found that he could achieve a truer image with a convex lens than with a concave mirror. The 'reflecting box' which Hooke had shown the Royal Society in March to May 1667 was not a reflecting telescope, but simply a means of shortening a conventional telescope by using two flat mirrors. Nevertheless, it is likely that Hooke was less than pleased to see credit for the reflecting telescope go to a newcomer, and he at once set about building a bigger and better instrument to establish his supremacy in the field.[1]

Meanwhile, Hooke found his reputation under attack (as he saw it) on another front. Newton's second letter to Oldenburg, describing his experiments on light and colours, was read to the Society on 8 February. Newton's experiments and conclusions were simple and compelling, and his 'ingenious discourse' was greeted (Oldenburg wrote) with 'singular attention and an uncommon applause'. For a more expert evaluation, Newton's paper was given to three Royal Society heavyweights, Robert Boyle, Seth Ward (now Bishop of Salisbury) and Hooke. Only Hooke, as far as we know, wrote a report. Its tone was coloured, no doubt, by his resentment at the rapturous reception Newton's little reflecting telescope had received a few weeks earlier.

Newton was already Lucasian Professor of Mathematics at Cambridge, but he was only twenty-nine, and a novice in the field of public scientific debate. Hooke, on the other hand, was an experienced and combative debater, and was regarded (not least by himself) as England's leading expert on light, refraction and optical instruments. It is unfortunate but not surprising that he treated Newton's first contribution to the subject (which did not contain the impressive mathematical exposition that Newton later provided) with a certain condescension. He heard Newton's letter read at the meeting of 8 February, and spent about four hours reading it through and writing a reply. After claiming, in typical Hooke fashion, that he was already familiar with Newton's observations ('having by many hundreds of trials found them soe') he recapitulated his argument in *Micrographia* that colours were the result of the disturbance or distortion of white light by refraction, and that they did not exist until the light passed through the prism. He accepted that Newton's explanation of what happened to the beams of light as they passed through the prisms was a plausible hypothesis, but not that it was the only possible one, as Newton had claimed:

> I do not therefore see any absolute necessity to beleive his theory demonstrated, since I can assure Mr Newton, I cannot only salve [solve] all the Phaenomena of Light and colours by the Hypothesis I have formerly printed ... but by two or three other ... Nor would I be understood to have said all this against his theory as it is an hypothesis, for I ... esteem it very subtill and ingenious ... but I cannot think it to be the only hypothesis; not so certain as mathematicall Demonstrations.

Hooke also picked up Newton's suggestion that light was a substance, a stream of particles or corpuscles, rather than a motion, and saw that this concept of the nature of light, the corpuscular theory, would fit very well with Newton's argument that white light was a compound of many colours. 'Tho yet, methinks, all the colourd bodys in the world compounded together should not make a white body; and I should be glad to see an expt of the kind.' He took the opportunity to repeat his wave or pulse theory, and to point out that it would also be possible for waves of white light to be a compound of many different wave motions. It was his view, though, that there was no more need to assume that the colours produced by the prism were components of white light, than there was to assume that the various notes produced by an organ or the plucked string of a musical instrument were separate components of the string or the air in the bellows.[2]

Hooke read his comments on Newton's paper to the Royal Society on 15 February, and Oldenburg sent a copy to Newton, as Hooke knew he would, a day or two later. The tone of Hooke's critique struck the Royal Society as hostile and disrespectful to Newton, and it was decided not to publish 'so sudden a refutation' alongside Newton's letter in *Philosophical Transactions*. In fact, it was not published until 1757. Hooke dwelt on his own accomplishments too long, as anyone who knew him would have expected, and his tone in places was patronizing, boastful and insensitive. From April to June, whenever the weather was sunny enough, Hooke repeated Newton's prism experiments for the Royal Society, confirming all the observations Newton had made. He stuck to his original opinion, nevertheless, that other explanations for the separation and reunification of colours were possible, and that Newton's demonstration was not an *experimentum crucis*. No doubt Hooke did not recognize the genius of the young man he was dealing with, and he was certainly unaware of Newton's neurotic sensitivity to criticism or opposition of any kind. But his central argument, that Newton's experiments as described in the first letter were not conclusive, and that other explanations for the phenomena he reported were possible, was not unreasonable, as nineteenth-century work on light and colour eventually demonstrated.[3]

Newton's immediate reply to Hooke's February paper was mild, but four months later, prompted by Oldenburg, he sent the Royal Society a very full, powerful and personal response, which clarified his original paper and answered Hooke's key points one by one. He reprimanded Hooke for misunderstanding his main argument, for failing to recognize the power of his prism experiments and for using the occasion to reiterate theories from *Micrographia*. He rebuked Hooke for urging him not to give up work on improving refracting telescopes: 'it is not for one man to prescribe Rules to ye studies of another, especially not without understanding the grounds on wch he proceeds'. He was surprised and irritated that Hooke had taken his tentative suggestion that light was a body or substance for his fundamental supposition, and that he had implied that his work on colours depended on it. He rose to Hooke's challenge to provide experiments to show that a mixture of many colours would produce white. Rainbow colours turning to white on a spinning top or wheel, the effects of coloured light in a church with stained-glass windows, the white effect of the multi-coloured bubbles of soapy froth or of a roughened metal mirror, the greyness of household dust (assuming that dust is a compound of many coloured particles, and grey is a less luminous form of white), all helped to make his point.[4]

Although Newton dismissed the wave or corpuscle question as a side issue, he found it much easier to envisage the separation and reconstitution of white light if it was composed of particles of different colours, and unlike Hooke he could not accept that a light wave might behave in the same way. 'I can easily imagine how unlike motions may crosse one another, yet I cannot well conceive how they should coalesce into one *uniforme* motion, & then part again & recover their former unlikenesse.' Furthermore, he regarded it as impossible that light, which apparently travels in straight lines, could be composed of waves or vibrations, which would spread or bend round corners as sound waves do. In passing, he suggested a fundamental improvement to Hooke's wave theory, by proposing the idea that vibrations of different colours might be of different 'depths or bignesses' or (as we would say) wavelengths, and thus refract in different ways. 'That the largest Vibrations are best able to overcome the resistance of a refracting superficies [surface], and so break through it with least refraction: Whence ye vibrations of severall bignesses, that is, the rays of severall colours, wch are blended together in light, must be parted from one another by refraction, & so cause the Phaenomena of Prisms & other refracting substances.' It is ironic that Newton, whose authority enabled the corpuscular theory of light to triumph over the wave theory for 150 years, should have proposed such an important refinement to the theory he rejected.

In his covering letter to Oldenburg, Newton said that (in implied contrast to Hooke) he had 'industriously avoyded ye intermixing of oblique & glancing expressions'. In fact in draft after draft of the letter he had made his references to Hooke more numerous and more offensive. In the final version Hooke's name was mentioned over thirty times. As Newton's biographer says, he 'virtually composed a refrain on the name Hooke'.[5] For Hooke, who had to listen to

Oldenburg reading out parts of Newton's offensive rebuttal in a Royal Society meeting on 12 June, and then see it printed – unanswered – in Oldenburg's *Philosophical Transactions*, the experience must have been deeply wounding. His response to it, which was apparently written for Lord Brouncker rather than Oldenburg or Newton, shows that he was surprised and stung by the venom of the attack. He tried to undo the damage done by his initial critique with apologies and disclaimers: 'if there be any thing therein that any ways savors of incivility or reproach I doe heartily begge his pardon and assure him twas innocently meant ... I was soe far from imagining that Mr. Newton should be angry that I cannot yet beleive that he is', since in a philosophical dispute 'a freedome & liberty of Discoursing and arguing ought to be Tollerated.' Hooke protested (not very convincingly) that he would be only too happy to see his own hypothesis on colours disproved, and that he did not insist on its truth: 'Mr. Newton is much mistaken if he thinks that I opposed his hypothesis for the sake of asserting my owne, for possibly if that were the hypothesis asserted I could produce more objections against it then it may be he hath yet thought of. In short I will assure him that I doe as little acquiesce in that for a reality as I doe in his.'

Once these pleasantries were over, Hooke returned to the argument. He concentrated his discussion on the nature of light, and wondered why Newton now seemed 'to be afraid of saying what a Ray of light is ... he seems now unwilling to stick to his first supposition that it is a body, I suppose because he sees there will be more objections against that than against the supposition of Light a motion'. He tackled Newton's argument that if light were composed of waves there would not be distinct shadows by describing his experiment with a thin beam of light in a darkened room. 'By placing the Edge of a razor in the cone of the Suns radiation at a pretty distance from ye hole ... and by holding a paper at some Distance from the razor in the shadow thereof, your Lordship plainly saw, that the Light of the sun Did Deflect very deep into the shadow'. It was an experiment like this, a demonstration of diffraction, that was used by Augustin Fresnel in 1818 to establish that light did indeed travel in transverse waves (with the waves at right angles to the rays), just as Hooke had argued in *Micrographia*. Hooke stuck to his claim that the crucial questions, whether white light was homogeneous or a mixture, and whether colours were created by refraction or already existed in unrefracted light, were still unresolved, despite Newton's so-called *experimentum crucis*. The distinctive behaviour of a coloured ray after refraction did not prove that it had had a distinct existence before refraction.[6]

Since this letter was found in Hooke's papers, not Newton's or Brouncker's, it is quite likely that he decided not to send it, and there is certainly no evidence that it ever reached Newton. The letter's tone and content were not especially conciliatory, and even if Newton had read it there is no reason to assume that it would have softened the animosity that had begun to develop between the two men. It is often said that the first dispute between Hooke and Newton was fomented by Oldenburg, who is depicted as a meddler and a breaker of confidences. But in this debate (as the editors of his correspondence have pointed out) Oldenburg simply carried out the orders of the Royal Society, and acted as an intermediary between scientists as he had always done. In his second letter Hooke protested that his first one was 'never intended . . . for Mr. Newton's perusall', but he must have known that papers read before the Royal Society were always sent to those affected by them. Even if he had wished to do so, Oldenburg could hardly have prevented a debate between two such sensitive and self-regarding men from becoming acrimonious. No doubt Hooke's hasty critique was largely responsible for getting the relationship off to such a bad start, but Newton was not an easy man to deal with, and had a succession of bad-tempered disputes with other scientists later in his career. Even the great Huygens found his letters on Newton's paper on light and colours answered in a sharp and reprimanding tone, driving him to reply: 'seeing that he maintains his doctrine with some warmth, I do not care to dispute.' Newton so disliked the rough and tumble of public debate that in 1673 he threatened to resign from the Royal Society and tried to cut himself off altogether from other scientists.[7]

Newton's dramatic triumph threatened Hooke's primacy as an inventor and maker of optical instruments as well as his reputation as a scientist. One of the issues that had caused greatest heat in the Hooke–Newton correspondence was the question of the value of refracting telescopes, of the kind that astronomers had been using since 1609. Hooke had spent a great deal of his own time and other people's money in trying to improve these telescopes, and his recently completed observations of Draco and stellar aberration had been based on them. Unlike most of his contemporaries, Hooke would not accept Newton's conclusion that refracting telescopes were bound to suffer from the dispersion of light into its component colours. Since the mid-sixties he had been working on a new type of compound lens which would overcome some of the problems of chromatic aberration that Newton mistakenly believed were ineradicable in refracting telescopes. He had dropped many hints about what he was doing, but kept some of the details frustratingly secret. Probably fearful that Newton would snatch this prize too, Hooke chose the Royal Society meeting of 18 January 1672, a week after Newton's reflecting telescope had been seen, to announce that he had invented a new way of making lenses for microscopes, telescopes and burning glasses 'by which the light and apparent magnitude of bodies may be most prodigiously and regularly increased; and whatever almost hath been in notion and imagination, or desired in optics, may be performed with great facility and truth'. Rather than describe his device in plain English Hooke lodged it in code, but told the secret to the Society's President, Lord Brouncker, and to Wren. The next week Wren and Brouncker declared their preliminary approval of Hooke's method, but the Society, wary of Hooke's tendency to promise more than he could perform, urged him to do something 'that might convince the world of the reality thereof'. The solution to Hooke's cipher is lost, but apparently what he had in mind was a compound lens filled with fluid. He had been studying the refracting power of fluids since the early 1660s, and in *Micrographia* he said that he had made microscopes with 'Waters, Gums, Resins, Salts, Arsenick, Oyls, and with divers other mixtures of watery and oyly Liquors'. These were interesting ideas, but there was a better solution to the problem of making an achromatic compound refracting lens, and it was almost under Hooke's nose. When Chester Moor Hall made the first such lens in 1729 he cemented a convex lens of conventional crown glass to a concave lens of flint glass. Hooke knew the first English manufacturer of flint glass, George Ravenscroft, and visited his works in 1673, but apparently did not investigate the optical properties of his glass.[8]

By March the shine had gone off Newton's little reflecting telescope. Its metal speculum had tarnished, and even Newton confessed that the 'instrument hath its imperfections both in the composition of the metall & in its being badly cast'. The key to producing an effective and long-lasting reflecting telescope, which Newton had not really done, was to make a concave metal mirror, or speculum, which was not distorted and would keep its shine. Instructed by the Royal Society, but driven by his own desire to reassert his supremacy as a master of optical instruments, Hooke had started working in January 1672 to make a better telescope than Newton's, using 'a Metall not obnoxious to tarnishing'. By 25 January, helped by London's leading optical craftsman, Christopher Cock, he was able to show the Royal Society a four-foot reflecting telescope with a partly polished metal speculum, which 'did pretty well, but was undercharged'. It was shown again the next week, when Hooke was told 'to see it perfected as far as it was capable of being'. In March he came up with some new ideas for using silver in reflecting lenses and for making concave lenses in large numbers by stamping them out between two dies, one concave, the other convex. His very first diary entry, on 10 March 1672, mentioned these two ideas: 'I told Cox how to make Reflex glasses by Silver and hinted to him making them by printing'. In April Cock produced a new steel concave, but his next effort, on 8 May, was pronounced falsely polished by Hooke, and in July Hooke declared that his refracting lens still gave better results than his reflecting one. His diary entry for 28 July shows how effective he thought his best telescopes could be: 'viwed Mars and saw a spott in it & I believe a satellite'. Since Mars' two small satellites were not seen again until 1877, it is unlikely that Hooke had seen one of them with his imperfect telescope.

Over the summer of 1672, when the Royal Society was in its long recess, Hooke worked with Cock to make a perfect steel speculum. Nearly every day in early August he either saw Cock or polished his seven-inch speculum. Diary entries for 9 and 11 August suggested good progress: 'Gave Lord Brouncker Reflex Speculum . . . Fitted my Newton'. A week later Hooke started polishing a larger speculum for a nine-foot telescope in a new workshop he had built in Gresham College cloister, and had his carpenter, Coffin, make a frame for it so that it could be tried out. On 23 August he saw the Moon with it, 'very big and distinct', but in early September an observation of Mars went less well. For the rest of September he was preoccupied with his work on the Fleet Ditch, the Royal College of Physicians, Bridewell, the Thames wharves, St Michael Cornhill, and other London duties. At the beginning of October the Royal Society committee that had been meeting weekly in his rooms through the summer urged him to finish the telescope, and his friend Sir Robert Moray gave him 'a metall of excellent Reflection that would not tarnish'. He had made the task more difficult by increasing the size of the speculum, and it was well into November before Cock had made the grinding tool that he required to finish it. On 21 December Cock spoiled the tool in some way, and when the metal began to warp a week later he began to lose heart. 'Cox the Glasse grinder thinkes that neither [Hooke's] Devices nor his Tellescope will obtain repute in the World, the Metall suddainly tarnishing', the mathematician and prolific scientific correspondent John Collins wrote to James Gregory.[9]

Hooke was seriously ill in December 1672, suffering from giddiness and sinus infection, or perhaps something worse. The word in London was that he was dying from tuberculosis. John Collins, who kept his friends well informed, reported that 'Mr Hooke is in a consumption and unlikely to recover'. But in January 1673 his new assistant Harry Hunt arrived, and Hooke taught him how to grind a speculum with a lead mallet. Hunt worked on it in his room, and on 22 January they carried it to Arundel House, to prove to the Royal Society that they were making progress. Hooke and Hunt kept grinding and polishing, and by the beginning of February the speculum seemed true, if not fully polished. But several trials in February were disappointing, and Hooke turned to other matters. It is possible that a letter to Collins from James Gregory, which was read at the Royal Society in March 1673, helped to change Hooke's thinking about making a speculum. Gregory suggested that the problems of metal speculums could be overcome by the use of mirrored glass, so designed that the glass surface and its mirrored rear would have the same focal point. On 29 July Hooke visited George Ravenscroft's new Savoy glassworks with Wren, and two weeks later he told Cock, who had made a new speculum, that he had found a 'new way of pollishing glasses by a small gage or tooth'. Between August 19 and 23 Hooke and Cock worked on a glass speculum with a mercury-foiled back, helped by the glass-grinder Hugh Bolter, or 'Dog Rogue Bolter', as

Hooke called him. No more is heard of the glass speculum in the diary or the Royal Society minutes for nearly six months, but a letter was found in Hooke's papers, probably written around this time, describing and picturing his new reflecting telescope. Hooke's telescope followed Gregory, and differed from Newton's, in that it allowed the observer to look directly at the object being viewed. The eyepiece was set in the middle of the large speculum, and the observer saw an image reflected from this object speculum onto a smaller concave mirror set in the middle of the tube.[10] Finally, on 5 February 1674, almost exactly two years after he had set himself the task, Hooke presented his reflecting telescope to the Royal Society, meeting in his own rooms. It was, as the minutes of the meeting acknowledged, the first practical Gregorian telescope, in which the observer looked directly at the object, that had ever been constructed.

Hooke's extraordinary fertility of mind, and his continuing need to satisfy the Royal Society's appetite for novelties, involved him in a wide range of scientific and technical projects even at the height of his dispute with Newton. Some of these were comparatively trivial, but in others he was working at the frontiers of knowledge. To begin with the trivial: Hooke had a long-standing interest in unorthodox methods of communication, and on 29 February 1672 he proposed a new method of long-distance signalling, using large letters or secret symbols, high ground and a telescope. The Royal Society tried the idea out the following week, using foot-high letters and a man with a two-foot telescope in a boat anchored across the Thames from Arundel House. Long-distance signalling had been used since ancient times, but Hooke's scheme was more elaborate and ambitious than the simple signalling with flags, shields or fires that had preceded it. However, nobody rushed to find a practical use for his idea. In 1793 the French government adopted an optical telegraph system much like Hooke's, combining semaphore and telescopes, and the British Admiralty introduced a signalling system based on shutters with black and white faces to coordinate coastal defence in 1795. The Royal Navy did not use semaphore or flashing lamps until the mid-nineteenth-century.[11]

Hooke pursued old interests alongside his newer one. In February and March 1672 he presented experiments to illustrate two phenomena that had fascinated him for years, combustion and congruity. To demonstrate the puzzling effects of congruity (the tendency of particles of one or more substances to cling together) he blew bubbles in soapy water. The experiment suggested more questions than answers:

> It is pretty hard to imagine, what curious net or invisible body it is, that should keep the form of the bubble, or what kind of magnetism it is, that should keep the film of water from falling down, or the parts of included and including air from uniting. The experiment, though at first thought it may seem one of the most trivial in nature, yet as to the finding out the nature and cause of reflection, refraction, colours, congruity and incongruity, and several other properties of nature, I look upon it as one of the most instructive: of which more hereafter perhaps.[12]

In his combustion experiment Hooke hoped that by 'looking steadfastly' at a candle flame magnified by a lens or concave mirror he would be able to see the process by which the candle reacted with the air to produce smoke, flame, light and heat. He watched the stream of liquor or 'jet d'eau' rising from the wick, and pronounced that this was 'nothing else but the mixture of the air with the parts of the candle, which are dissolved into it in the flame; ... the action of dissolution in most bodies producing heat and light'. Unlike Boyle, Hooke believed that only the 'nitrous' part of the air was involved in burning, and he was interested in finding out whether this part of the air disappeared in the chemical process of combustion. This would not be unexpected, in the light of the loss of volume he had found when two 'congruous' liquids were mixed. Later that year, he devised an apparatus which would indicate whether air was consumed during combustion, as he thought. The details of the experiment are unclear, but it probably involved placing an upturned glass vessel over water, the method used by John Mayow at around the same time, and by Lavoisier a hundred years later. Lavoisier's advantage over Hooke and Mayow was his possession of phosphorus, which burns the oxygen in air more completely than a candle, and which was not available until 1677. Hooke used a burning glass to set his experiment going and had trouble with leaks. From November 1672 to February 1673, he tried the experiment repeatedly, both in his rooms and at the Royal Society, but the apparatus always failed. At last, on 5 March 1673, the experiment seemed to succeed, and the volume of air decreased. Several failures followed, but on 18 March he recorded this historic moment in his diary: 'tryd the Experiment of fire by the help of burning glasse and found air decrease.' The next day he told the Royal Society that the air had lost a twentieth of its volume. What had happened, unknown to Hooke, was that the candle had gone out when about a quarter of the oxygen in the air had been converted into carbon dioxide, which was then dissolved in the water.[13]

The word 'gas' was coined in the early seventeenth century by the eccentric chemist or alchemist J. B. Van Helmont, who collected the carbon dioxide produced by burning charcoal and called it 'gas sylvestre'. Van Helmont mixed up his useful chemical ideas with so much supernatural nonsense that their meaning was lost, and as far as Hooke, Boyle and their contemporaries were concerned there was no gas but air. Yet Hooke was fairly sure, on the strength of his recent experiments, that air contained a particular 'nitrous' component, at least 5 per cent of its volume, which was necessary to respiration and combustion. What he and Boyle needed to find out now was whether other gaseous 'exhalations' that they could create were 'true air', able to sustain life and fire. A month after his successful combustion experiment, Hooke put aside the liquor penetration experiments he had been working on, and joined Boyle in creating and testing gases. On 23 April, at a Royal Society meeting, they mixed aquafortis, or nitric acid, with crushed oyster shells (calcium carbonate) in a bottle with a bladder fixed to its top to collect the 'air' that would be produced by their reaction. The swelling bladder kept the Society interested over the next few weeks, and on 14 May, in a well-attended meeting, Hooke tried to test its effects on a burning candle. The 'factitious air' in the bladder, which was carbon dioxide, was fed 'by a certain contrivance' into a jar, but mixed with some true air in the process. A candle which had burned for fifty-five pendulum swings with true air now went out after forty-five swings. Two weeks later he tried the experiment again, this time using an open-topped vessel. When the candle was held near the top of the glass it burned well, but when it was lowered into the 'produced air' it went out at once. It was clear, as Hooke recorded in his diary, that the 'damp air' actually quenched the candle's flame. He celebrated his success by going to Garraway's with Daniel Colwall and

Lord Moray, and then going home to have sex with Nell. A week later he tried the same experiment with the gas from bottled ale, and got the same result. Hooke had come a long way in a few weeks. He had demonstrated the existence of gases that were not air, shown that an ingredient making up a small proportion of the volume of air was used up in combustion, and provided clues that combustion was a chemical process, not just a physical mixing. But he took his study no further, and switched his attention back to optical matters (reflecting telescopes and precision sights for quadrants) and to his work on rebuilding London. It was left to his friend John Mayow to pursue the experiments and publish his work on the role of 'nitro-aerial' matter in respiration in combustion in 1674. Mayow was only thirty-six when he died in 1679, after marrying a penniless woman in the belief that she was wealthy. With his death, interest in promoting the ideas that he and Hooke had pioneered faded, and the way was left open for the triumph twenty years later of the altogether misguided phlogiston theory of Georg Ernst Stahl. A hundred years on, when Lavoisier and Priestley overthrew Stahl's theory, they did so in part by revisiting the work of Mayow and Hooke.

Hooke's workshop and papers were full of unfinished or forgotten projects, which were only made public when he feared that a rival was about to claim authorship of the same idea. This habit of keeping his inventions secret, and then declaring, as soon as another scientist announced a particular invention, that he had made a better one, or that he had had the same idea years ago, earned him a reputation as a braggart who was almost incapable of conceding another's superiority in any field. Beyond his own intimate circle Hooke's unpublished or forgotten achievements were often doubted, and he was forced to rummage through old Royal Society records to find evidence that he had made an announcement or performed an experiment at a meeting five or ten years earlier. His production of a micrometer in 1667 to beat Richard Towneley's was a recent example of his practice, and his reaction to Newton's paper on light and colours was another. There was a third example in 1673, when the brilliant young German philosopher and mathematician Gottfried Wilhelm Leibniz arrived in London with his new arithmetical calculating machine. Leibniz demonstrated the machine, which used a set of revolving geared discs to add and multiply, to the Royal Society on 22 January. At this meeting, as Leibniz later complained to Oldenburg, Hooke took an almost indecent interest in the workings of the machine:

> he removed the back plate which covered it, and absorbed every word I said; and so, such being his familiarity with mechanics and his skill in them, it cannot be said that he did not observe the machine. That he did not distinctly trace out its wheels I readily admit. But in such cases it is enough for a man who is clever and mechanically-minded to have once perceived a rough idea of the design, indeed the external manner of operation, and then for him afterwards to add to that a little of his own . . .'[14]

If Leibniz had been worried by Hooke's interest in his machine, he was right. Between 2 and 4 February 1673 Hooke worked on an engine that could multiply to twenty places, and on 5 February he told the Royal Society that he could make a simpler and more reliable machine than Leibniz's. An afternoon spent with Hooke on 8 February did not calm Leibniz's anxieties, and he wrote privately to Oldenburg two weeks later expressing the hope that Hooke, who had shown no previous interest in arithmetic machines and who had so many inventions to his credit, would not 'demand possession of that which had been publicly proposed already by someone else'. But if Hooke thought something could be done better, such scruples did not hold him back. During late February and early March 1673, when he was also busy with the speculum and weather clock, he worked with Richard Shortgrave, Harry Hunt and the mathematician Dr John Pell to produce a working calculating machine. On 5 March, at the Royal Society meeting in which he showed his fire experiment successfully ('air decreased') for the first time, Hooke demonstrated his new machine with satisfying results: 'all understood and pleased'. After that his optimism seems to have faded. In a paper he gave to the Royal Society on 7 May 1673 he took a more realistic view of the value of mechanical calculators. For addition and subtraction, no machine could beat pen and paper for convenience and speed, and for squares, cubes and square and cube roots John Babington's or Dr Pell's mathematical tables were unrivalled. Multiplication and division were best done using Lord Napier's metal or parchment rods ('Napier's Bones'), and Leibniz's machine, with its 'wheels, pinions, cantwrights, springs, screws, stops, and truckles', was just too complicated, bulky and expensive to be of any practical use. Its results could only be checked by repetition, and its only novel quality was its ability to multiply and divide. Hooke ended on a familiar note: 'I have an instrument now making, which will perform the same effects with the German, which will not have a tenth part of the number of parts, and will not take up a twentieth part of the room, that shall perform all the operations with the greatest ease and certainty imaginable'. What happened to Hooke's machine after this is not clear. In a Cutler lecture written shortly afterwards he claimed to have 'a mechanical way of calculating and performing Arithmetical operations, much quicker and more certainly than can be done by the help of Logarithms', but he did not present it to the Royal Society, or mention it again in his diary. Perhaps the most significant result of his attempt to make a calculating machine was that he earned the hostility of one of the rising stars of European natural philosophy. Leibniz wrote angrily to Oldenburg in February 1673 that any 'right-minded and decent' man would yield credit to another 'rather than incur the suspicion of intellectual dishonesty and want of true magnanimity should they chase after falsehood with an unworthy kind of greed'. Hooke's claims to priority were 'unworthy of his own estimate of himself, unworthy of his nation, and unworthy of the Royal Society'.[15]

13. Measuring the Heavens

(1673–1676)

In November 1673 the Council of the Royal Society decided to move its weekly meetings from Arundel House back to Gresham College, 'considering the conveniency of making their experiments in the place where Mr Hooke, their curator, dwells, and that the apparatus is at hand'. Hooke prepared for the Society's return by buying six new chairs, and on 1 December the Joint Gresham Committee threw a party in the Committee Room, and 'with wine and sweetmeats welcommed the Society thither'. The Committee was doing its best to make the College a place of peace and scholarship again, but it was not easy. In November a sub-committee resolved to remove 'families now residing there; and unfitt meetings which were kept there', and on 7 December the Joint Committee heard a report from Hooke on the state of the college. He told them that since the summer a garden to the south of the college, and right next to his rooms, had been converted into a coaching inn and stables. There was 'a great & continuall stench and noyse from the said new-made hostelry they having made a horse Pond and a Laystall or Dunghill where all the Dung of the said Stables is heaped up'. As a solution to this problem, Hooke asked the Committee to agree

> that a new way may be made into the Long Gallery out of the roome where in the Royal Society meets in Gresham Coll that thereby he may be inabled to make himself a study with light into the Quadrangle of the College; and soe free himselfe from sitting as now he doth in the suffocating stink of the stables hors dung and Jakes [privy], which hath bin a great cause of his late illnesse.[1]

The Committee agreed that he should make these changes 'at as Cheape a Rate as may be', and left him to fix the details with their clerk, John Godfrey. Hooke celebrated by going to see Lord Rochester's new five-act verse tragedy, *The Empress of Morocco*, at the Duke's Theatre, a new Wren theatre on the river front west of the Fleet.

Sitting in his close and foul-smelling study that November and December, Hooke read Johann Hevelius's new book, *Machina Cœlestis* ('The Celestial Machine'), a stout defence of the traditional astronomical methods he had inherited from Tycho Brahe, and an implicit rejection of Hooke's ideas on observational accuracy. The book struck Hooke as being obstinately – even dangerously – old-fashioned. Hevelius's instruments seemed to be hardly more accurate than those used by Tycho Brahe in the 1590s, and it seemed tragic and almost incredible that progress in astronomy should be threatened by the influence of one man, however great his past achievements. He seized the opportunity to challenge Hevelius, and to produce a manifesto for the application of the best modern technology in astronomy and the other sciences. He spent the morning of 10 December writing a lecture criticizing Hevelius's book, and delivered it as a Cutler lecture to an audience of leading natural philosophers the following day. Warned by Oldenburg (who was in Hooke's audience) of Hooke's impending attack, Hevelius repeated his claim that telescopic sights were prone to error, and appealed again to his long record of practical observation.

To strengthen his case against Hevelius, Hooke conducted an experiment on the members of the Royal Society on 15 January 1674. Using a marked ruler held at an appropriate distance, he showed that not a single man at the thinly attended meeting could distinguish a minute of a degree with his unaided eyes, however 'earnestly and curiously' he tried. A week later a second experiment 'convinced all', and Hooke declared that for under £10 he could make a quadrant with which an astronomer could measure angles to a second.[2] All through 1674 Hooke worked with Tompion on his new quadrant, demonstrating his improvements one by one to his friends in Garraway's. One of those he kept in the picture was the ambitious young astronomer John Flamsteed. On 7 May Flamsteed gave Hooke some advice on making metal reflectors for the quadrant, but Hooke was not impressed: 'My way much better thus which I told not'. At the end of the month they met again, but Flamsteed's reaction to the quadrant evidently displeased Hooke, who noted in his diary that

Flamsteed was 'a conceited cocks comb'. Flamsteed could have been Hooke's strongest ally in the battle for precision instruments, but personal animosities, for which both men were probably to blame, were already driving them apart.

Hooke's attack on Hevelius was finally published as *Animadversions on the First Part of the Machina Coelestis of Johannes Hevelius* in the middle of December 1674. *Animadversions* started with a sustained and convincing exposure of the inadequacy of Hevelius's astronomical instruments. Hooke pointed out that on the arc of a brass quadrant or sextant with a six-foot radius, a minute measured only a fiftieth of an inch, and a second a three-thousandth of an inch, which no man living could distinguish with his naked eye. It was conceivable that a man of exceptional eyesight might distinguish angles to half a minute, but it was 'not possible to make any observation more accurate, be the instrument never so large.' Plainly, he argued, instruments could not be more accurate than their least accurate component, and all the care that Hevelius had devoted to marking the divisions on his instruments, refining his plumb lines, improving his sights and increasing the scale and rigidity of his quadrants counted for little, 'since the power of distinguishing by the naked eye is that which bounds and limits all the other niceness [precision]'. In short, Hevelius's instruments were no more accurate than those used by Tycho Brahe almost a hundred years earlier, and all his achievements were undermined by his use of plain sights. Nothing that Hevelius had written could explain his strange resistance to telescopic sights, whose value Hooke had painstakingly explained to him in the 1660s. It was as 'if he had some dread of making use of Glasses in any of his Sights'.

> Whether it were, that he supposed Glasses to have some hidden, un-intelligible, and mysterious way of representing the Object, or whether from their fragility, or from their uncertain refraction, or from a supposed impossibility of fixing them to the Sights, or whether from some other mysterious cause, which I am not able to think of or imagine, I cannot tell.

Hooke tempered his attack on Hevelius with praise of his contribution to the progress of astronomy: 'he hath gone as far as it was possible for humane industry to go with Instruments of that kind, and . . . he hath calculated them with all the skil and care imaginable, and delivered them with all the candor and integrity.'

> But yet I would not have the World to look upon these as the bound or *non ultra* of humane industry, nor be perswaded from the use and improvement of Telescopic Sights, nor from devising other ways of dividing, fixing, managing and using Instruments for celestial Observations, then what are here prescribed by *Hevelius* . . .

Hooke ended his attack with an eloquent, if somewhat boastful, appeal to natural philosophers to keep up the search for greater precision in scientific observation, to move on from the achievements of Tycho Brahe, and to take his own approach, rather than Hevelius's, as their model:

> I have myself thought of, and in small modules try'd some scores of ways, for perfecting Instruments for taking of Angles, Distances, Altitudes, Levels, and the like, very convenient and manageable, all of which may be used at Land, and some at Sea, and could describe 2 or 3 hundred sorts, each of which should be every whit as accurate as the largest of *Hevelius* here described, and some of them 40, 50, nay 60 times more accurate . . . These I mention, that I may excite the World to enquire a little farther into the improvement of Sciences, and not think that either they or their predecessors have attained the utmost perfections of any one part of knowledge, and to throw off that lazy and pernicious principle, of being contented to know as much as their Fathers, Grandfathers, or great Grandfathers ever did, and to think they know enough, because they know somewhat more than the generality of the World besides: . . . Let us see what the improvement of Instruments can produce.[3]

To illustrate his point Hooke proposed the construction of a huge brass semicircle of thirty feet radius, mounted on a firm stone wall built exactly along the meridian, and marked in degrees, minutes and seconds. The moving limb of the semicircle should be a counterweighted thirty-foot telescope, fitted with a convenient reflex eyepiece and a seat from which the operator could scan the whole celestial arc by turning the handle of a windlass. 'By this means (in one Nights

Observation) the Declinations of some hundreds of Stars may be taken to a Second by one single Observator . . . the Observations will be nearly 30 times more accurate, the charge not a quarter, and the labour not a tenth part so much as the other wayes made use of by *Ticho* and *Hevelius*'. With his eye on Charles II, Hooke argued that this 'mural quadrant' deserved the support of 'some Prince, whose Name and Honour will thereby be Registred among those glorious Celestial Bodies to all Posterity'.

A few pages later Hooke described another remarkable invention, which brought together some of his most ingenious devices in one extraordinary machine. He described a large iron quadrant, with a radius of about five feet, in which accuracy was achieved not only by telescopic sights and cross-thread micrometer eyepieces but by the clever use of a screw adjustment on the movable limb of the quadrant. Hooke had first described this perpetual screw device, which is still used in marine sextants, in a Royal Society meeting in March 1666. It could be used as a dividing engine, to mark the divisions of a quadrant with much greater precision than existing geometrical techniques. In explaining how a circle could be divided mechanically, using gearwheels instead of geometry and approximation, Hooke anticipated Jesse Ramsden's work on dividing engines and other machine tools in the 1770s, but Ramsden's precision engineering techniques gave him much higher levels of accuracy than Hooke could have achieved. Hooke's screw device also allowed an observer to make more exact readings than before. To move the limb of a five-foot quadrant through one degree, or one inch along the arc, might take thirty screw turns, with each turn representing two minutes. Each turn of the screw could itself be divided into many parts, so that the final adjustment would be 'so exceeding exact, and withal so Mathematically and Mechanically true, that 'tis hardly to be equallized by any other way of proceeding.'[4]

Hevelius had several assistants, but Hooke's aim was to create an instrument which could be used by a single observer. To this end, there was a long turning rod running from the adjusting screw to the observer's position, enabling one man to move the telescopic limb of the quadrant without help. Another improvement, also derived from Hooke's work on quadrants and sextants in 1666, was to use oblique reflecting eyepieces to enable an observer to look through a small hole above the point at which the two telescopes were joined and see objects through both at once, whatever the angle between them. This was possible because Hooke placed the eyepieces at the centre of the quadrant, where the two telescopes met, rather than at the circumference, which was more usual. As soon as the two images touched, the observer stopped turning the long-handled micrometer screw, and the angle could be accurately established. Hooke also designed his fixed sight as a two-way telescope with two object lenses, enabling the observer to measure angles of up to 180 degrees by adjusting the eyepiece and thus look in the opposite direction. To match the accuracy of the calibration and the eyepiece, he added a water-filled bubble-level, refined but not invented by him, to ensure that the quadrant could be placed exactly perpendicularly for the measurement of azimuths. A short glass tube, made with care by a good glassmaker, was more accurate and convenient than a long plumb line.

Hevelius's large quadrant had to be moved by turning several hand screws, a task which needed at least two experienced observers, and which had to be repeated whenever the objects under observation moved, or appeared to move. Hooke's solution to this problem of continuous readjustment was bold and (literally) revolutionary – he designed an 'equatorial' quadrant, which would revolve in a plane parallel to the equator. His design mounted the quadrant on one end of a long and stout sloping wooden pole of 'very dry and strong Dram-Fir', the other end of which was steel-tipped, and resting in a hardened steel socket. Roughly in the middle of the pole was a large clock driven by a circular pendulum, which produced a steadier motion than a conventional swinging pendulum. A long screw or 'worm' turned by the clock would be connected to a toothed wheel fixed around the pole, so that the pole turned in time with the clock. An arrangement of joints connecting the quadrant to the pole meant that the quadrant moved too, 'to keep even pace with the seeming motion of the fix'd stars' without any effort on the part of the observer. To make this arrangement work Hooke had to invent a universal joint which would transmit the rotation of the wooden axis to the quadrant equally, whatever the angle at which it was set. The universal joint consisted of two shafts ending in two metal semicircular forks held at right angles to each other by a rectangular metal cross, with each fork joined to (and swivelling on) two opposite points of the cross. The two shafts could be turned in almost any direction, but when one shaft rotated it turned the cross, forcing the other shaft to rotate at the same speed. Look in any illustrated car repair manual today, and you will find Hooke's 'yoke and spider' universal joint pictured as part of the typical transmission system. The name of its inventor is virtually forgotten, but Hooke's joint still plays a vital part in modern engineering.

Thus Hooke brilliantly exploited the unrivalled breadth of his mechanical and scientific experience, his much-criticized versatility, to devise an instrument that none of his contemporaries could have produced, and which anticipated features that would not become standard for 150 years. His equatorial quadrant brought together his work on conical pendulums and clockwork, his knowledge of spirit levels, his refinement of micrometer screws and eyepieces, his experience of telescopic sights and his invention of the reflecting quadrant, and combined these with two new and valuable mechanical devices for the transmission of rotary motion, the universal joint (or 'Hooke Joint') and the worm-wheel. Though Hooke did not have the time or resources to make his new quadrant, his description and pictures made it possible for others to do so, and he even recommended a craftsman who had mastered the intricacies of the micrometer screw: 'if any person desire one of them to be made, without troubling himself to direct and oversee a Workman, he may imploy Mr. *Tompion*, a Watchmaker in *Water*-Lane near *Fleetstreet*'. As we will see, the instruments installed in the new Greenwich Observatory two or three years after the publication of *Animadversions* owed much to the inventive genius of Hooke and the skills of Tompion.

Animadversions struck some of Hooke's colleagues as an unnecessary and ungentlemanly attack on the reputation of a great astronomer, but Hooke believed that if the battle for precision instruments was not won then progress in every branch of science would be jeopardized. 'To what purpose is all this curiosity [accuracy]? . . . it is of infinite value, to any that shall design to improve Geography, Astronomy, Navigation, Philosophy, Physicks, &c.' With his new quadrant scientists could measure the refraction of the air, the better to understand the weather and astronomical observations. Its precision would allow astronomers to discover the movement of stars, to establish longitude, to examine the influence of other planets on the Earth, to measure degrees of latitude with the utmost accuracy, to establish distances on land and sea, to take exact measurements of the Sun, Moon and planets, of the height of hills and the 'Rotundity of the Earth'.[5]

Hevelius, who seems to have been appalled by Hooke's assault, counterattacked by questioning his opponent's ability to perform in the observatory what he had promised on paper, and mocking his habit of claiming that he was too busy to put his ideas into practice. In August 1675 he wrote to Oldenburg (and thus to the Royal Society): 'whereas he ought to demonstrate his discoveries by observations already made by himself, he does the business in an abundance of mere words and grandiloquent reasons . . . in most cases it is usual with him for tasks that are to be completed to be interrupted by other things, as though he were wholly destitute of leisure'.[6] The best way to remove doubts about the practical value of the quadrant was to make it, and this is what the Royal Society urged Hooke to do when he described it to them on 3 December 1674. At the same time Ward, Petty, Wren and Jonas Moore, all friends of Hooke's, were told to meet as a committee and evaluate his finished instrument.

It was hardly realistic to expect Hooke to make either of his proposed quadrants in his own rooms, or even in the new turret observatory the Royal Society had built for him at Gresham College. But just at the right time another opportunity presented itself. Because precision astronomy offered the chance of a solution to the longitude problem, Charles II was tempted to provide money for a new royal observatory, equipped with the best modern instruments. On 23 September 1674 Hooke took a boat with his wealthy friend Sir Jonas Moore, the Surveyor-General of the Ordnance, to look at Chelsea College, an abandoned theological college which the Royal Society had been struggling to find a use for ever since the King gave it to them in the late 1660s. Having recovered from the cold he caught on the boat, Hooke told the Royal Society Council that Moore was ready to spend £250 building an observatory in Chelsea. The Society, delighted to have found a wealthy patron at last, immediately elected Moore, an able mathematician, to its membership. The man who aspired to run the new observatory was John Flamsteed, Moore's protégé. Flamsteed had been writing to Moore in a tone that was very critical of Hooke's vanity and secrecy, but as soon as he realized that Hooke was in a position to influence Moore's choice he changed his tone. A firkin of ale arrived for Hooke at Gresham College in October 1674, and Flamsteed sent a letter of retraction to Moore: 'I know you will not let him understand any thing by my letters that may be prejudiciall to my concerne and for the futur my opinion of him shall be much more charitable'. Flamsteed's manoeuvres were successful, and in March 1675, thanks to Moore, the King appointed him as his 'astronomical observator'.[7]

The Society's delight over the conversion of Chelsea College was premature. A committee which had been set up by the King to consider the claim of a Frenchman, Le Sieur de St Pierre, to have solved the longitude problem eventually decided that Greenwich, not Chelsea, was the best site for a royal observatory, and that the new establishment should be independent of the Royal Society. The fact that Wren and Hooke were on this committee did the Royal Society no good, because both were hostile at this time to the Society's leadership, and had been thinking of setting up a rival club of their own.[8] The Royal Society's exclusion did not mean Hooke's. The royal warrant for the Observatory named Wren as its designer, but as usual Hooke played a shadowy but significant part in his friend's work. On 22 June 1675, the day the warrant was issued, Hooke noted in his diary: 'At Sir Ch. Wren order to view spittlefields for Title, and to direct Observatory in Greenwich park for Sir J. More. He promised money.' Hooke was at Greenwich with Flamsteed and Halley a week later, and on 2 July, in company with those two, he drew or 'describd' an observatory. On 28 July he spent most of a day at Greenwich and 'set out' the observatory, apparently to his own or Wren's plans. After that the diary is silent, and it seems that the observatory was completed over the next ten months under the supervision of Moore and Flamsteed. Hooke, as we shall see, had other problems on his mind in 1675.

Animadversions had made Hooke the greatest public exponent of the instrumental and observational accuracy that the new observatory was intended to exploit, and he was deeply involved in designing and equipping the Greenwich Observatory. In accepting his new post Flamsteed committed himself to working closely with a man he disliked, and to filling his observatory with instruments he distrusted.

In July 1675, when work on the Observatory had just begun, he complained that Hooke, who would not have to use the instruments himself,

> will needs force his ill-contrived devices on us. hee talks of such things as none but those that understand them not can esteeme possible or probable: that an instrument of no more than 18 inches Radius should measure an Angle to less than 6 seconds, that hee has an instrument or Quad[rant] of 36 foot radius that weighs not a pound and which he can put into his pocket, and severall things of the same sort but larger far and the good successe of his [spring-driven] watch . . . has so swelld him that he is not to be persuaded that his unreasonable discourses are in the least erroneous: I know not how to deale with him but if nothing else will make him tractable have resolved to excerpte a Catalogue of his errors from his workes and represent them fairely to him to bring him into a freindlyer behavior.[9]

Flamsteed apparently never presented Hooke with this catalogue of errors, and if he had done so it is hard to imagine it easing the relationship between these two difficult and self-regarding men. Their relationship never improved, and in the 1680s, mutual distrust ripened into enmity and contempt.

Hooke and Tompion built a smaller version of the thirty-foot mural quadrant described in *Animadversions*, ten feet in radius, for the new Observatory. At Moore's insistence and expense it was installed on the east wall of the Quadrant House in July 1676. Its special feature was that the arc was marked at five-degree intervals by brass studs, and there was a fully graded subsidiary five-degree arc which could be moved and clamped into any position on the quadrant to allow accurate readings to be made. Flamsteed was unhappy with it from the start. In November 1677 he told Towneley that he was 'busy correcting our ill contrived Quadrant', and two months later he complained that he could hardly make Hooke's device serviceable, and that its performance would never achieve the accuracy Hooke had bragged about. 'But I am not much dissatisfied, since this has shown the difference between boasts and performance'. The double index (the five-degree movable arc) was too awkward, he told Moore in July 1678, and was impeding his work: 'I tore my hands by it and had like to have deprived Cuthbert [an assistant] of his fingers'.[10]

About six months after this incident Hooke's mural quadrant disappeared from the Observatory and from the historical record. Flamsteed's new mural quadrant, which was also unsatisfactory, was not installed until 1681, and in the meantime he had to make do with two smaller instruments, one designed and the other inspired by Hooke. In 1676 Flamsteed installed an equatorial sextant of nearly seven feet radius, with two telescopic sights and (at Hooke's suggestion) a perpetual screw adjustment on the limb. Flamsteed claimed to have designed this sextant himself, but he had watched the progress of Hooke's equatorial quadrant throughout 1674, and it was made by Tompion, who had also made Hooke's instruments. In 1677 Flamsteed counted the turns of the perpetual screw to calibrate the arc into minutes and seconds, just as Hooke had suggested in *Animadversions*. In January 1677, at a time when Hooke's influence in the Royal Society was particularly weak, the Society's Council ordered that all its astronomical instruments stored at Gresham College should be loaned to the Greenwich Observatory, 'and that Mr Hooke's new quadrant be forthwith finished at the charge of the Society'. This was Hooke's marvellous three-foot equatorial quadrant, the one he had described in *Animadversions*, with its perpetual screw adjustment, double reflecting eyepiece for seeing two images at once, two-way telescope for angles up to 180 degrees and water level for exact perpendicular and accurate altitude observations. Its most ingenious and advanced feature, diurnal rotation powered by clockwork, was not included, and was not really a practical proposition with seventeenth-century (or even eighteenth-century) materials and manufacturing techniques. The problems of distortion associated with constant movement were not overcome until the nineteenth century, and the clock-powered equatorial quadrant was eventually produced in the 1820s by the astronomer Richard Sheepshanks.

Hooke's quadrant was forward-looking and ingenious, but ingenuity did not always mean practicality. Flamsteed reported that Hooke's equatorial quadrant 'was so ill-contrived that I could not make it perform better than my first'. In Hooke's defence, it should be said that most new instruments needed a period of adjustment, and that Flamsteed was notorious for blaming his own failures, especially his lack of published results, on the shortcomings of other people's instruments. He sang a different tune in January 1678, when the Society tried to have its astronomical instruments returned to Gresham College, where it thought Hooke would make better use of them. Flamsteed fought hard to keep the instruments, and managed to retain them until after Moore's death in August 1679.[11]

*

MEANWHILE, HOOKE'S DISPUTE with Hevelius rumbled on rather inconclusively. In March 1676 Hevelius wrote to the Royal Society (via Oldenburg) complaining about Hooke's attack and promising to answer it, and in January 1677 the Society wrote to assure him that *Animadversions* had been written 'without any approbation or countenance from the Society'. Then, at Oldenburg's prompting, Flamsteed entered the debate. Hevelius's first letter to Flamsteed, written in Latin in June 1676, showed the depth of his anger with Hooke. He wrote, he said, 'as a defence against my many detractors ... who seek to destroy my good name and reputation, and utterly despise everything I do. Yet ... they themselves have achieved nothing in that field'. Hooke 'boasted that he could distinguish everything sixty times more accurately with a certain tiny little instrument', and claimed credit for a system of sextant mirrors Hevelius had used on a polemoscope (a sort of periscope) forty years earlier, but when had these boasts ever been tested?

> To be honest, I repeatedly have serious doubts as to whether, with his telescopic lenses and polemoscope mirrors, or with any other devices whatsoever, that man (I repeat, *that* man) will be able to produce results more definitive or consistently more complete than ours.[12]

Since Flamsteed shared this opinion of Hooke, he could reply to Hevelius with warmth: 'do not be anxious about your defence against him, since he has, even now, still not made an observation with his quadrant, as I hear, nor do I believe that he will achieve very many, because of heavy commitments, in which he is kept busy from day to day, and a throng of business.' The barrier to a full alliance between Flamsteed and Hevelius, as the latter realized, was that Flamsteed was as strong an opponent of open sights as Hooke was: 'You are undertaking with all your strength and in fact with much better instruments exactly the same thing that Hooke is trying to show with lesser ones'. This letter, which Hevelius sent at the end of 1676, persuaded Flamsteed that he could not continue his correspondence with him. Instead, he used an intermediary, Oldenburg, to deny that he was involved in a conspiracy with Hooke:

> You may please to inform him that there is no such familiarity or plot betwixt us as hee supposes. That I know not of one observation that Mr Hooke has made with any of his small Quadrants, and am of opinion that hee will never make any by reason of his multitude of businesse and that I thinke it scarcely possible to make observations to neare that preciseness he imagines and asserts with instruments much larger than either his or any of those hee discourses of.[13]

In 1679 the Royal Society tried to settle the dispute by sending Edmond Halley, an able astronomer and friend of Hooke's, to Danzig to see Hevelius at work. Halley became friendly with Hevelius and even friendlier (gossips said) with his much younger wife, and the reports he wrote from Danzig were equivocal. Hevelius had extraordinary instruments and he could distinguish angles between stars to half a minute (the limit conceded by Hooke), but Halley could not. In earlier correspondence Hevelius had claimed that he could measure angles to a tenth of a minute or even to a second.[14] Hevelius wrote a full response to Hooke in his *Machinæ Cœlestis* (part 2) in 1679, but most copies of the book were lost in a catastrophic fire which destroyed his observatory and its wonderful instruments that September. Hevelius managed to repair much of this damage, but he never quite recovered from the disaster, and when he died in 1687 the tradition of naked eye astronomy died with him.

14. The Coiled Spring

(1674–1675)

It is virtually impossible to give an orderly account of Robert Hooke's crowded social and intellectual life in the 1670s, when his diary records almost every meeting, every building or surveying task, every new idea he had. A daily narrative of his life in these years would be impossibly complicated, but concentrating on a few scientific projects or a couple of famous disputes ignores the wonderful vigour and variety of his career, and diminishes his story. If the dominant theme of 1674 was his work on his new quadrant and his critique of Hevelius's astronomical methods, the year's subplots included brief sexual affairs with his servants Doll Lord and Bette Orchard, his negotiations with his brother and Sir Thomas Bloodworth over the position of his niece Grace, a frustrating battle with Sir John Cutler over his unpaid salary, the formation of a new scientific club and an attempt to revitalize the Royal Society, his design and building work on the Royal College of Physicians, Bethlem Hospital, Montagu House, the Thames Quay, the Fleet Ditch, and the Royal Exchange, his collaboration with Wren on the Monument and several City churches, his continuing duties as a City Surveyor, a series of painful or dangerous accidents on the City streets, his cooperation with John Ogilby on the mapping of London and his world atlas, and his never-ending search for scientific, medical and technical novelties, some trivial, others of lasting significance.

In the first week of January 1674 Hooke met Cutler in Tooley's coffee house ('promised money but I know not when'), made love to Doll, spoke or experimented on double writing, sailing chariots, memory, putting mills in the arches of London Bridge and the accuracy of the human eye, and ate five larks. The following week he inspected and reported on the serious damage caused to the Fleet

River and its brick arches by a breach in the Turnmill Sewer, the Fleet's filthy headwater. He also settled a dispute over a fallen wall in the City, organized the demolition of part of St Antholin's Church, Watling Street (one of the churches ruined in the Great Fire), bought some new dancing shoes and 'Tumbled down'. During the rest of the month he devised a pedometer ('pacing saddle with waywiser') for surveyors and map-makers, decided to start a new club, presented Christ's Hospital (of which he was a governor) with the badge and motto for their new mathematical school, lectured on Hevelius and the quadrant, wrote his tract on the movement of the Earth, kept up the pressure on Sir John Cutler, drew a plan for the Merchant Taylors' School, explained his new reflecting eyepiece for two telescopes ('to see forwards and backwards') to the Royal Society's Operator, Richard Shortgrave, fell out with Oldenburg and Lady Ranelagh (Boyle's sister), devised 'a watch like a celestial globe' to show the movement of the heavens, received £1,500 for his work for Sir Christopher Wren, invented a new code for writing the date, and 'fell into dirt at Lad Lane'.

In February 1674 Hooke demonstrated his reflecting telescope through which 'the observer looked directly at the object' – the first practical Gregorian telescope – to the Royal Society, and investigated the magnetic field of a spherical lodestone or terella in the hope of explaining the Earth's magnetic field and magnetic variation. He also collected information on several new manufacturing processes – hardening lead by mixing it with iron and antimony, Dwight's secret way of making china (or 'porcelain') in his new Fulham pottery, Reine's method of joining plates of glass, Harriott's cheap technique of soldering and gilding with silver, tin and brass – and examined an 'extravagant' plan to run eighteen-inch pipes from the London Bridge waterworks to provide water for fighting fires. He continued to work on Merchant Taylors' School, reported to the Lord Mayor on the state of the Thames quay, walked through Wren's great model of the new St Paul's, and met his cronies – Wild, Aubrey, Hoskins, Hill, Lodwick, Petty, Godfrey, Blackburn, Haak – at Garraway's coffee house.

In March John Aubrey, one of his closest friends, was arrested for debts of £200. He was free again after a few days, and continued to rely on Hooke's generosity to keep him out of trouble. Hooke's scientific work concentrated on the magnetic field of lodestones and his reflecting telescope, and checking the proofs of *An Attempt to Prove the Motion of the Earth*, which was published in May. On 23 March, a day on which Hooke's City Surveyor duties took him to Newgate, Fleet Bridge and Holborn Bridge, Wren's little son Gilbert died. Upset by this news (or perhaps by eating strong cheese), Hooke slept badly that night and 'dreamt of viragoes and other strange phenomena'. Hooke's work as an architect and construction manager was expanding. On 7 and 8 March he estimated the cost of the Royal College of Physicians' new operating theatre, a building that would occupy him for four years. And at the end of the month he started working on two naval buildings, the Victualling Office and the Navy Office in Seething Lane, which had been destroyed by fire in 1673. At the same time he paid off Doll Lord and hired Bette Orchard for a trial period, at a salary of £3 a year. She started work two weeks later, and became Hooke's mistress (if that is not too grand a description of their brief liaison) on 26 July.

In April 1674 Hooke was working on the clock mechanism and paving for the new Royal Exchange, and in the middle of the month he took on one of his biggest building projects, the design and construction of a new Bethlem Royal Hospital for the insane. Bethlem (or 'Bedlam') was built on Moorfields (now Finsbury Circus), only a few minutes' walk from Gresham College. Hooke was involved in every detail of the huge project, and he worked on it every week, and sometimes every day, for the next two years. On 15 April he joined a royal commission with Lord Brouncker, Seth Ward and others to examine the claim by Henry Bond that he could calculate longitude by establishing the pattern of compass variations around the world. Hooke immediately saw that Bond's claim, which was based on the idea that the magnetic poles were 'somewhere in the air and . . . left behind by the motion of the earth', was the work of an incompetent amateur: 'His way then was this – he from two observations jumped at random upon an hypothesis and at random calculates tables to it'. Later in the month Hooke exchanged letters with Hevelius over the virtues of open and telescopic sights, reported to the City on the three-foot layer of mud and refuse that had been deposited in the Fleet Ditch by the January floods, bought goose grease suppositories, 'exceeding good for the piles', and resolved never to go to coffee houses again. He kept to his resolution for a single night.

On 2 May 1674 one of the most important associations in his career – and also a significant one in the history of scientific instruments – began. Hooke visited Thomas Tompion, the watchmaker, in Water Lane near the Tower, and talked to him about watches, his wheel-making engine (for clockwork) and the shape of arches. In Tompion Hooke found a craftsman with the skill to follow his most complex and delicate designs, and in Hooke Tompion found a practical scientist who was able to feed him ideas that would earn him a supreme reputation among English watchmakers. Their cooperation began with work on the reflecting quadrant with telescopic sights and micrometer screw adjustment, a version of which Hooke was showing to Brouncker, Petty and Flamsteed only a few days later. On 24 May his old servant and lover Nell Young, who still met him often, brought him an aborted foetus, which he preserved in alcohol and added to the Royal Society's repository of curiosities.

In June 1674 Hooke visited Fish Street Hill to watch the Monument being built, and ordered the demolition of the tower of St Bartholomew-by-the-Exchange, one of the City's fire-damaged churches. He met Richard Towneley, the Lancashire virtuoso, to swap information on making watches and telescopes, and the great Thomas Hobbes, whose gaze had disconcerted him ten years before. He hurt his eye looking at the Sun through a helioscope on 15 June, but had recovered enough to go to the playhouse with Wren and Sir John Hoskins to see *The Tempest* (probably Thomas Shadwell's new version) a few days later. Perhaps the most important development in June was Hooke's growing friendship with Tompion. They met at Salisbury Court coffee house on the 25th, and two days later, at Child's coffee house, Hooke told Tompion about his idea for a 'poysed watch floating in water' and his plan for a two-cistern water-clock. They met almost every day in early July to work on Hooke's new quadrant. Whether his ideas prospered or failed, Hooke had a happy habit of expecting great things from them. At the end of June he 'invented the way of printing with the common press pictures made with Pinns'. This, he was sure, would be 'an invention of Great use'.

At the beginning of July Hooke's Bedlam model was accepted by the Bridewell authorities who ran the hospital, and he spoke to Sir Christopher Wren about St Martin within Ludgate, one of the City churches the Surveyor-General was charged with rebuilding. He joined the leading parish officials on 4 July in the Greyhound tavern in Fleet Street, and later in the month he met one of them again to discuss a five-guinea fee for designing the new steeple. Over the next few years he often visited the church or its site, and clearly played an important part in its rebuilding. St Martin within Ludgate escaped the Blitz almost undamaged and it remains an excellent example of a Wren–Hooke church, only a few minutes west of St Paul's. On 17 July, between these two meetings, Hooke was almost killed by a careless coachman on a City street. 'Sir J. Frederick would have sent him to Bridewell [prison], I reprieved him. I owe my rescuer much.'

By this time, the middle of 1674, the Royal Society had lost much of its initial vigour and enterprise. Some of its early leaders, including Moray and Wilkins, were dead, and others, including Wren and Lower, were no longer active experimenters. Hooke, whose endless flow of ideas, devices and experiments had kept the Society's meetings alive in the 1660s, was now more interested in making new instruments, and in his work for the City, than in performing scientific experiments for the amusement of a dwindling audience of amateurs. A reflection of the Society's decline was the fact that it only met once between the end of April 1674 and its return from summer recess on 12 November. Hooke, who knew the Royal Society better than almost anyone else, turned his mind to its problems in the summer of 1674. In late June or early July he wrote a paper on the Society, and on 16 July he discussed the running of the Society with Lord Brouncker (still its President), Hoskins and Colwall. An undated paper by Hooke in the Royal Society's archives may date from this time, though it is also possible that it was written later in the 1670s. In either case, it probably expresses his ideas about the aims and membership of the Society pretty well. What he wanted was a Society of men truly committed to the pursuit of natural philosophy, without the ballast of idlers and gossips that it presently carried. Members should undertake to work for the Society in an orderly way, reading books, consulting fellow scholars, testing hypotheses and presenting their work in meetings as part of a systematic programme of study. The 'newfound world' of natural philosophy had to be conquered, Hooke said, just as Cortés had conquered Mexico, with an 'army well Disciplined and regulated though their number may be small'. Hooke had no doubt been flattered once by the idea that he was indispensable, but now he was tired of his weekly obligation to perform for the Society. He wanted an association in which every member was obliged to display his findings, and in which there were twenty or more paid curators to keep the work of the Society going. Restricting membership to active scientists would also help to achieve his second aim, that members should be bound by an oath of secrecy not to communicate to outsiders things they had learnt through the Royal Society. He was tired of seeing his own ideas spread across Europe, often to be used (as he saw it) by others without proper recognition of his priority. In an exclusive Society bound by an oath of secrecy, he could be true to his own nature by being boastful and secretive at the same time. He already saw Oldenburg, the Society's efficient and highly communicative secretary, as the worst source of leaks, though in fact Oldenburg's correspondence was nearly always with Royal Society fellows who were not resident in London. This is why on 9 July 1674 Hooke showed the new quadrant that he had been working on with Tompion 'to all but Oldenburg'. Hooke ended his paper with one of his characteristic lists, weighing the burdens of membership of a reformed Royal Society against its many benefits. The pleasures he drew – or hoped to draw – from Society membership were these:

Suitable Acquaintance
Delightful Discourse
Pleasant Diversion by Experiment
Instructive Observation by tract
Considerable Intelligence by Letters
New Discoverys by Inventors
Solution of Doubts and Problemes
An easy way to know what is already know[n]
A Liberty of Perusing repository
  of seeing and using the Library
  of Perussing Modules and Instruments
  Of Perusing Letters & Discourses

of being present at tryalls – mechanick, Optick, chymick
Anatomick Astronomick –
An opportunity to shew their Abilitys with advantage
to Get a character of the Society
To have an account sent to him every fortnight of all is doing.[1]

In late August and early September 1674 Hooke accepted a major private building commission. His client was Ralph Montagu, a courtier and diplomat, who had recently married Elizabeth Wriothesley, the extremely wealthy daughter of the Earl of Southampton and widow of the Earl of Northumberland. To celebrate this unhappy but highly profitable union Montagu wanted Hooke, who was by now regarded as one of London's leading architects, to design and build him a palatial mansion in the French style on Southampton Fields. These fields had been a part of the Bloomsbury estate of the Earl of Southampton, one of the most active developers of Restoration London, who had divided his lands between his three daughters when he died in 1667. Hooke visited the site on 2 September, finished his first draft of the new house by the 13th, and had a model of the building ready by mid-December. Building began the following spring, and Hooke was intimately involved with every stage of the work, from marking out the plot to setting out the garden, until 1680.

At the same time he continued his work as a City Surveyor and as Wren's associate in rebuilding the City churches. On 19 September he agreed a price for the east and south sides of St Stephen Coleman Street with the mason, Joshua Marshall. This church, which was built on medieval foundations, possibly to Hooke's design, was destroyed in 1940. On 15 September he had another narrow escape as he scurried around the City streets, when he was 'almost run through by the pole of a coach'. Perhaps Hooke's disabilities made him prone to accidents, but London's streets were notoriously hazardous for walkers, with their potholes, puddles, overhanging shop signs, open cellar flaps, and piles of dirt and rubbish. Pavements were not separated from the roadway, and pedestrians who did not 'keep to the wall' (which had its own dangers) risked being hit by carts and carriages. Except on the darkest nights, the streets of the City were unlit, and even when candle or oil lamps were displayed on moonless winter nights they were dim, infrequent and unreliable. Hooke was himself working on ways of making oil lamps whose wicks would be fed by a steady flow of oil, but practical results were some way off. In the meantime he, like others whose business took them onto the City streets, had to get used to falling over.

Hooke's friendship with Tompion seems to have revived his interest in spring-driven watches, a subject he had apparently neglected since 1670. They talked about double-spring watches on 4 October 1674, and Hooke gave Tompion (who may not have needed it) advice on how to keep the pallets of a watch moving. He also returned to another youthful interest, the secret of human flight, in this conversation with Tompion. On 7 October he tried an experiment 'about the artificiale strength by water, air, fire, by which flying is easy and carrying any weight', and felt happy enough about its results to tell Sir Robert Southwell, a Royal Society Council member 'that I could fly, not how'. Another of Hooke's interests in the autumn was a new kind of horizontal mill. His idea, which he sketched in his diary, involved a horizontal revolving wooden cross with adjustable sails on the end of each limb. He invented the mill on 26 September, tried it out two days later ('Very good'), and discussed its construction and uses with Tompion, Wren and several craftsmen during October. By the end of October a model mill had been made, and on 19 November Hooke and Shortgrave, the Royal Society's Operator, prepared the model for a demonstration before the Royal Society, which took place on 17 December.[2]

Hooke's ideas for revitalizing the Royal Society were shared by some of the Society's most influential members. On 27 August 1674 the Council (which did not include Hooke) discussed plans for paying for weekly experiments by chasing up unpaid subscriptions, and for 'the ejection of useless members'. Hooke's paper was not discussed, but Sir William Petty made proposals on 29 September which roughly echoed his views. Petty's plan was that the Society should identify a number of fellows who would undertake to entertain meetings with a 'considerable experimental discourse' at least once a year, or pay a £5 fine. On 15 October the Council agreed on a legal declaration for these volunteers to sign, and took steps to make the Society a more secretive organization. Non-members were to be excluded from meetings, and members were to sign a promise not to reveal experiments or information which the President or Vice-president had declared confidential. This did not go far enough for Hooke, who drafted another paper on 'Secresy and Secretary' in November. The shortage of fellows who were ready to show experiments or read papers at Royal Society meetings was not solved. There was a good series of lectures in the autumn of 1674 and the following spring, but the virtuosi could not sustain their efforts, and by March 1676 the Council was discussing the same problem all over again.

With his time and energy mostly devoted to his work on the Monument, Bedlam, Montagu House, the Fleet Ditch and the City churches, and the preparation of his attack on the observational methods of Johann Hevelius for publication in December, Hooke's contribution to Royal Society meetings in the winter of 1674–5 was small. In December 1674 he told the Society about his new quadrant and explained his horizontal mill, and in January he talked about the nature of springiness and read his Cutler lecture on helioscopes, which also included accounts of short telescopes, the universal joint, the dial-making instrument and an eclipse of the Moon. At the same meeting, on 28 January 1675, Oldenburg read out a letter from Christiaan Huygens which had profound implications for Hooke's future.

Huygens's letter to Oldenburg gave notice of 'a new invention of watches by himself the secret of which he conceals in an anagram'. To judge from his diary, Hooke did not notice the significance of the letter. Over the next three weeks he produced new designs for Montagu, persuaded Sir John Cutler to admit before a witness that he owed him four and a half years' wages, organized labour and materials for the foundations of Bedlam on Moorfields, supervised the erection of his new turret in Gresham College, ordered work to begin on the frontage of the Physicians' Theatre, and appeared before a hearing in Westminster Hall on the new steeple for St Martin Ludgate. In the Royal Society meeting of 11 February Hooke was drawn into a discussion with Dr Croone about the flight muscles of birds, and into revealing his own work on flight: 'Declared that I had a way of making an artificial muscule and to command the strength of 20 men. Told my way of flying by vanes tryd at Wadham. Told Dr Wrens way by kites, of the unsuccessfulness of Powder for this effect, and what tryalls and contrivances I had made.'

In the meeting of 18 February 1675 Oldenburg read Huygens's second letter, in which the secret of his new spring-driven pocket watch was revealed and the anagram was deciphered into Latin. In English it ran as follows: 'The axis of the movable circle [the balance wheel] is attached to the centre of an iron spiral.' In other words, Huygens was using the regular vibrations of a spiral spring in place of a pendulum, enabling him to construct pocket watches and, he thought, marine clocks that would keep good time in rough seas. Hooke was not taken by surprise, because Boyle had told him about the watch the day before. He immediately appealed to the Society's journals and to Sprat's *History of the Royal Society* to prove that he had invented and made watches in the same way over ten years earlier. The meeting was not convinced, and its reply to Huygens suggested that Hooke's spring watch had not been a success. Hooke, on the other hand, was sure that he had made an accurate spring-driven watch before Huygens (or 'Zulichem', as he always called him, using the latter part of his title), and set out to persuade others – and especially the King – that this was so. On 19 February he showed his spring watch to Sir Charles Scarborough, the King's physician, and the next day, convinced 'Zulichems spring not worth a farthing', he searched the Royal Society register book and found a reference to his spring watch in 1666. He was so sure of his priority over Huygens that he told the Council of the Royal Society that he had solved the longitude problem, and that he would take his prize as an annual income of £150, not a £1,000 lump sum.[3] In late February and early March 1675, still busy with Montagu House and Bedlam, he found time to see Tompion, the man who could turn his ideas into working reality, and to have a long discussion on spring watches with Wren.

Hooke's sense of urgency was intensified on 6 March, when his friend and ally at Court, Sir Jonas Moore, gave him the shocking news that Huygens had offered Oldenburg the English patent on his spring watch if one could be secured from the King. If Oldenburg were successful, Hooke would be prevented from developing his own marine timekeeper on the principles he had understood and talked about for more than a decade. From this moment he was convinced that Oldenburg was using his position as Secretary of the Royal Society to fill his own pockets, and that Huygens had learned the secret of Hooke's original spring watch from him. It seemed to him that Oldenburg was cheating the Royal Society, too, because Huygens, who could not hold an English patent because he was foreign, had offered the patent to either Oldenburg or the Royal Society. As it happened, Oldenburg was also an alien and ineligible to hold an English patent, but neither Hooke nor Oldenburg knew of this rule. Moore's news seemed to confirm Hooke's suspicion that Oldenburg was working against him in the Royal Society, and that the body which he had served for over twelve years would take Huygens's side in the priority dispute. Because of this Hooke decided to ignore the Royal Society's claim to act as an impartial arbiter in matters of scientific priority, and to take his case – and his watch – to the King, by whose favour patents were granted.[4]

Ever since the 1670s Hooke's critics and defenders have argued over the merits of his case. It is certain that Hooke had thought and spoken about spring-regulated watches in the 1660s, though it is not clear whether the springs he had in mind were spiral or straight. The fact that he was so sure, in private as well as in public, of his own priority makes it seem likely that he thought of the spiral spring before the idea dawned on Huygens in January 1675. But it is equally clear that in early 1675 he did not have an efficient working example of a watch regulated by a spiral spring, and there is no firm evidence that he had ever made one. This accounts for his frantic activity between March and October 1675, when he and Tompion struggled to produce an accurate and reliable spiral-spring watch for the King.

Hooke had written in detail of his double spring balance watch in his draft letter to Charles II in 1664 (see pp. 35–7), but his public statements about spring watches in the 1660s had been generalized or deliberately vague, and could not help him sustain his claim to priority over Huygens. The reference in Sprat's *History* that he valued so much simply mentioned 'several new kinds of Pendulum watch for the Pocket, wherein the motion is regulated, by Springs, or Weights, or Loadstones, or Flies moving very exactly regularly'.[5] The little that Hooke had revealed to the Royal Society or in Cutler lectures was of course transmitted to overseas members like Huygens, though in this case the informant was not Oldenburg but Hooke's admired friend Sir Robert Moray, who had choked to death in 1673. Moray had written to Huygens in August 1665 to tell him about Hooke's

> altogether new invention, or rather twenty of them, for measuring time exactly as your pendulum clocks, as well on sea as on land, for, according to him, they cannot in any way be disturbed by changes in position, or even the air. It is, in a word, to apply to the balance, instead of a pendulum, a spring, which can be done in a hundred different ways; and he even went so far as to tell us that he has undertaken to prove that one can so adjust the oscillations that small and large will be isochronous [equal in time]. It would take too long to describe these in detail, and he claims to be publishing the whole thing in a little while.

There are no grounds here for the claim that there was a plot between Moray, Oldenburg and Huygens to rob Hooke of his just reward. Hooke's decision to speak about his watch in a Gresham College lecture, delivered in 1664, freed Moray of his obligation to secrecy, and clearly his intention in writing to Huygens was not to undermine Hooke's claim to priority but to assert it. Huygens, replying to Moray, was unimpressed. 'Mr Hooke speaks a little too confidently of this invention of the Longitude in [*Micrographia*], as in several other things.' He had already seen springs used instead of pendulums in clocks in Paris, but he thought that a ship's motion and perhaps changes in temperature were likely to make them inaccurate at sea.[6]

Although it seems likely that Hooke's 'hundreds' of ways included watches with spiral springs there is nothing in the written record to prove that he actually made a watch like this before 1675. The secret watch he had described to Boyle, and then to Moray and Lord Brouncker, in the early 1660s certainly had a spring attached to the spindle (or arbor) of the balance wheel (because Moray told Huygens so in August 1665), and although the shape of the spring was not specified its position makes it most likely that it was a spiral. Because his three confidants could not offer him the terms he wanted Hooke did not reveal its mechanism to them or anyone else, and though his secrecy made it possible for him to claim in 1675 that he had perfected the spring-regulated pocket watch around 1660 it also prevented him from calling witnesses to substantiate his claim. The fact that he was still working on a variety of marine timekeepers, using pendulums and lodestones as well as springs, later in the 1660s makes it fairly clear that he had not achieved a satisfactory solution to the problem in 1658 or 1662, as he later claimed to have done. The pocket watch seen by Magalotti at the Royal Society in February 1668 was 'regulated by a little spring of tempered wire', which sounds like a hair-spring, not a spiral.

Hooke's friend and biographer Richard Waller explained his habit of leaving inventions unfinished until a rival threatened to come up with the same idea:

> It must be confess'd that very many of his Inventions were never brought to the perfection they were capable of, nor put into practice till some other Person . . . cultivated the Invention, which, Hooke found, it put him upon the finishing that which otherwise possibly might have lain 'till this time in its first Defects: whether this mistake arose from the multiplicity of his Business which did not allow him a sufficient time, or from the fertility of his Invention which hurry'd him on, in the quest of new Entertainments, neglecting his former Discoveries when he was once satisfied of the feazableness and certainty of them, I know not.[7]

This seems to be what happened when Hooke got news of Huygens's pocket watch in 1675. He had been toying with ideas for various longitude timekeepers for fifteen years or more, but had never made a perfect example of what he had in mind, or even established which of his 'twenty ways' would really work. Late in 1674 his friendship with Tompion revived his interest in watches, and when the Oldenburg–Huygens conspiracy (as it seemed to him) came to light in March 1675 he worked frantically to produce a watch that would block his opponents' bid for a royal patent and win one for himself. Because he had suggested in the early 1660s that there were a hundred ways to apply a spring to a watch he felt that he had established his claim to the whole territory, and he felt justified in crying 'plagiary' when anyone else suggested a spring-regulated watch. In the end, though, the winner of the contest would have to demonstrate *one* way to keep time at sea, not a hundred. Flamsteed, whose letters to Towneley described the unfolding saga, had already heard Hooke's complaints by 16 March 1675. In Huygens's watch, so he had heard,

> the axis of the ballance wheele playes upon and is moderated by a gentle spring which gives it its æquality, but here Mr Hooke

> proclaimes that the invention was his and done some yeares agone and by an English Gentleman imparted to Monsieur Huygens. however tis certaine hee had a watch or two made that had the balances moderated by a spring but it went so ill that it was esteemed inferior to the usual contrivances. but now hee has one in the makeing and tis thought will be perfect before Monsieur Huygens' come over, that hee sayes shall be excellent.[8]

On 8 March 1675, two days after hearing of 'Oldenburg's treachery', Hooke told Tompion how to fix two curved springs like spokes to the inside of a balance wheel, and Tompion set about producing a spring watch. Hooke's sense of urgency was increased on 3 April, when he was told that Oldenburg had procured a patent for Huygens, and four days later, thanks to Sir Jonas Moore, Hooke and Tompion had an audience with the King and presented him with a preliminary version of the spring watch. Hooke was happy: 'The King most graciously pleasd with it and commended it far beyond Zulichem's. He promised me a patent and commanded me to prosecute the degree [longitude].' This conversation hints that the King already had one of Huygens's watches, but this is not so, since Huygens's first watch did not reach Lord Brouncker (*via* Oldenburg) until June. The next day Hooke and several Royal Society members met at the Bear tavern, where Sir Robert Southwell told him that after meeting Hooke and Tompion the King had refused to grant Oldenburg a patent for the Huygens watch. Discovering that Oldenburg and Lord Brouncker were working together for Huygens's patent, Hooke gave them a piece of his mind and 'told them of Defrauding'. But his battle was far from won. On 10 April Moore warned Hooke that the King had said that Oldenburg would get the patent 'unless we made hast with the watch', and over the next few days Hooke and Tompion searched for the best way to use a spring to produce an accurate watch that was not upset by movement. They tried a 'perpendicular Spiral spring' (a helix or jack-in-the-box spring) on 13 April and a better 'double perpendicular spring' the next day.

For the rest of the month Hooke was apparently too busy to write his diary, but he met Wren several times 'about Oldenburg's Fals information'. He was furious to discover that Lord Brouncker had spoken against him at Court, probably denying the value of the watches he had made in the 1660s: 'Brouncker a Dog for belying me to the King.' Early in May he met Tompion almost every day, and finally settled on the 'thrusting spring' as the best for the King's new watch. On 17 May Hooke took the new watch to the King, who received it kindly and locked it in his closet. Hooke met the King again on the next two days, first in St James's Park, where Charles assured him that the new watch was 'very good', and then at Whitehall, where the King returned the watch to him, presumably for adjustment or repair. This was probably the watch that carried an inscription which summarized Hooke's claim to have invented the spring-regulated watch in the 1650s: 'R. HOOK invenit 1658. T. TOMPION fecit 1675' (Invented by R. Hook 1658. Made by T. Tompion 1675).

Producing a good spring watch was not proving to be as easy as Hooke had hoped, and he reverted to some earlier ideas. He showed the King a lodestone watch on 21 May 1675, and a double pendulum sea clock a week later. His diary for June and July gives the impression that he and Tompion were still searching for the best mechanism, and beginning to fall out over Tompion's inability to produce the definitive spring watch in time to beat Huygens. Flamsteed's letters give a different impression. Writing to Towneley on 8 June, he thought that Hooke's new watch was regulated by the vibrations of 'a payre of fine springes such as are used in Childrens toyes to force a puppet out of the box when the lid which holds it downe is removed. these as far as I can understand lay hold upon the Axis of the ballance and their vibrations with equall force being nearly æquitemporaneous keepe the motion equall.' Four days later, on 12 June, Hooke added a spring to an old watch of Sir Jonas Moore's, and gave the watch to Moore and Flamsteed for observation, but not inspection. The catch that released the back of the watch was filed down so that they could not open it and see how the spring was fixed to the watch mechanism. Flamsteed, who was not a particularly friendly witness, found that once the spring had settled it kept time with a pendulum to within a minute a day for ten days. 'I found it goe well to admiration for keepeing proportionable pace with the pendulum', he wrote to Towneley in June. 'Yet I can give no account of it for tis so contrived that wee cannot open it, and Mr Hooke is so proud of the Contrivance that the King's owne (if we may beleive him) is still a secret to him: as for Mr Tompion's clock it goes not forward hee being otherwayes employed at present.' Over the next month Moore's watch kept excellent time, but stopped twice.[9]

Hooke was playing a cunning game, deploying the skills of a craftsman and a wheeler-dealer alongside those of a natural philosopher. Science was his livelihood not his hobby, and he was not bound by the aristocratic moral code that Boyle, Huygens and Hevelius claimed to follow.[10] Nobody but Tompion and perhaps the King knew how his watches were constructed, and even Tompion (his apprentice George Graham said much later) was often asked to make components without knowing how Hooke was going to use them. Huygens, acting like a gentleman, found himself outmanoeuvred on all fronts. In France his own watchmaker, Thuret, claimed in January 1675 that he had invented the spiral spring watch, and though he soon withdrew the claim it was unlikely that the alliance between the two men was as close and effective as that between Hooke and Tompion. Although Huygens had already had experience of Hooke's tactics in an earlier dispute, he was slow to get the measure of his opponent. In May Oldenburg warned him that he needed to get a working example of his watch to England to beat Hooke, and on 7 June Oldenburg, who was now under direct attack from Hooke for giving away secrets and deliberately failing to register Hooke's inventions and demonstrations in Royal Society journals, begged Huygens to send a note affirming that he had been told nothing about Hooke's watch before sending the anagram in January 1675.

Huygens's watch soon arrived, and by chance Hooke was at Court when Lord Brouncker, who was plainly acting as Huygens's agent, brought it to the King on 21 June. The watch did not work properly and had neither a minute nor a second hand, and Huygens, pressed by Oldenburg and Lord Brouncker, set to work on an improved version to impress the King. The new watch did not arrive until December 1675. Huygens was distressed to find himself in such a race, and made the point several times that it was easy enough to vary the construction of an original invention a little and then claim it as your own. In doing so, he said, Hooke was behaving just like a Parisian watchmaker. Hooke used the same argument against Huygens in October, and in fact it was this argument – '*facile inventis addere*' (it is easy to add to an invention) – that had led him to keep his imperfect spring watch secret in the early 1660s.[11]

Hooke and Tompion worked on the King's watch in June and July 1675, with Hooke chastizing Tompion from time to time for his slowness. By the end of July the royal watch was ready again, and Hooke, who had just discovered the wonderful purgative powers of spirit of sal ammoniac, was in high spirits. Then the spring on the King's watch became loose, and it was not finally presented to the King until 26 August. The watch impressed the King's brother and Prince Rupert, and kept good time for more than a fortnight. When Hooke went with Wren to see Charles on 15 September the King told him that it was accurate to a minute a day. At about the same time Oldenburg sent Huygens the distressing news that Hooke's watch not only worked well, but that it beat seconds, something that Huygens believed was almost impossible with his design.[12]

Hooke and Tompion had several watches on the go at the same time, some for wealthy customers and others for experiment. Working on one of these, an oval watch, at the end of August 1675, Hooke perfected a mechanism with two linked spring-driven balances, the idea that had occurred to him in the early 1660s and to Huygens in January 1675. He wrote a description of it using Wilkins' invented language, the 'Universal Character'. At the same time he seems to have hit upon the law of springs that still carries his name. On Thursday 2 September he went to Tompion's: 'told him of my way of opposite springs which I had fully experimented before. All springs at liberty bending equall spaces by æquall increases of weight.' The next day Hooke turned his discovery to practical use, and made some 'philosophicall Scales to shew the King', in which a weight would stretch a vertical coiled spring ('equall spaces with equall weights') and move a pointer down a marked scale. This was also the day on which Wren's wife died of smallpox. Two days later Hooke successfully tested his new watch with opposite springs, whose double action would, he hoped, cancel out the irregularities caused by the movement of a ship or pocket. On 18 September Hooke told Tompion how to adjust the spring to keep the watch running accurately. This was a secret that Flamsteed had been trying to coax from Hooke for a long time, he told Towneley, 'but could never get any other answer from him then that hee know the whole Theory of Springs, and would discover [reveal] it on a good occasion, hee makes a mighty secret of it'.[13]

His mastery of this new subject, the theory of springs, helped him on 5 October, when the King met him in Whitehall and told him that the weather had altered the speed of his watch. In fact, springs were much more vulnerable to temperature changes than pendulums, because they were affected not only by expansion but also by a weakening of the strength of the spring in heat. Bypassing this difficult (and at the time insoluble) problem of temperature, Hooke told him instead about his new law of springs. Charles was interested, and the next day Hooke was invited into the royal closet to show the King 'the Experiment of Springs. He was very well pleasd. Desired a chair to weigh in'. The conversation turned to shipping, and Hooke tried to interest the King in his ideas for the more efficient design of warships, using a 'movable keel' or centreboard. Hooke worked on this device the following week, but Charles was much more inclined to trust natural philosophers with his watches than with his navy.

By the autumn of 1675 the danger that Oldenburg would win a royal patent for the Huygens watch had passed, and he wrote to Huygens reporting Hooke's success at Court and the likelihood that he would win the royal patent. In some distress at Hooke's public attacks on his honesty, he urged Huygens to defend him against the charge that he had trafficked in information for personal gain, and begged him to send a better watch with a minute hand without delay.[14] Huygens's new watch, a gold one with a minute hand and an improved mechanism, arrived just before 13 December, but after this a serious illness forced him to withdraw from the contest. Hooke continued to work on the design of an accurate and reliable watch in the autumn of 1675. In mid-October he sketched a rotating 'circular fly' in his diary, and designed a watch based on this principle for Tompion to make. But Tompion was apparently losing interest in the contest, and Hooke's diary entries in the week before Christmas 1675 suggest that their partnership was temporarily at an end: 'Fel out with Thompion . . . Calld on Tompion, who will doe nothing . . . Tompion a Slug'. Hooke was not finished with watches, but for the moment he and Huygens had fought each other to a standstill. The real winners were the balance-spring watch and the London watch industry. Whether we credit Huygens or Hooke and Tompion with its invention, the contest between them improved the spring-regulated pocket watch to such a degree that within a year or two it was established as a commercial product, and London (especially Clerkenwell) watch-making was set on the path to European dominance.

Access to the printing press allowed Hooke and Oldenburg to conduct their increasingly bitter and personal dispute in public. In the summer of 1675 Hooke was preparing for publication an expanded version of a Cutler lecture on helioscopes he had delivered in January, and he took the opportunity to publish his charges against Oldenburg and Brouncker, and to tell his story of the origins of the longitude watch. On 3 September, when the early pages of *A Description of Helioscopes and some other Instruments* were with John Martyn, the Royal Society's printer, Hooke began writing a postscript attacking Oldenburg and asserting his claim to the longitude watch. The piece was read and (according to Hooke) 'approved' by Wren three weeks later, and on 10 October Hooke went to the printer's to correct the final pages. There he met the unfortunate Oldenburg, who told him (perhaps having seen the sheets attacking him) that the matter should have been settled in the Royal Society, not in print. Unmoved, Hooke started distributing copies to his friends and supporters the next day.

Helioscopes was an account of Hooke's work on scientific instruments over the previous few years. The helioscope, which had featured regularly in Hooke's Royal Society performances in 1675, was a telescope that used flat black reflecting lenses to dim the Sun's image so that 'the weakest eye may look upon it, at any time, without the least offence'. Hooke went on to describe his reflecting telescopes and precision quadrants, and to give fuller details (with a fine clear illustration) of the universal joint he had mentioned in *Animadversions*. The essay ended with a typically confident (some would say boastful) announcement of some of the inventions he had already published ('discovered'), and of others yet to be revealed. The latter were described in the familiar anagrammatic code, to conceal their details and to add to the aura of mystery and hidden knowledge that he liked to surround himself with. To read the whole paragraph gives some idea of why such men as Flamsteed, Huygens, Oldenburg and Hevelius found Hooke insufferable, and why others have found him so fascinating:

To fill the vacancy of the ensuing page, I have here added a *decimate* of the *centesme* [i.e., a thousandth] of the Inventions I intend to publish, though possibly not in the same order, but as I can get opportunity and leasure; most of which, I hope, will be as useful to Mankind, as they are yet unknown and new.

1. *A Way of Regulating all sorts of* Watches or Timekeepers, *so as to make any way to equalize, if not exceed the* Pendulum-Clocks *now used.*
2. *The true Mathematical and Mechanical form of all manner of Arches for Building, with the true butment necessary to each of them.* A Problem which no *Architectonick* Writer has ever yet attempted, much less performed. abcccddeeeeefggiiiiiiiill mmmmnnnnnooprrsssttttttuuuuuuuuux.
3. *The true Theory of* Elasticity *or* Springiness, *and a particular Explication thereof in several Subjects in which it is to be found: And the way of computing the velocity of Bodies moved by them.* ceiiinosssttuu
4. *A very plain and practical way of counterpoising Liquors, of great use in Hydraulicks.* Discovered.
5. *A new sort of Object-Glasses for* Telescopes *and* Microscopes, *much outdoing any yet used.* Discovered.
6. *A new* Selenoscope, *easie enough to be made and used, whereby the smallest inequality of the Moons surface and limb* [edge of its disk] *may be most plainly distinguished.* Discovered.
7. *A new sort of* Horizontall Sayls *for a* Mill, *performing the most that any Horizontal sayls of that bigness are capable of; and the various use of that principle on divers other occasions.* Discovered.
8. *A new way of* Post-Charriott *for travelling far, without much wearying Horse or Rider.* Discovered.
9. *A new sort of* Philosophical-Scales, *of great use in Experimental philosophy.* cdeiinnoopssttuu.
10. *A new Invention in* Mechanicks *of prodigious use, exceeding the chimera's of perpetual motions for several uses.* aaaaebccdd eeeeegiiilmmmnnooppqrrrrstttuuuuaaeffhiiiillnrrsstuu[15]

Item 2 was Hooke's new catenary (hanging chain) theory of arches, which introduced the principle that the shape of a stable arch had to include the inverted shape of a rope or chain hanging freely from two points. His anagram translated in Latin as '*Ut pendet continuum flexile, sic stabit contiguum rigidum inversum*', which means 'As a pliable continuum hangs down, so a contiguous rigid [structure] will stand the other way up'. The mathematics of this theory were not understood until the eighteenth century, but the principle was correct, and it may have guided Christopher Wren in his work on St Paul's and the other London churches. On 5 June 1675 Hooke spoke to Wren about St Paul's: 'He was making up of my principle about arches and altered his module by it'. We should remember (as Hooke did not) that when Hooke demonstrated his early ideas on arches to the Royal Society in January 1671 Oldenburg mentioned that Wren also had a demonstration on the subject. Item 4 was developed next year in *Lampas* and items 3 and 9 relate to the work on springs and spring balances that Hooke did in August and September 1675. The anagram ceiiinosssttuu, as Hooke revealed in his Cutler lecture on *Potentia Restitutiva or Spring* two years later, stood for *Ut tensio sic vis*, 'the power of a spring is proportional to its extension', by which he meant that a spring is stretched in proportion to the force or weight pulling it. This was the principle that Hooke had applied to the spring balance he had shown the King in September, which is also described in the anagram cdeiinnoopsssttuu – *Ut pondus sic tensio*, or the weight is proportional to the extension or stretch of the spring. The anagram describing the mysterious 'new Invention in *Mechanicks* of prodigious use' was later translated by Hooke as '*Pondere premit aer vacuum quod ab igne relictum est*', or 'The vacuum left by fire lifts a weight'. This principle, which was restated by Jean de Hautefeuille and Huygens in the late 1670s, was regarded by some of Hooke's admirers as the principle behind the invention of the steam pump by Thomas Savery in the 1690s. In fact the principle described by Hooke here is not the one applied by the makers of steam engines, and the partial vacuum created by combustion would not have provided the power they needed. Their vacuums were produced by the creation and condensation of steam using an external heat source, rather than by the partial consumption of air by internal combustion, as Hooke suggests. Of course, their inventions built on the earlier work by Boyle and Hooke on the creation of vacuums and airtight pistons, but they still had to overcome the problem of harnessing the power of steam, and in this they owed as much to Hooke's protégé and amanuensis, Denis Papin, as to Hooke.[16]

Helioscopes finished with an account of Hooke's new watches, written in John Wilkins' 'Universal Character', the new 'international' language which Hooke wanted to publicize and bring into wider use. Towneley translated the passage for Flamsteed in late November 1675, but could not quite understand what Hooke meant without seeing inside the King's watch. In brief, the passage argued that the best way to make a pocket watch that resisted gravity or shaking was to have two balance wheels 'moving constantly contrary, the first to the second', with their motion regulated by springs or some other force.[17]

On the title page of *Helioscopes* Hooke had quoted a few of the famous words used by the poet Virgil when he denounced a plagiarist: '*Hos ego, &c. Sic vos non vobis*'. Their meaning only became clear when readers came to the Postscript, the section of *Helioscopes* that had the greatest impact in his small world. Here he told the whole story (as he remembered it) of his invention of the spring-regulated watch in 1658, his aborted negotiations with 'several Persons of Honour' (Boyle, Brouncker and Moray) in 1660, and the reason he had kept the watch secret for fifteen years since then. 'Upon this I was told, *That I had better have then discovered* [revealed] *all, since there were others that would find it out within six months*; to which I answered, that *I would try them one seven years*; and it is now above twice seven, and I do not find it yet found out.' The trial of Huygens's pendulum clock at sea in 1662 had given him a momentary alarm, but the clock failed to keep good time. In 1664 Hooke had given several Cutler lectures in Gresham College, to an audience that included 'many strangers unknown to me', in which he revealed twenty ways of applying springs to the balances of watches, and of regulating their vibrations. He had given discourses and shown models to the Royal Society, and made some spring watches in 1664 or 1665, 'though I was unwilling to add any of the better applications of the *Spring* to them, as waiting a better opportunity for my advantage'. Huygens must have known of Hooke's achievements in this field, Hooke said, but in this as in the case of the circular pendulum he had presented the inventions as his own. Huygens's spring watch, Hooke argued, showed that he knew no more than Hooke had known and published more than ten years before, and that he did not know 'the other Contrivance' (probably the double balance) 'without which the first part of the Invention is but lame and imperfect, and doth but limp on one leg, and will some time hobble, and stumble, and stand still. And the said Watches will not be *tres Juste*, nor shew the Longitude at Sea or Land, but on the contrary, they will be subject to most Inequalities of motion and carriage, and with many of those motions will be apt to stand still'. This was all true enough, but it is clear from Hooke's diary that he had not finally settled on the double-spring idea himself until early September 1675, when he explained it to Tompion, though he had described a double-spring device in 1664.

Hooke's attack on the honesty and competence of a fellow member of the Royal Society went beyond the conventions of philosophical dispute that the Society expected its members to follow, and confirmed that however fine a scientist Hooke was, in his conduct he was more a mechanic than a 'gentleman'. Huygens was shocked: 'I had noted for some time that he was vain and foolish, but I did not know that he was as malicious and insolent as I see he now is'. He advised Oldenburg to recruit some well-placed friends to match Hooke's allies at Court, and appealed to Lord Brouncker to defend Oldenburg against the accusations of 'such a man as Mr Hooke', who was motivated by self-interest and an 'egotistic pretension to have invented everything'.[18] On 21 October the Royal Society's Council considered Hooke's ejection, but his friends on the Council, or perhaps his evident value to the Society, saved him. In any case, Hooke seems to have been ready to break with the Society over its support for Huygens, and especially over Oldenburg's role. He told John Aubrey in late August 1675 that he dreamed of 'setting up a select clubb, whether 'twill take or not I know not. As we are, we are too much enslaved to a forreine spye, and think of nothing but that, and while 'tis soe I will not doe any[thing] towards it. I have many things which I watch for an opportunity of publishing, but not by the R.S.. Oldenb. his snares, I will avoid if I can.'[19]

Oldenburg, who correctly saw Hooke's attack on Huygens as an attack on his own part in the dispute, replied through his own publication, the *Philosophical Transactions*, on 25 October. Reviewing *Helioscopes*, he pointed to the contrast between Huygens's openness and Hooke's secrecy, and said that when Oldenburg had brought a diagram of Huygens's watch to the Royal Society Hooke had not let him take it away 'without permitting him first to copy it'. So, the review implied, Hooke was the spy and the plagiarist, not Huygens and Oldenburg. None of the watches Hooke had made in the 1660s had worked, Oldenburg said, and it was only since Huygens's letter had arrived that his interest in the subject had revived. Offering Hooke a chance to save face, Oldenburg conceded that 'pregnant and inventive Heads . . . may, and not seldom do, fall into the same Discoveries and Inventions about the same time'.[20]

Hooke had indeed seen 'Zulichem's watch scheme and transcribed it' on 18 March 1675, but this might have been just the decoded message, which did not give much away. When he saw 'the Lying Dog Oldenburg's *Transactions*' on 8 November he 'resolved to quit all employments and to seek my health'. Like his other resolutions this was soon forgotten, and three days later he was at the Royal Society lecturing about his new self-regulating oil lamps ('Oldenburg viewd and took notes') and checking the Secretary's minutes to discover, as he expected, that many of his earlier experiments before the Society had not been recorded. Three weeks later he demonstrated lamps and waterpoises again, and talked at length about watery damp and fiery damp, drawing on his experience in the plague year with the well-shafts on Banstead Down.[21] Hooke saved his written response to Oldenburg's attack until September 1676, when he added a postscript to his published Cutler lecture, *Lampas*. His argument was that he had not told Oldenburg about his successful watches, 'since I looked on him as one that made a trade of Intelligence', but that details of his invention had been revealed to thousands, who heard his Cutler lectures in 1665. Hooke denied any underhand motive in reading Huygens's description of his watch, and instead accused Oldenburg of seeking a patent for himself 'thereby to defraud me'.

Hooke's ferocious attack on Oldenburg now seems misguided, or even paranoid. To describe Oldenburg's vast multilingual scientific correspondence, which kept members of the Royal Society, including Hooke, in touch with the best natural philosophers in England and Europe, as a trade in intelligence, was quite unjust. When Oldenburg made a proper record of Hooke's experiments Hooke called him a spy, and when he did not Hooke thought he was suppressing his achievements. In fact, Oldenburg's published correspondence shows that he promoted and praised Hooke's work whenever he could, and did his best to prevent Hooke's unfortunate tone from causing offence. For instance, in February 1675 he wrote to Huygens with a copy of Hooke's *Animadversions*, in which Hooke had accused Huygens of publishing an account of the circular pendulum without acknowledging his own prior invention of it. The excuse he made for Hooke reveals much about his thoughts on his brilliant but graceless colleague: 'There are people who, not having seen much of the world, do not know how to observe that decorum which is necessary among honest folk. This is between you and me'. In 1675 Oldenburg was already sixty, and he could not supplement his modest and hard-earned Royal Society salary from building and surveying fees as Hooke did. If he seized upon Huygens's offer of a patent as a way of providing for his wife and two children, who (apart from Hooke) could blame him? No one on the Council of the Royal Society seemed to share Hooke's condemnation of him, and the enmity between the two which persisted until Oldenburg's sudden death in September 1677 was deeply unfortunate for both men.

The events of 1675 damaged Hooke's position in the Royal Society, and reinforced his reputation as a braggart and a rough fighter. The collection of distinguished scientists who disliked or despised him was growing dangerously large, and now included Newton, Hevelius, Leibniz, Huygens and Flamsteed. Even his old friend and patron Robert Boyle seems to have lost patience with him by November 1675: 'At Boyle's. He would doe nothing. Try him noe more'. But the year also demonstrated Hooke's irrepressible self-confidence, and his impressive ability to cope with several stressful and exhausting projects at the same time. Like one of his own springs, he got stronger under tension: '*ut tensio, sic vis*'. Hooke's dispute with Huygens overlapped at its beginning with his quarrel with Hevelius and at its end with a new dispute with Newton, but at the same time Hooke was responsible for at least four important buildings, managed much of Wren's work on the City churches and kept up a flow of new scientific ideas. Design and building work on the Bethlem Hospital, the Royal College of Physicians, Montagu House and the Monument continued throughout the year, and in the summer Hooke took on two new commissions, the Greenwich Observatory and a house for Sir Richard Edgecombe that was never built.

The old Bethlem (or Bethlehem) Hospital stood outside the City wall, between Bishopsgate Street and Moorfields, and therefore it had not been destroyed in the Great Fire. But in January 1674 the Bethlem and Bridewell Court of Governors decided that the fourteenth-century buildings were 'old weake & ruinous and too small & streight' to meet seventeenth-century needs. Hooke was already involved in rebuilding Bridewell Prison, which was administered with Bethlem, and he was well known for his skill and competence as City Surveyor. On 14 April 1674, just after royal approval for a new building on Moorfields had been given, Hooke was asked to view the site, and the same day he agreed to design the hospital. He brought drawings to a governors' meeting in July, and it was decided that Bethlem should be a 'single pile' building, with one row of rooms on each floor, rather than the usual double pile building, with two rows of rooms divided by a wall or corridor. This would enable the governors to achieve their two main aims, healthy and well-aired rooms for the inmates, and a long and grandiose frontage to impress the citizens of London.[22]

Hooke may not have been an architect of the highest inspiration, but he was efficient, fast and innovative. He designed and supervised the whole building, from the sewers and gardens to the turrets and clocktower, between 1674 and 1677. The ground was levelled in January 1675, the foundations were set out in March, building of the stone-fronted pavilions began in July, and work on their turrets started in September. At end of that month Hooke inspected the statues made by the mason Thomas Cartwright (probably the lion and unicorn) for the Bethlem gateway. The famous statues of Melancholy and Raving Madness, which are now in the Bethlem Royal Hospital Museum, Beckenham, were made by Caius Cibber, the Danish sculptor who worked on the Monument and St Paul's. On 29 August 1676 the hospital was almost finished, and ready for the King's inaugural visit.

Although Bethlem was built quickly and at a reasonable cost (£17,000), it did not look cheap. The main body of the hospital was built in brick, but in the centre and at each end there were magnificent stone-faced pavilions, each crowned with a large cupola and decorated with ornate Corinthian pilasters. The decision to build the pavilions in stone caused 'a great Huff with treasurer and Chase' on

19 July 1675, but drinks at Hooke's expense in the Rose after the meeting smoothed things over. Carved swags in the Dutch style separated the first-floor pavilion windows from those in the attic, and at the base of the fine French roofs there were fashionable decorative balustrades. Most of the building was surrounded by fourteen-foot brick walls, but at the front, where there was no danger of the inmates escaping, the walls were lower, and broken by wide iron gratings to give the public a better view. Inside, there were immensely long galleries, perhaps the longest in England, lined with twelve-foot by nine-foot cells. At first it was intended to divide the galleries with iron gratings to separate 'the distracted men from the distracted women', but the governors judged that gratings would be too ugly, and decided to keep the lunatics in their cells.

Hooke's building played a significant part in the development of the asylum. It was the grandest hospital that had ever been built in England, and seemed to set a new standard in the humane custody of the insane. In the words of a modern specialist in the development of asylums, Hooke's Bethlem 'not only determined the next century and a half of England asylum design, it was immediately and pointedly celebrated as offering accommodation for lunatics that was spacious, airy and light'.[23] Its design influenced Wren's Chelsea Hospital in the 1680s and George Dance's St Luke's Hospital in the 1750s. Even the new Bethlem in Lambeth, built in 1815 and now altered and occupied by the Imperial War Museum, repeated much of Hooke's plan.

Londoners and visitors to the city admired Bethlem's stately facade and its innovative cosmopolitan style, as they were meant to do. A poem in 1676 set a tone that was followed for more than a century:

> So Brave, so Neat, So Sweet it does appear,
> Makes one Half Madd to be a Lodger there.

Ned Ward, the 'London Spy', took a different view, and saw the building itself as a symptom of the City's madness, equalling the Monument in its crazy ostentation. All through the eighteenth century Hooke's Bedlam and its lovely gardens were a favourite place of promenade and amusement for Londoners. Of London buildings, only Greenwich and Chelsea hospitals were pictured more often, and the building looked so much like a palace, wits said, that those inside who believed they were kings would have their delusions reinforced.

Hooke's ability to break the monotony of a 550-foot facade with projections, recessions, pediments and balustrades was widely admired and imitated. The great hospital remained a famous London landmark until 1810, when serious structural faults – perhaps the consequence of its hasty construction – were discovered, and it was demolished.[24]

In Fish Street Hill the Monument reached a critical phase in July 1675. The great hollow pillar was finished, and decisions had to be made about the statue or emblem that was to sit on its top. Wren had favoured a brass statue and toyed with the idea of a phoenix, but in July the King decided that 'a large ball of metall guilt would be most agreeable', and the City Lands Committee ordered Hooke to negotiate with honest and able City workmen to get a nine-foot globe made at the lowest possible price. Hooke was fond of great golden globes, which he also used on Bedlam and the Physicians' theatre. He talked it over with Wren as they walked in the King's Privy Garden on 3 August, and then used his excellent knowledge of London craftsmen to find cheap and skilful founders, turners and braziers to make the ball, which was to be a wooden core covered in finely chased copper, brass and gilt, and also to produce iron railings to make the viewing platform safe. A blacksmith, William French, was contracted to make a strong iron balcony 'according to the Modell Agreed upon by Mr Hooke the City Surveyor', and Hooke chose a master brazier, Robert Bird, to make the ball, which was now to take the shape of a flaming urn. On 20 November Hooke discovered that Bird had 'bungled' the urn, and spent the day 'much disturbd in health and spirit'. A month later Bird's progress was better, and by the end of January 1676 the urn, which weighed over 1,450lb, was apparently ready. Hooke must have been happy with the finished work, because eight months later he employed Bird to make a golden ball for the top of the Royal College of Physicians' theatre, and Bird was the craftsman chosen to make copper weathervanes for several City churches – a dragon for St Mary-le-Bow, a pelican for St Mary Abchurch and a giant 'A' for St Anne and St Agnes. Later, according to his diary, Hooke used Bird as a supplier of domestic pots and pans. In April 1676 Hooke went to the top of the Monument several times and his fellow City Surveyor, John Oliver, arranged for the construction of an iron frame to hold the urn. In June and July

Hooke and Oliver supervised the making of ornamental copperwork to surround the urn and its cradle.[25] Work continued that summer, and in October Hooke watched as the wooden scaffolding around the huge 202-foot pillar was taken down. The Monument was one of the wonders of the city, standing almost as high as the tallest City churches. From its gallery Hooke could see Harrow and Shooter's Hill, over five miles away. Nearly a hundred years after its construction James Boswell found climbing to the gallery a terrifying experience. Boswell need not have worried. The Monument, which is still the largest free-standing stone pillar of its kind in the world, more than fifty feet higher than Nelson's Column or the column of Marcus Aurelius in Rome, withstood the Blitz and the thunder of heavy modern traffic, and survives today, alone of all Hooke's major secular structures in London. Perhaps inevitably, it was almost universally regarded as the work of Sir Christopher Wren, at least until the publication in 1935 of Hooke's diary, which revealed the extent of his work on it.

The Monument provided a high platform for the sort of experiments on weight and atmospheric pressure that had taken place at the top of old St Paul's before the Fire, but Hooke's hope that it might be used as a huge telescope tube were dashed when it was found to vibrate, just as his own Gresham College turret did. On the other hand, Londoners who decided to end their lives in style discovered that the upper gallery, despite Hooke's railings, provided an excellent launching pad. The first known victim, a pickpocket, jumped or fell from the platform – or perhaps the scaffolding – on 24 April 1676.

To judge from his diary, Hooke hardly ever took time off from his work. There was an occasional walk with Grace or his friend John Godfrey, a game of chess with Theodore Haak, a ride to Acton with Dr Sydenham, or a trip to the theatre with John Hoskins to see an 'Atheistical wicked play' by Thomas Shadwell, but most of his leisure time at home or in coffee houses was spent in conversations about natural philosophy or craftsmanship. Even when he 'drank rare but heady wine' with the successful court painter Peter Lely, who had briefly been his master when he first came to London, they talked of improving sight and of the 'picture box', or camera lucida.

Hooke's ability to perform so well in so many fields was not simply a matter of energy and hard work. His great skill, intelligence and knowledge enabled him to complete tasks in a day when others would have taken a week. Time and again his diary records apparently very time-consuming tasks finished in a morning or afternoon: 'Drew Sir R. Edgecomb's House', 'Set out Observatory', 'Invented Aetherial Baroscope [to measure air pressure] by helicall [spiral] spring'. 'Riddle of arch ... Invented Candlestick to keep the flame at the same height'. Even when it is not clear what he is doing, his diary communicates his immense energy as he scuttled around London: 'Monday, August 9th. – At Sir J. Mores. Pye, Aldgate, Brattle, 15sh. To Tompions and Montacues and Youngs and Boyles and Tompions &c. Invented flying chariot.' His solid grounding in the mechanical theory of matter enabled him to propose convincing, and often correct, explanations of natural phenomena without laborious investigation, and to make incisive and impromptu contributions to Royal Society meetings. Commenting on his fellow curator Nehemiah Grew's discourse on the tastes of different plants on 25 March 1675, for instance, Hooke declared his notions about flavours, 'that they were produced by the Dissolution of tastable bodys by the saliva, and such as were not Dissoluble by the saliva were tast lesse'.

Also, Hooke had help. Grace, freed at last from her obligations to the Bloodworth family, lived in the turret and kept Hooke's domestic accounts. Another family member, Tom Gyles ('a pretty boy'), the son of Hooke's Isle of Wight cousin Robert, joined the Hooke household in July 1675, but as a pupil – and sometimes a lazy and dishonest one – as well as a servant. Richard Shortgrave, the Society's Operator, often did jobs for Hooke, and Harry Hunt, who shared Hooke's rooms, had become an assistant of the highest ability. The two had their squabbles ('Harry surly and proceeded [to] putt on hat – he shall march'), but Hunt was Hooke's indispensable colleague for thirty years. Much of Hooke's scientific and building work involved preparing accurate and attractive drawings, and Hunt (like Hooke) was a fine draughtsman. Many of the ground plans of City church sites prepared by Hooke's office for Wren, including fourteen delivered on 12 April, were Hunt's work. And Hunt provided Hooke with a copy of the 'naked woman picture' they picked up in Islington in August 1675, which Hooke kept until his death in 1703.

As Hunt's skills developed and he seemed likely to move on,

Hooke needed a new apprentice, 'a sober virtuous young man and diligent in following such things as I shall imploy him about'. The position would be unpaid, but beneficial to the boy, Hooke told John Aubrey: 'though he do me service yet I shall assist him much more, and therefore I think it will be enough for me if I take him with nothing if I find him meat, drink, lodging washing & instruction'. Aubrey recommended George Snell, the brother of a coffee-house acquaintance, as a lad of ingenuity. But when Snell turned up with his trunk at four thirty on the afternoon of 21 October 1675 Hooke found that he 'Stutterd Intollerably'. Whether the unfortunate lad moved in at all is unclear, but he was obviously gone by 5 November, when Hooke 'Sent Snell to his stuttering Brother'.

15. 'Oldenburg Kindle Cole'

(1675–1676)

Although Hooke continued to speak and demonstrate at its meetings in 1675, his disillusion with the Royal Society was intense, and his attendance was irregular. On 25 November there was criticism of his salary, and five days later he recorded 'Base doings' when the Society, packed with over fifty members, elected its officers for the year. As usual, Hooke was not chosen for the Council. Even in his troubled sleep, he thought of breaking away from the Royal Society: 'drempt of Lord Brounker Slipping down and my scape'.[1] He needed a forum in which he could display his knowledge and ingenuity, and he tried to form a new club in which he could do so away from Oldenburg's prying disapproval. On 10 December he agreed to meet regularly at Joe's coffee house with a group of Royal Society friends, Aubrey, Hill, Lodwick, Wild, Sir John Hoskins and Sir Jonas Moore. Joe's was a coffee house in Mitre Court off Fleet Street, and became more famous later as the Mitre tavern, Dr Johnson's favourite inn. The club met a few times in December, with Hooke (if his diary is a true record) contributing most of the ideas, but then seemed to merge into the casual coffee-house conversation groups that were the mainstay of his social life. He tried again at the very end of the year, this time involving Wren and some heavyweight Royal Society members, Thomas Henshaw and Dr William Holder, Wren's brother-in-law. The members of the New Philosophical Club met on 1 January 1676 and bound themselves to absolutely secrecy, even about the existence of the club. The talk, which Hooke recorded like a society secretary in his longest diary entry, ranged over the wave motion of light, the existence of water on the Moon (which Hooke had seen, he thought, through his new selenoscope), devices for walking on ice, making lightning and thunder effects for the stage, the prevention of decay by vacuum-packing, and the fertilization of plants. Hooke asserted 'that all plants were femalls. that their wombs were impregnated by insects either bred on them or otherwise flying in the air at their season of blowing'. Next day he started a journal for the club, and for two months the group met on occasional Saturdays, generally at Wren's house.

Hooke used the meetings to develop his interest in music and sound. For some years he had been thinking about music, which attracted him because of its universal system of notation, its basis in vibrations and mathematics, and its insights into the workings of the senses. In November 1674 he had propounded his 'Musick theory' in Garraway's, first to a Dr Fulwood, and then to Garraway and John Godfrey, who both made fun of his ideas. The Philosophical Club gave him a more sympathetic hearing. In the first meeting he proposed a discussion of his new musical notation, and on 8 January he spoke of his 'way of improving Musicall Instruments'. He told them of devices for changing the pitch of a virginal ('to raise it higher and lower') and the volume of an organ, of his 'improvement of the clavicimbalum' or harpsichord, of 'the way of the sounding of a bell', and of how he could make a string vibrate without sound, 'but told them not how, meaning the sympathetick motion'. A week later, at Wren's house, he showed his new notation to Dr Holder, a clergyman and composer of church music who had gained a brief celebrity by teaching a deaf-mute to speak. Dr Holder had also published a book on speech and hearing, and spoken to the Royal Society on the structure of the ear. The three men discussed the science of musical sounds, and Hooke gave them in confidence the benefit of his thoughts on the subject:

> I told him but *sub sigillo* [under seal, in secret] my notion of sound, that it was nothing but strokes within a Determinate degree of velocity. I told them how I would make all tunes by strokes of a hammer. Shewd them a knife, a camlet coat, a silk lining. Told them that there was no vibration in a puls of sound, that twas a pulse propagated forward, that the sound in all bodys was the striking of the parts one against another and not the vibration of the whole. Told them my experiment of the vibrations of a magicall string without sound by symphony that touching of

> it which made the internall parts vibrate – caused the sound, that the vibrations of a string were not isochrone but that the vibration of the particle was. Discoursd about the breaking of the air in pipes, of the musick of scraping of trenchers, how the bow makes the fidle string sound, how scraping of metall, the scraping the teeth of a comb, the turning of a watch wheel &c., made sound. Compard sound and light and shewd how light produced colours in the same way by confounding the pulses.

Stimulated by this conversation, and perhaps by Holder's information on the ear, Hooke spent the next morning writing about sound. The piece was not published, but after his death an essay on the effects and nature of music was found among his papers, and perhaps this was what he wrote that morning. He began with an account of the emotional power of music, and went on to consider why some vibrations were audible, pleasing and harmonious to the human ear, and others not. He sought the answer in the nature of vibrations and in the construction of the ear itself, which 'by means of several little bones, nerves and muscles . . . can be soe tuned, as it were, or stretch't, that it becomes harmonicall or unison to whatsoever sound is heard'. A string slipping out of tune sounds disagreeable 'because the eare hast not time to tune itself all the while', while other sequences or combinations of sounds seem harmonious because their vibrations are consonant or coincident, and thus the ear can easily attune itself to their combined effects. In the paper, he also looked briefly at the effects of different sorts of music on the emotions. 'Wherefore if the time be very swift and short, it rouses & quickens the spirits & facultys, as being drawn into a concert with the nimble & busy acustick spirits; whereas if the notes are playd or sung with slow time, the spirits are made thereby dull, sottish & heavy.' Hooke was deeply interested in musical vibrations, both in their own right and as a metaphor for light, congruity and other phenomena, but he never published a full study of the subject. Like so many of the interests of this busy and mercurial man, it was cast aside unfinished. But two months later, on 28 March, he decided to make a machine in which a revolving toothed wheel would strike a piece of metal at various speeds to make different musical notes, illustrating the point he had made to Wren and Holder, that the pitch of a particular sound was determined by the frequency of its vibrations. On 28 March he 'directed Tompion about sound wheels. The number of teeth'. The first public performance of this ingenious machine took place, as far as we know, in a Royal Society meeting in July 1681.[2]

The essential institution in Hooke's social and intellectual life, the coffee house, was briefly under threat in the winter of 1675–6. At the end of December Charles II issued a proclamation that coffee houses should be 'put down and suppressed' because 'in such houses divers false, malitious and scandalous reports are devised and spread abroad to the Defamation of His Majesty's Government and to the Disturbance of Peace and Quiet of the Realm'. Luckily for Hooke and his friends the proclamation proved impossible to enforce, and on 10 January 1676 Hooke recorded that it had been suspended for six months. Without the serendipity of coffee-house company, would Hooke that month have talked till nearly midnight with a 'tall soldier of fortune', taught a Quaker to make cantilevers, learnt that Captain Wood could shoot through thirty deal boards set a foot apart, or struck up a friendship with a one-eyed travelling painter?

Coffee houses, most of all Garraway's, were Hooke's favourite places for meeting friends and craftsmen, and for picking up, showing off and passing on pieces of scientific and mechanical information. But whether he liked it or not, the Royal Society was the forum in which the greatest scientific debates took place. One in which he was deeply involved, the disagreement with Newton over the nature of light and colour, sprang into life again that winter. Newton had withdrawn into isolation after his bruising disputes with Hooke and Huygens in 1672–3, but he made contact with the Royal Society again early in 1675. In London about a Cambridge matter, he attended three Royal Society meetings, the first of which was on 18 February, the meeting at which the secret of Huygens's watch was revealed. Three weeks later he had the satisfaction of hearing Hooke defend his famous prism experiment against Francis Linus, an old scientist who had written a letter claiming that the separation of colours did not take place as Newton described it. Perhaps because of Newton's visit, Hooke's own interest in light and colours revived at this time. On 11 March he repeated his opinion that light was 'a vibrating or tremulous motion', that colours were produced 'by the proportionate and harmonious motions of vibrations intermingled', and that colours were determined by the 'lenth of the pulse'. He had adopted the wavelength theory of colours that Newton had offered him in June 1672, but still clung to most of the explanation of colours that he had advanced in *Micrographia*. The following week he read a paper on 'several new properties of light', in which he explained a phenomenon he called 'inflexion', but which is now called diffraction, in which a round beam of sunlight allowed into a very dark room could be seen to spread into a cone and produce a circular image on a piece of paper with a faint ring or penumbra around its edge. If the paper is the one printed in Hooke's *Posthumous Works*, which it appears to be, he went on to describe the effects of placing the edge of a razor across the beam of light, which produced a shadow which was not entirely black, but which had 'a certain faint light' cast into it, especially along its edge. After many tests and trials, he concluded that 'the effect was ascribable wholly to a new Propriety of the Rays of Light, and not at all to any Reflection or Refraction, or any other common Propriety of Light'.[3] This was an account of the experiment he had mentioned in his letter to Lord Brouncker (for transmission to Newton) in mid-1672, but since the letter was unsent this Royal Society discourse was apparently his first public description of the phenomenon of diffraction. Newton had not heard of this experiment before, but he was not unduly impressed by it. According to his own account he told Hooke (wrongly) that his new property of light was only a new kind of refraction, and Hooke, irritated or hurt, had replied that 'though it should be but a new kind of refraction, yet it was a *new one*.'

The matter rested there until the end of 1675. The warmth of his reception in February and March persuaded Newton to resume his contact with the Royal Society, and on 7 December he sent Oldenburg a 'Hypothesis' and 'Observations' on the Properties of Light to be read at a Society meeting. This very long paper was written as a response to Hooke's arguments against him in 1672, and Hooke's name featured prominently – and sometimes unfavourably – in it. The first part of his paper, which dealt with refraction, reflection and the nature of hypothesis, was read by Oldenburg on 9 December. The second part, which was read a week later, dealt with the causes of colours and the appearance of coloured rings in thin plates, the two topics over which Newton and Hooke had clashed in 1672. To make things worse, Newton's paper described the 18 March meeting, in which Hooke had defended diffraction as a 'new kind of refraction', with a hint of mockery: 'What to make of this unexpected reply, I knew not'. Worse still, Newton added that he had discovered that Hooke's experiment was not new, but that it had already been described by the Italian physicist Francesco Grimaldi, who died in 1663, and repeated by another writer. Newton softened this by adding 'I make no question but Mr Hook was the Author too', but for his own reasons Oldenburg deleted this line and did not read it out in the meeting. Perhaps he wanted to sharpen Newton's criticism of Hooke, even hoping that this would revive the animosity between the two men. Hooke was hurt and surprised, because he really thought he had found something new. His reaction, as recorded by Oldenburg, was to assert that most of Newton's discourse 'was contained in his *Micrographia*, which Mr Newton had only carried farther in some particulars'.[4]

Oldenburg immediately wrote to Newton with his version of Hooke's reaction to his paper. He had good reason now to make trouble for Hooke, and he was doing so. But Newton regarded Oldenburg as an entirely trustworthy informant: 'I have no means of knowing what's done but by you, I hope you will continue that equitable candor'. Newton had been inclined to give credit to *Micrographia* for its contribution to the understanding of light and colour, and in his 'Observations', which were not read to the Royal Society until 10 February 1676, he acknowledged that *Micrographia* had 'delivered many other very excellent things concerning the colours of thin plates, and other natural bodies, which I have not scrupled to make use of so far as they were for my purpose'.[5] But Hooke's reported remark angered Newton, as Oldenburg probably meant it to, and led him to reconsider the originality of *Micrographia*, and the extent to which he was prepared to acknowledge his debt to it. In December and January he sent Oldenburg two angry and effective letters in which he argued that very little in *Micrographia*'s treatment of light and colour had been original to Hooke, and that nothing that Hooke thought he had taken from *Micrographia* had in fact been Hooke's own work. He conceded that Hooke was the first to say that light was a vibration of the aether, but mentioned that the idea that colours arise from the size of the vibrations had been suggested by him to Hooke in 1672. Hooke's account of refraction and the creation of colours was mostly taken from Descartes, he said, and though his work on the colours in thin plates was new his explanation of these colours had been overturned by Newton's own work. While it was true that he had quoted some of Hooke's observations on thin plates it had been him, not Hooke, who had managed to measure the thickness of the plates. Therefore, he concluded, 'I suppose he will allow me to make use of what I tooke the pains to find out'. This letter was sent on 21 December 1675, and he followed it with a second, sent on 10 January 1676, in which he emphasized again that most of the ideas in *Micrographia* that he had used were Descartes's, not Hooke's. 'I desire Mr Hook to shew me therefore, I say not only ye summ of ye Hypothesis I wrote, wch is his insinuation, but any part of it taken out of his *Micrographia*: but then I expect too that he instance in [it] what's his own.' By the end of the letter Newton's anger had abated, and he finished with a friendly postscript: 'If you have opportunity pray present my service to Mr. Hook, for I suppose there is nothing but misapprehension in wt has lately happend.'[6]

Oldenburg used Newton's letter of 21 December to cause Hooke the greatest possible discomfort. He read the first part of the letter, which was concerned with an experiment with static electricity in rubbed glass, to the Royal Society on 30 December, but kept the attack on Hooke secret. The second part, belittling the originality of *Micrographia* and rejecting the notion that Newton's work on light and colour rested on it, was read out without warning on 20 January, taking Hooke completely by surprise. Oldenburg did not record Hooke's reaction to the letter, but Hooke's diary shows that he believed that Newton's animosity had been aroused by Oldenburg's false account of his reaction to the reading of Newton's hypothesis on 16 December. Newton was plainly a formidable scientist and a dangerous opponent, and Hooke could not allow a troublemaker, 'Oldenburg kindle cole', to fan the embers of an old argument into a new outbreak of hostilities. So that evening Hooke wrote a letter to Newton, and the next he tried it out in Garraway's on his friend Abraham Hill. Despite its occasional Hooke-isms, it was a persuasive and finely phrased offer of reconciliation and mutual support. This is the letter he sent to his 'much esteemed friend, Mr Isaack Newton':

The Hearing of a letter of yours read last week in the meeting of the Royall Society made me suspect y[t] [that] you might have been in some way or other misinformed concerning me and this suspicion was the more prevalent with me, when I called to mind the experience I have formerly had of the like sinister practices. I have therefore taken the freedom wch I hope I may be allowed in philosophicall matters to acquaint you of myself, first that I doe noeways approve of contention or feuding and proving in print, and shall be very unwillingly drawn to such kind of warr. Next that I have a mind very desirous of and very ready to imbrace any truth that shall be discovered though it may much thwart and contradict any opinions or notions I have formerly imbraced as such. Thirdly that I doe justly value your excellent Disquisitions and am extremely well pleased to see those notions promoted and improved which I long since began, but had not time to compleat. That I judge you have gone farther in that affair much than I did, and that as I judge you cannot meet with any subject more worthy your contemplation, so I believe the subject cannot meet with a fitter and more able person to inquire into it than yourself, who are every way accomplished to compleat, rectify and reform what were the sentiments of my younger studies, which I designed to have done somewhat at myself, if my other more troublesome employments would have permitted, though I am sufficiently sensible it would have been with abilities much inferior to yours. Your Designes and myne I suppose aim both at the same thing wch is the Discovery of truth and I suppose we can both endure to hear objections, so as they come not in a manner of open hostility, and have minds equally inclined to yield to the plainest deductions of reason from experiment. If therefore you will be pleased to correspond about such matters by private letter I shall very gladly imbrace it and when I shall have the happiness to peruse your excellent discourse (which I can as yet understand nothing more of by hearing it cursorily read) I shall if it be not ungrateful to send you freely my objections, if I have any, or my concurrences, if I am convinced, which is the more likely. This way of contending I believe to be the more philosophicall of the two, for though I confess the collision of two hard-to-yield contenders may produce light yet if they be put together by the ears of other's hands and incentives, it will produce rather ill

> concomitant heat which serves for no use but ... kindle cole. So I hope you will pardon this plainness of your very affectionate humble servt
>
> Robert Hooke.[7]

Hooke did not resist the temptation to blame Oldenburg for misrepresenting his opinions, or to claim that pressure of work had prevented him from perfecting his own work on light. His claims that he did 'noeways approve of contention or feuding', and that he was happy to see his own ideas overturned in the cause of truth would not have sounded very convincing to Hevelius or Huygens, or to the many lesser scientists he had bruised in the course of his career. It was not fighting that he disliked, but losing. But this was a graceful peace-offering, and Newton accepted it. His reply, sent on 5 February 1676, praised Hooke's 'true Philosophical spirit', and welcomed his proposal that they should write privately to each other, rather than fight in public for the entertainment of the Royal Society. 'What's done before many witnesses is seldome without some further concern then that for truth: but what passes between friends in private usually deserves ye name of consultation rather than contest, & so I hope it will prove between you & me.' Newton said that 'pertinent Objections' would revive his interest in the subject of light, and that no man was better able to offer them than Hooke. He went on, in one of the most famous passages in English scientific correspondence:

> I hope you will find also that I am not so much in love wth philosophical productions but y[t] I can make them yield to equity & friendship. But, in ye meane time you defer too much to my ability for searching into this subject. What Des-Cartes did was a good step. You have added much several ways, & especially in taking ye colours of thin plates into philosophical consideration. If I have seen further it is by standing on ye sholders of Giants ... I have reason to defer as much, or more, in this respect to you as you would do to me, especially considering how much you have been diverted by buisiness.[8]

Perhaps Newton's 'shoulders of Giants' compliment was more conventional than heartfelt. The first-century Roman poet Lucan had used the image of a pygmy on a giant's shoulders, and so had the twelfth-century French philosopher Bernard of Chartres and Robert

Burton, in his *Anatomy of Melancholy*, which was first published in 1621. Newton, naturally enough, left the pygmy out of the metaphor, but the idea that he used the phrase as a malicious joke on Hooke's deformity is inconsistent with Newton's character and with the general tone of the letter. Newton did not enjoy controversy, and wanted to end his feud with Hooke. It is a sign of his sincerity that he wrote to Oldenburg on 15 February suggesting that the passage in which he had compared the work of Hooke and Grimaldi should be omitted when his 'Hypothesis' was registered in the Royal Society's records. Newton's letter to Hooke ended with a proposal that the two should cooperate in observing a particular star, but the correspondence did not continue, and the collaboration that both men said they wanted did not take place. Newton and Hooke were eloquent in propounding the ideal of fruitful collaboration between scientists, but neither had the temperament for putting it into practice. If Robert Hooke was proud, sensitive, suspicious and intolerant of opposition he had met his match (in this as in other respects) in Isaac Newton.

After his demonstration of Newton's prism experiment on 27 April Hooke did hardly any work for the Royal Society in 1676, and often missed its meetings. Without him, many Royal Society meetings were rather dull and poorly attended affairs, with readings from Oldenburg's extensive correspondence and an occasional botanical demonstration from Nehemiah Grew to occupy the afternoon. In a council meeting on 6 March Lord Brouncker introduced a new scheme to pay members for performing experiments, but Hooke took this as an attack on his own work ('Councell Resolved against me. Brouncker, Oldenburg, Colwall and Croon, busy bodys'), and his hostility to the Society deepened. Perhaps Hooke was also angry at his exclusion from the committee that was set up that day to find a custodian for the Society's Repository, which had at last been moved out of his rooms and established in the west gallery of Gresham College. But it is likely that he had found caring for this growing collection of rarities a nuisance, and he had recorded its removal from his rooms on 7 February with apparent indifference. In May, at the suggestion of Daniel Colwall, the Treasurer of the Royal Society, Hooke met Lord Brouncker to discuss an increase in his experimental demonstrations, but nothing came of their conversation. Hooke probably missed the last meetings of the spring session (the last formal meeting was on 15 June), and by the time the Society reconvened after its long recess on 26 October his relations with it were worse than ever.

Hooke's withdrawal from his Royal Society work in 1676 reflected his anger with Oldenburg, and his loss of faith in the system by which the authorship of ideas or inventions revealed in Royal Society meetings was recorded and protected. In all his recent disputes, with Hevelius, Huygens, Oldenburg and Newton, the Society's role had been unhelpful or hostile. He could use the demands of his architectural practice as an excuse for his neglect of Society business. Bethlem and the Monument were in their final stages in the first half of 1676, and at the same time he was much involved with work on the roof and gardens of Montagu's palatial mansion in Bloomsbury. In January he was working on various buildings adjoining the Merchant Taylors' Hall (one of the few to survive the Great Fire), and negotiating the mason William Hammond's contract for building the Royal College of Physicians' theatre. In March he arranged with the mason Thomas Cartwright for the construction of Wren's magnificent steeple for St Mary-le-Bow, the first classical steeple in London. This was a busy time for the City church builders, with four or five churches nearing completion, and work on another ten just beginning. At around the same time Hooke undertook to make drafts and models for the rebuilding of the Tangier Mole (the breakwater protecting the port of Tangier, then an English colony), which had been damaged by storms in 1673–4. In the spring he designed a house for the Earl of Oxford and his wife in the Whitehall Privy Garden (near the Horse Guards) for a £100 fee. There were two new commissions later in the year: repaving the choir of Westminster Abbey for his old teacher Dr Busby in July, and planning a garden for Londesborough Park, Lord and Lady Burlington's mansion in the East Riding of Yorkshire, in November. In the past, though, work such as this had not stopped him from performing his Royal Society duties to the full.

His scientific interests were still as varied and enthusiastic as ever. He took up new subjects, and the old ones, like the air pump, better ways of polishing lenses, improving the design of carriages or 'chariots' and developing new dyes, still interested him. He was still telling Wren, Boyle and Harry Hunt of his intention to fly, and on 11 January 1676 he 'contrived flying by pulleys without wheels'. This conjures up a ludicrous image, but the design of the contraption is unclear. Hooke was always picking up new ideas, either from scientists (especially Wren) or from craftsmen. Ideas sown in his fertile mind usually flourished, especially when they had a practical value in the building world. On a coach journey to Bloomsbury in January, Wren told him about a way of painting white marble. He noted all the details of the method, and when Oldenburg showed an example of Prince Rupert's marble stained in two colours at the next Royal Society meeting Hooke claimed that he could do it better, with more colours. By 16 March he had mastered the technique, and was ready to show it off: 'Staind marble with cochineel, Logwood, Saffron, Litmus. Shewd it to the Society.' A week later he made a long and careful record of information on Indian dyes given him by another Royal Society fellow, and in May he was told of a way of making white varnish with linseed oil, vinegar and potash. On 10 July he found out, probably from his friend Wild, about making black varnish and how to turn white marble black and black marble white. Hooke was a disseminator as well as a hoarder, and in December he passed on to Elizabeth Pepys a recipe for making white varnish from mastic and turpentine that he had got from Thomas Henshaw.

In January and February 1676 he was working and writing on waterpoises, devices that would accurately measure the specific gravity (weight compared with a similar volume of water) of different liquids. He was particularly interested in the weight and movement of liquids in 1676 because he was working on the problem of making an oil lamp which would burn steadily with a regular supply of oil to its wick. Hooke's interest was in heat rather than light, but in the candlelit world of post-Fire London, oil lamps were the public and domestic lighting of the future, and ways of making them safer, brighter and more reliable were of interest to everyone. If natural philosophers, with all their knowledge of lenses, reflectors, combustion and liquids could contribute nothing useful to the development of better lamps, their frequent claims that their work would benefit the whole community would ring hollow.

Boyle and Wren were both interested in the question of the oil lamp, and Hooke first demonstrated his design for a lamp with a steady oil supply in a Royal Society meeting in March 1667. He returned to the subject in November 1675, and on 2 December 1675 he showed a Royal Society meeting one of his new lamps. In his lecture he explained that his device promised steady heating for hatching eggs and for working with glass and iron, and that the principle could also be used where a steady flow of water was needed. He pursued his theme in a Cutler lecture on 3 February 1676. The lecture, which was published in a revised form as *Lampas, or Descriptions of some Mechanical Improvements of Lamps and Waterpoises* in September, attempted to show that theoretical science was the best way to solve practical problems. So he began with an explanation of two of his favourite scientific concepts, congruity (the tendency of different substances to penetrate into each other, as oil rises up a wick), and the shape and nature of flame. A man who tried to solve problems without scientific knowledge, he said, was groping in the dark, and 'after long puzling himself in vain attempts and blind trials . . . may at length stumble upon that which had he been inlightned by the true Theory, he would have readily gone to at the first glance.' Hooke went on to describe not one but eight ways of making a lamp with a steady flame, none of which depended on either of the two theories he had discussed, and two of which were simply lamps whose wicks floated on their reservoir of oil. A more impressive device involved a hemispherical bowl of oil, with a sealed hemisphere above it which pivoted into the lower bowl as the fuel was consumed, keeping the oil level steady. There were practical problems in creating self-regulating oil lamps, but Hooke was sure that their many scientific and manufacturing uses would make the effort worthwhile. A steady current of water from a cistern made on the same principles could be used 'for sawing or grinding stones by an Engine; for gauging of Glass Tools, or grinding glasses by an Automaton, . . . for washing and Fulling of Cloth; . . . for maintaining any slow and constant motion, as that of a Jack, or Clock; an Engine for continually stirring of a liquid body, or shaking, tumbling, and turning of dry Solids and powders, of which sort there are a great number of uses in Chymistry'.[9]

Hooke followed this lecture with another delivered a week later (on 10 February 1676) defending the branch of natural philosophy on which his lamps were based, hydrostatics, from a strange attack by a Cambridge theologian, Dr Henry More. In a recent book More had argued that the behaviour of liquids could not be explained by the force of gravity, and had introduced a new force, the 'hylarchic spirit', to explain this and other puzzling physical phenomena. Hooke seized the opportunity to denounce the intrusion of mystical and spiritual concepts into the study of the physical world, and to proclaim the primacy of rational scientific enquiry and the rules of mechanics in explaining natural phenomena. He showed how the flow of water from a vessel, which More had explained by invoking the 'hylarchic spirit', could be easily understood if it was realized that the upper parts of a fluid exerted pressure on the parts below, and cited his January 1669 experiments to show that the speed of the flow of water from a hole in a vessel was proportionate to the square root of the height of the liquid above the hole. Thus the motion of liquids obeyed the 'duplicate proportion' rule that had been established for the motion of all projectiles, pendulums and falling bodies. Even where phenomena were not yet understood, Hooke went on, we should not 'perplex our minds with unintelligible Idea's of things, which do no ways tend to knowledge and practice, but end in amazement and confusion'.

> This Principle therefore at best tends to nothing but the discouraging Industry from searching into, and finding out the true causes of the Phenomena of Nature: And incourages Ignorance and Superstition by perswading nothing more can be known, and that the Spirit will do what it pleases ... Whereas on the other side, if I understand or am informed, that these Phenomena do proceed from the quantity of matter and motion, and that the regulating and ordering of them is clearly within the power and reach of mans Industry and Invention; I have incouragement to be stirring and active in this inquiry and scrutiny, as where I have to do with matter and motion that fall under the reach of my senses, and have no need of such Rarefied Notions as do exceed Imagination and the plain deductions of Reasons therefrom.

This passage was characteristic of Hooke's attitude to scientific knowledge and superstition. Some things may still be unexplained, and some may resist our efforts to understand them for many years to come, but the introduction of mysterious or magical concepts that discouraged the search for rational understanding could never be helpful. Alchemy, astrology, the philosopher's stone and other superstitious remnants still held a place in many seventeenth-century scientific minds, but not in Hooke's. His belief that every mystery in the natural world would eventually yield to the power of experiment, effort, argument and rational thought guided his life as well as his writing. His conversations in the middle of May 1676 illustrate his insatiable interest in mechanical and natural problems of every description. In Garraway's on 12 May the talk was of witches and the tricks they were supposed to use in causing injury to farm animals. The following day he walked in the park with Wren and talked about 'flying, about concave shell of the earth, about divers magneticall poles and meridians', and then went on to Man's coffee house, where he talked with Sir John Hoskins, and later with Wren, Ent and Aubrey, about the scotoscope (for seeing in the dark), refraction, anatomy, 'of the French menagirie, of Petrifactions, of the Helicall muscule of the gutt, of the shell of the Earth, of Pidgions of the Cliff loving salt'. A week later Hooke and Wren talked, apparently with mutual enthusiasm, about Hooke's plan for a 'flying chariott by horses'. And it was not just talk. Hooke was still working on the theory and manufacture of springs, aiming to produce a spring-regulated watch that would keep good time in rough conditions. On 22 March he put together a 'round sea watch', and two days later, despite the onset of severe fits ('increasing and rising hideously'), he met Tompion at Man's coffee house. 'Told him of my way of springs by a hammer and anvill, like making pins which he approved of. I conceive it the best in the world.' Tompion was kept busy in the spring repairing the watches he and Hooke had made for Lady Tillotson, Lord Bedford and the King, and preparing the Greenwich sextant and quadrant.

On 25 May, passing the evening with three of his usual coffee-house companions, John Aubrey, Sir John Hoskins and Abraham Hill, Hooke heard about *The Virtuoso*, a new play by Thomas Shadwell, which made fun of a natural philosopher who seemed to bear a strong resemblance to him. The Duke's Company were performing the play in their theatre behind the church of St Bride. The King, whose goodwill was so important to Hooke and the Royal Society, had been in the audience. Hooke spent much of the next few days with Grace, watching the terrible fire that had broken out in Southwark, just across the bridge from his new column, but on

2 June, a Friday, he went to see the play with his friends Tompion and Godfrey. His City Surveyor colleague John Oliver was also there that night.

The play's central character, Sir Nicholas Gimcrack, was a composite figure based on Shadwell's quite detailed knowledge of recent scientific writings and the work of the Royal Society. By trivializing the work of serious scientists like Hooke and Boyle, Shadwell was able to present them as laughable and pathetic figures, indistinguishable from the amateurs or 'virtuosi' who collected natural curiosities and paltry pieces of information for their own sake. Gimcrack was a cuckolded husband rather than a bachelor, but some of his other circumstances and characteristics were like Hooke's. He was the guardian of two nieces whose romantic entanglements formed one of the play's main plot lines, he was a braggart who often claimed to have improved on other scientists' work, and he had his hands in almost every scientific pie, from the weight of air and microscopic eels to blood transfusions and flying to the Moon. Shadwell drew his information from the writings of Boyle, Thomas Coxe (a transfusionist) and others, but one of his main sources was *Micrographia*, and his versatile natural philosopher resembled no one so much as Hooke.

The central joke of *The Virtuoso* was that the work to which Hooke and the rest of the Royal Society had devoted their lives was of no practical use, and had given them no understanding of the human world. Gimcrack was 'one who has spent two thousand pounds in microscopes to find out the nature of eels in vinegar, mites in a cheese, and the blue of plums . . . One who has broken his brains about the nature of maggots, who has studied these twenty years to find out the several sorts of spiders, and never cares for understanding mankind'. In his first appearance in the play he was learning to swim with the encouragement of a fawning swimming master, lying on a table in his laboratory with a thread between his teeth, the other end of which was tied around the belly of a frog. Not only would he soon be swimming better than any man in England (though never in water) but he was also 'so much advanc'd in the art of flying' that he could already outfly a bustard. Having exposed Hooke's laughable aspiration to flight, Shadwell went on to mock his work on the respiration of dogs, and his (and others') efforts to transfuse blood between dogs and from sheep to man. In authentic scientific language,

Gimcrack and his friend Sir Formal Trifle made ludicrous claims about how they had turned a bulldog into a spaniel and a spaniel into a bulldog, and put sheep's blood into a madman: 'The patient from being maniacal or raging mad became wholly ovine or sheepish: he bleated perpetually and chewed the cud; he had wool growing on him in great quantities ... I shall shortly have a flock of 'em. I'll make all my own clothes of 'em.' Later, Gimcrack entertained his friends with an account of his studies of ants and spiders, and fell into the trap of describing the habits of the 'tumbling spider', a species invented as a trick by one of his nieces' suitors, through whom Shadwell commented on the folly and dishonesty of natural philosophers. 'As there is no lie too great for their telling, so there's none too great for their believing.'

As the play went on, the parallels between Gimcrack's work and Hooke's multiplied. Gimcrack devised questions for travellers to Lapland and Russia, performed experiments with an air pump, collected and weighed bottles of air from all over England (including Banstead Downs), found millions of eels in a saucer of vinegar, and used technical language lifted straight from *Micrographia*: 'it comes first to fluidity, then to orbiculation, then fixation, so to angulization, then crystallization, from thence to germination or ebullition, then vegetation, then plantanimation, perfect animation, sensation, local motion, and the like.' Experiments and ideas which seemed strange enough to unscientific men when Hooke explained them, were made stranger still through Shadwell's wit. The virtuoso released bottles of country air into his room so that he could enjoy fresh air from his armchair, made a tarantula dance to music, read the Geneva Bible by the light emitted by a rotting leg of pork, planned a study lit only by glow-worms, and watched military campaigns taking place on the surface of the Moon. In a scene that must have convinced Hooke that he was Shadwell's victim, Gimcrack explained that with his new speaking trumpet ('not invented by me, yet ... improv'd beyond all men's expectations') one parson could preach to the whole country, and princes could talk to each other without the expense of ambassadors.

Between farcical sexual and marital scenes which presented Gimcrack and his fellow virtuoso Sir Formal Trifle in the worst possible light, the play's climax was an attack on Gimcrack's house by a mob of ribbon weavers who believed he had invented a weaving engine that would put them out of work. The virtuoso was forced to declare what the audience and most of the cast had known all along, that none of his work had the slightest practical value: 'Hear me, gentlemen, I never invented an engine in my life. As Gad shall sa' me, you do me wrong. I never invented so much as an engine to pare cream cheese with. We virtuosos never find out anything of use, 'tis not our way.'

The play was funny and well received, the best satire on a scientific theme since Ben Jonson's *Alchemist*, sixty-six years earlier. But Hooke took his work and his dignity very seriously, and was very sensitive to mockery and insult. Although the leading actor did not (as far as we know) imitate his physical peculiarities, Hooke was sure that the title character was based on him, and he was mortified to see his life and work lampooned before a laughing audience. Even if Shadwell's target was English natural philosophy in general, nobody was more closely identified with the scientific work of the Royal Society than Hooke, and many in the audience must have recognized him. When he got home he wrote furiously in his diary: 'Damned Dogs. *Vindica Me Deus* [God avenge me]. People almost pointed'. Coffee-house friends who had seen the play seemed to smile derisively at him, and Hooke was still talking about it a month later, when he bought a copy of the play.[10] By that time he had begun, on an occasional basis, to share his niece Grace's bed, a practice which even Shadwell might have hesitated to dramatize.

In fact, Hooke shared some of Shadwell's contempt for aimless collectors of trinkets and trivia, and felt completely out of step with the leaders of the Royal Society. A diary entry for 22 June 1676, describing a lecture (perhaps a Cutler lecture) and a tavern gathering afterwards, captures his sense of estrangement and unrewarded effort: 'Wrote lecture of noyse and sound. Read it. It seemd to please. At Crown, Brouncker and Oldenburg. None drank to me.' During the summer recess of 1676 he tried to form a new scientific club on the lines of the New Philosophical Club that had met in January that year. If he could recruit most of the Royal Society's most active members, he might be able to replace the enfeebled society with a smaller and more serious organization. Boyle and Wren seemed interested, and on 2 July he contrived a 'new Decimall Society of Boyle, Wren, Hoskins,

Croon, King, D. Cox, Grew, Smethwick, Wild, Haak, for chemistry, anatomy, Astronomy and opticks, mathematicks and mechanicks'. He spent much of July talking to prospective members, with mixed success. Cox, Wren, Petty, Hill and Hoskins seem to have been keen on the idea, but Boyle, whose support might have given the club greater credibility, was hard to pin down. Groups of four or five met at Wren's house or in Garraway's in July and August, and there were occasional meetings on this scale for the rest of the year.

Hooke's interests in the early summer ranged from windmills to sunspots. In July he proposed a shipboard windmill which could be used to raise water or an anchor, to hoist sails or goods, and to wind a ship up a creek against the wind with an anchor and cable. And in June and July he observed a series of sunspots, and added a short section to his forthcoming book on oil lamps, *Lampas*, suggesting further research on the relationship between the appearance of spots and the unusual heat of that summer. In August he completed the circular fly watch mechanism he had invented the previous October, and went to Greenwich to supervise the installation or adjustment of his ten-foot mural quadrant for Flamsteed, a man who now seemed to Hooke to be 'proud and conceited of nothing'. He also considered another way forward, suggested to him by Ralph Montagu, whose Bloomsbury palace he was still building. He could give up his work in London and accompany his wealthy client, who was going to France as Charles II's ambassador to Louis XIV. He had 'resolved' to do this in early July, and perhaps it is what he had in mind on 15 August when he 'resolved to leave all imployment'. Both resolutions (like all his resolutions) were soon forgotten, and he stayed in London, working as frantically as ever.

In the autumn of 1676 Hooke's relations with the Royal Society took another turn for the worse. The last sheets of *Lampas* were printed in early September 1676, and later in the month Hooke started giving copies to friends and leading members of the Royal Society. It was not the text of *Lampas* but its postscript, which accused Oldenburg, 'one that made a trade of Intelligence', of attempting to defraud him of the rewards for his work on spring watches, that aroused interest. Oldenburg used his September *Philosophical Transactions* to denounce Hooke's 'immoral postscript', and the Royal Society Council (of which Oldenburg was a member) met on 3 October to decide what to do about it. On 12 October Hooke was summoned to a Council meeting, which insisted that the printer of *Lampas*, John Martyn, should publish a statement dissociating himself and the Society from the attack on Oldenburg, if he wished to retain his position as the Society's official printer. Hooke was not reprimanded, but he 'resolved to leave' the Society, nevertheless. The matter was not yet settled, and on 2 November the Council set two of its members, Dr Croone and Abraham Hill, to prepare a statement in defence of Oldenburg's and the Society's integrity for the next edition of *Philosophical Transactions*. The same meeting, to Hooke's surprise, appointed his friend and assistant Harry Hunt to replace Richard Shortgrave, who had died at the end of October, as the Society's Operator. In the meantime Oldenburg, who was much embarrassed by Hooke's attack, made several drafts of a statement for *Philosophical Transactions* trying to explain Huygens's offer of the patent, in which he argued that the two watches were so different it would have been possible for them to be patented separately, enabling both Oldenburg and Hooke to make their fortunes.[11] Oldenburg's central argument went to the heart of Hooke's case against him:

> if he had put his investigation timely enough in print, these contentions might have been prevented or at least more easily determined.
>
> But he alleges that he had publikly read of it before thousands as well Forrainers as English. If so, why then is he so angry, that it came, as he will have it, to the knowledge of Monsieur Huygens and that by the Publishers [Oldenburg's] communication. If he urges that Monsr Huygens should have acknowledged him to be the Inventor; Tis easy to answer, that he should not quarrel with the Publisher, but with M. Huygens . . . this bold man spares not the R. Society itself, who have so highly obliged him, since he sticks not slightingly to call that intelligence which he doth and must acknowledge to be theirs, a *Trade*.[12]

Hill and Croone, who were old friends of Hooke's, brought their report to the Council on 20 November. Hooke was sure that the Council had been rigged against him ('Lord Sarum, Dr. Holder, Sir J. More, &c. not warnd'), but the Council's statement, which defended Oldenburg's honest conduct of Royal Society intelligence, was not explicitly hostile to him. He hoped to win a seat on the Council when elections were held at the end of November, but though his friends Wren and Evelyn gained seats, along with Samuel Pepys, his own vote was not high enough. He was convinced that he had been cheated by Oldenburg and a cabal of the old Council: 'Much fowl play used in this choice . . . I had 15 votes to be one of the Councell, though Oldenburg reckonned but 13. Resolvd to Reforme these abuses . . . 41 papers given in though not soe many present'.

While his dispute with the Royal Society was going on that autumn, Hooke pursued his search for the perfect watch, passing his new ideas on to Tompion. In the middle of September they were working on a watch with an 'endlesse screw', a month after that, on 15 October, Hooke told Tompion 'the way of the single pallet for watches' and 'about laying ballances one over tother', and on 10 November he told him of his 'new striking clock to tell at any time howr and minute by sound'. Apart from an experiment on 7 December to demonstrate the impossibility of finding longitude from the degree of inclination of a dipping magnetic needle, Hooke did hardly anything for the Royal Society in the last months of 1676. Meetings were thinly attended, and usually had nothing more interesting than readings from Oldenburg's correspondence to amuse them. Hooke's Gresham lectures were poorly attended, too. On two successive Thursdays in November the hall was empty, and he had to wait around until one or two grumpy listeners turned up. By contrast, his club had some lively discussions. On 17 November, in Child's coffee house, Wren, Hooke, Hill, Hoskins and Aubrey discussed teaching children grammar by tables, and an inscription for the Monument, and two days later they talked about starting a correspondence with Paris and Amsterdam.

Although Hooke discussed the reform of the Royal Society with Ward, Wren and other friends, his position in the Society was weakened, and could hardly recover while Oldenburg and Brouncker dominated the Council. Within his own household things were not much better. Tom Giles, his young cousin, was turning out to be a sluggard and a pilferer, and Grace had become so tiresome that he was on the point of sending her back to his brother on the Isle of Wight. Hooke consoled himself in the company of his club and coffee-house friends, and in a growing sense of his own financial security. On the last day of 1676 he made a record of all his debts and credits. He was owed £325, more than six years' salary, by Sir John Cutler, £500 by the City Lands Committee for work on the Monument and other services since 1673, £200 for Bethlem Hospital, perhaps £75 by Ralph Montagu, £150 by Wren for his work on the City churches, his annual salary of £30 from the Royal Society, and £200 by his struggling brother John on the Isle of Wight. To set against these very substantial sums, Hooke's debts were domestic and small:

> I owe Mr Neile for 9 firkins of small beer and for 5 gallons of ale. I owe Mr Loach for velvet coat and lyning £5, my shoomaker for 2 pairs shoos and Goloshoos. 1 pair of shoos. Mr Berry for 1 Load of Coles, 12sh. Blagrave for wine 12sh. Much love to all my friends I owe. Lever for quadrant 20sh. Thomson, Bloomsberry for the same 12s. I owe none els a farthing to my knowledge.

16. In Two Worlds

(1677)

By early 1677 Hooke seemed to have gone the way of Sir Christopher Wren, and turned into a full-time architect and site manager with a part-time interest in science and mechanics. True, he still attended most Royal Society meetings, as he was expected to do, and joined the regular gathering at the Crown tavern in Threadneedle Street afterwards, but he made hardly any recorded contributions to their discussions and rarely performed experiments in the winter and spring session. Even the sensational demonstration of the qualities of phosphorus to the Royal Society in January 1677 could not entice him back into the scientific mainstream. A Hamburg alchemist, Hennig Brandt, had accidentally discovered phosphorus in 1669 while he was searching for a way of making the philosopher's stone. He sold the recipe to a chemist, Daniel Krafft, who kept it secret for eight years. Hooke, whose long-standing interest in shining or phosphorescent matter like rotting wood or putrid meat (the *Virtuoso*'s rotting pork reading lamp was only a mild exaggeration) might have given him an intense interest in this mysterious smouldering substance, recorded its appearance with something close to indifference. 'Oldenburg produced the *Phosphoros Baldwini.* Twas affirmd that it shined best. I could not perceive it. The manner was to open the box and expose the substance to the light whence it collected light as also it did from the candle but I could see neither.' Unlike Boyle, Hooke showed no interest in discovering a way of making phosphorus for himself, but since this involved boiling away over fifty buckets of putrefying urine the other residents of Gresham College had reason to be thankful for his inactivity.[1]

To be fair, Hooke was more active as a natural philosopher than the Royal Society minutes suggest. On 1 February 1677 Oldenburg read part of a letter from the Dutch microscopist Leeuwenhoek describing his discovery of vast numbers of minute creatures – single-cell organisms – in ditchwater and pepper water (an infusion of black pepper), using a powerful but tiny single-lens microscope. Hooke was no longer an active microscopist, finding (he said) that powerful single lenses hurt his eyes, and the Royal Society now relied on its second curator, Nehemiah Grew, for accounts of animal and plant structure, and to repeat the work of the leading European microscopists, Marcello Malpighi, Jan Swammerdam and Leeuwenhoek. But it was Hooke, rather than Grew, who devised a way of using the magnifying power of the water itself, in conjunction with a microscope, to duplicate Leeuwenhoek's observations. There was apparently not much interest when he announced this idea on 22 February, and Hooke's contribution was not minuted. But his diary recorded the moment: 'I told my way of small microscopes, of using the drop of water for a lens and seeing it'. His account of a one-wheeled chariot on 8 March was greeted with similar indifference, and when he joined a discussion on 10 May Flamsteed told him that 'he spake nothing to purpose'. Another of his interesting ideas, a 'wheelhorse' or 'double engine for the feet', was never presented to the Royal Society, though he told Boyle about it in April. If Hooke's planned vehicle had two wheels, not four, perhaps it was an early appearance of the pedalless bicycle, or velocipede, which was next suggested by de Sivrac in 1790, and made after 1815. Hooke was not a keen horseman. A ride on a lively horse out to the villages of Southgate and Edmonton with his friend Dr Scarborough in April 1677 left him exhausted, and a ride with Hoskins to Banstead the previous June had bruised his testicles. So a two-wheeled vehicle propelled by swinging the legs would have been a convenient way of getting him from one appointment to the next, if one discounted the laughter it would provoke on the London streets.

The weakness of Hooke's position in the Royal Society was epitomized by the meeting of 25 January 1677, which publicly disowned his criticisms of Hevelius and accepted the accuracy of Hevelius's work. On the same day the Council agreed to send his screw quadrant to Greenwich, and Lord Brouncker announced that he wanted Hooke to give up his key to the Arundel House library and return his books. Flamsteed needed Hooke's quadrant because the large wall-mounted quadrant Hooke had designed and built for him in 1676 did not work well. On 5 June 1677 Hooke and Tompion went to Greenwich to set the mural quadrant's arms properly, but Flamsteed was not satisfied, and a week later he arrived at Gresham College to collect the smaller quadrant the Royal Society had promised him, along with anything else that might be useful. Hooke was not one to give in without a struggle. He held on to his library key, searched the Society's records for evidence to support his case against Oldenburg and Brouncker, and did his best to build alliances with influential fellows who might take his side. On 8 June he 'contrivd about chusing Sir J. Hoskins new president and new moduling the Royal Society'. No doubt his sense of injustice sometimes made his company tiresome even to his friends. 'I had a great dispute about the Society's abuse to me with Hill and Hoskins', he wrote on 5 July.

Hooke's other public forum, the Gresham and Cutler lectureship (in term and vacation weeks respectively), seems to have been virtually defunct in 1677. Often he prepared his Thursday afternoon lectures, waited till three o'clock for an audience that never arrived, and went away with his lecture unread. Royal Society members who were once keen to hear his lectures now stayed away, and sometimes his few listeners seem to have wandered into the hall by mistake or on personal errands. On 25 October nobody came for the lecture, 'but a rusty old fellow walkd in the hall from 2 till almost 3'. Embarrassingly, on 5 July, when only Captain Panton, who had made a fortune playing cards with courtiers and was using it to develop Panton Street (off the Haymarket), turned up to hear Hooke's Cutler lecture on comets, a Cutler family servant, Thomas Axe, was there to 'spy' on Hooke's performance. It was the same the next year, 1678. Generally nobody turned up, and on the rare occasions that someone appeared they were either spying for Cutler or there by mistake. On 31 October 1678 'a fellow with a blue apron layd asleep all the time there should have been a lecture, I suppose a spy', and the following week 'onely one came, peepd into the hall, but stayd not'. Since Cutler seems to have been determined not to give Hooke a penny of what he owed him, his spies' reports made little difference.

Hooke had not completely abandoned his scientific work, although he had fewer opportunities to display it in public. On Saturday 21 April 1677 he rose before three o'clock to observe a comet which

Bates the carpenter had told him about in Garraway's the previous evening. Tom forgot to wake him on Sunday morning, but he watched it with two friends early on Monday, and used a six-foot telescope and a cross-staff to measure its size and position, and track its path across Andromeda and Cassiopeia towards the rising Sun. The next morning Aubrey, Wild and other Garraway's friends joined the watching party, but clouds disappointed them, and Hooke never saw the comet again. He dug out his old notes and lectures on the comets of 1664 and 1665, and began writing a Cutler lecture on the nature of comets and their relevance to the great problem of planetary motion. He worked quickly, stitching old and new material together to save time, and by 5 May he was discussing publication of the unfinished work with John Martyn, the Royal Society's printer. Martyn, whose reputation had suffered in the *Lampas* scandal, was unwilling to publish it without a licence, and *Cometa* was not published for a year.

With Wren's help, Hooke was beginning to turn his mind again to the fundamental question of planetary motion. The three basic laws they had to work from had been established earlier in the century by Kepler on the basis of Tycho Brahe's observations. First, planets moved in elliptical orbits, with the Sun as one focus of the ellipse. Second, planets moved more slowly in those parts of their orbits that were further from the Sun, in such a way that the area between the arc of the orbit travelled in a particular period of time and the Sun were always equal: short fat sectors had the same area as long thin ones. Third, in comparing one planet with another, the square of the planetary year was proportional to the cube of the average distance of the planet from the Sun. It was clear to Hooke that planets moved more quickly when they were close to the Sun because the Sun's gravitational attraction – one of the forces that propelled each planet – was more powerful at close range, and diminished with distance. But what was the mathematical relationship between gravitational force and distance? In the summer of 1677 the two talked about Kepler's laws and tried to derive the mathematics of the planetary system from them. In a conversation on 16 August which ranged from primitive superstition to modern science Wren told Hooke 'of curing his Lady of a thrush by hanging a bag of live boglice about her neck' and then 'explained his way of solving

Kepler's problem by the Cycloeid'. On 20 September, discussing Kepler's second law of planetary orbits, Wren 'affirmd that if the motion were reciprocall to the Distance the Degree of velocity should always be as the areas, the curve whatever it will'. From this account in Hooke's diary it seems that Wren had not yet hit upon the principle that made sense of Kepler's laws, that the gravitational force between two objects diminishes in proportion to the square of the distance between them, the so-called inverse square law. Unless Hooke misunderstood him (which is not very likely) Wren still believed, as he did in the 1650s, that gravitational force was inversely proportional ('reciprocall') to the distance between two bodies, rather than to the distance squared. It was possible for a good mathematician to derive the inverse square law from Kepler's three laws in conjunction with Huygens's statement of the same principle in centrifugal force, which he had published in his treatise on pendulums, *Horologium Oscillatorium*, in 1673. Wren and Hooke both owned this important book, and Hooke had read it in November 1675. In 1686, when the first discovery of the inverse square law was a matter of dispute between Hooke and Newton, Newton recalled that when he had met Wren in London in 1677 (probably in the spring) Wren already knew about 'duplicate proportion' – the inverse square law. Yet it appears that Wren did not know the law in September 1677, or did not explain it to Hooke. Thus Hooke did not incorporate the inverse square law into *Cometa*, which was published in 1678, or even in his republication of *An Attempt to Prove the Motion of the Earth* in 1679, in which he repeated that the degree to which the Sun's attractive power diminished with distance was 'not yet experimentally verified'. These may seem obscure issues, but they lie at the centre of Hooke's last great dispute with Newton, and the disappointment that clouded his final years.[2]

The scientific project that occupied most of Hooke's time in 1677 was the perfection of his spring-regulated watch, which was now quite widely used by his friends and customers. His most successful device was the paired or double balance, in which two interconnected sprung balance wheels vibrated in exact opposition to each other, to neutralize the effects of rough motion and to overcome the tendency of spring-regulated watches to stop. An undated document found among his papers explains what he had in mind: 'I made two ballances exactly alike in all particulars and soe placd them in a frame that the axis of them both lay in the same line, then by the help of two small beames & strings . . . I soe united them that look how much and how fast the one was movd one way soe much and soe fast was the other movd the other way . . . by this means I was able to rowle or tumble the instrument any wayes without altering in the least its regular motion.'[3]

We can trace the improvement and perfection of the double-balance watch through the diary in 1677, and see how Hooke spread the knowledge of its construction to a second watchmaker, John Bennett of Clerkenwell, perhaps with the intention of reducing his dependence on Tompion. On 1 April Tompion 'shewed his watch with 2 ballances', and two days later Hooke explained 'the new way for a ballance' to Bennett. By 27 May Tompion had made a working double-balance watch, and two weeks later, just before the new watch was delivered to Sir Jonas Moore, Hooke took it to Bennett's. Sometimes it is hard to distinguish Hooke's creative contribution from Tompion's, but it is generally clear that Hooke devised and tested the original ideas, leaving it to Tompion or Bennett to produce the finished working watch. On 24 June, for instance, he instructed Tompion 'about the King's striking clock, about bells and about the striking by the help of a spring instead of a pendulum, as also the ground and use of a fly and of the swash teeth'. The diary often shows him making new sorts of springs and trying them out in watches – a twisting spring in January, a reverse spring in July. At the beginning of July he showed Tompion a way of making a silent pendulum using his universal joint, and on 27 July he fitted a 'great watch with 1 loop spring for the two ballances'. This seems to have been an important day, and Hooke's diary entry ended: 'Double Ballance perfected.' The time it took for Hooke's watches to get from the laboratory to the market place was surprisingly brief. On 12 August 1677 Dr Scarborough, Hooke's friend and customer, bought a watch from him 'with two springs bended about cylinders and a third spring regulator moved by strings'. Hooke considered following Sir Jonas Moore's advice and applying for a new patent for his watch, but he never did so. At least, by defeating Huygens's and Oldenburg's bid for a monopoly he had kept the road open for himself, Tompion, Bennett and others to work towards a perfect pocket watch.

In July 1677 Hooke rediscovered his earlier interest in measuring, recording and predicting the weather, and picked up the threads of his work on the weather clock. On 7 July he fitted up a new device, an aneroid (liquid-free) barometer or 'otheometer' (pressure gauge), in which a spiral spring connected to a toothed wheel was used instead of a column of mercury to measure the weight of the air. For the next ten days he kept a close watch on four instruments which he thought might indicate changes in atmospheric conditions, a thermometer, mercury barometer, an otheometer and a 'magnetic meter' of some type. He soon found that the otheometer moved 'directly opposite to the barometer', and that the movements of the magnetic needle seemed to have no discernible relationship with atmospheric fluctuations. Nevertheless he had the confidence by 15 July, a wet St Swithun's Day, to set aside folklore and tell his friends in Jonathan's coffee house that there was fine weather ahead. Though the otheometer was taken down and apparently discarded five days later, his work on weather measurement continued, and produced some interesting results in 1678.

As his scientific work contracted, his work as an architect and construction manager grew. In the early 1660s the King granted first a lease and then the freehold of part of St James's Fields, the land to the north of his favourite London home, St James's Palace, to Henry Jermyn, Earl of St Albans, a courtier and protégé of the King's mother. After the disasters of 1665 and 1666 there was a booming demand for good houses in the new West End, and Jermyn began developing St James's Square and its surrounding streets as a fashionable and largely self-contained community, with its own Wren church, market and central square. He leased building plots to men who were able to raise the capital to build handsome houses for sale or rent to aristocrats or rich Londoners. Some of these speculative builders were themselves titled men with court connections, like Sir Thomas Bond and Sir Thomas Neale, some were financial fixers like Nicholas Barbon, and others were master builders. Hooke had been dealing with master builders, on behalf of Sir Christopher Wren, Ralph Montagu, the City authorities, the Royal College of Physicians, and the governors of Bridewell and Bedlam, ever since the Great Fire, and now his experience was in demand in the new West End. In January 1677 he began working for John Hervey, Treasurer of the

Queen's household, a substantial property holder in St James's and the Strand and a kinsman of Jermyn and Montagu. On Hervey's behalf, he evaluated a new house built by the master mason Abraham Storey on St James's Square (now number 6), and tried to knock £500 off Storey's £5,500 asking price. Hooke knew Storey well, since they had worked together on the City church of St Edmund the King in the early 1670s. Storey offered to do more work on the house if the price was held at £5,500, but in March, after two months of delay and negotiation, Hooke managed to get the house and the extra work for £5,000. He also viewed other houses in St James's on Hervey's behalf, including those built by Richard Frith, Nicholas Barbon and the Earl of St Albans. One success led to another. In November 1677 Lord Ranelagh (Lady Ranelagh's son) employed Hooke to buy a house in St James's Square from the developer John Angier, and he negotiated a price of £4,800, with payments spread over forty-two months.[4] Hervey was obviously pleased with Hooke's work, and began employing him as his agent in dealing with the rest of his West End houses. He asked Hooke to arrange with the master carpenter Roger Bates, who had worked on Bedlam, to demolish and rebuild two of his older houses, and for the rest of 1677 Hooke attended on him frequently. He directed the finishing of Hervey's house in St James's Square, and provided designs for seven houses in the Strand, five of which were built in 1678. Years later, Hooke would often illustrate his lectures on shells and fossils with examples he had found in brick pits that had been dug in St James's Fields in the 1670s.

Hooke's friendship with Boyle made it fairly easy for him to find work among the West End aristocracy. Boyle's family, the Burlingtons, had large estates in and near London, and his sister's family, the Ranelaghs, were active in the property market, too. In March 1677 Lady Ranelagh asked Hooke to organize the rebuilding of the back of her house in Pall Mall. Hooke designed the building, set out the ground and introduced Lady Ranelagh to the builder Roger Bates. Contractors like Bates and Storey naturally valued their relationship with Hooke, and kept it alive by offering him discreetly wrapped gifts. He refused a gift from Bates in May, but apparently not as a matter of principle, since he accepted twenty guineas from Storey's wife in July. In May he went down to Lady Ranelagh's country house in Chiswick to supervise work on the 'kitchen, Great Stairs, railes, Gates, floors, doors, &c.', but he was much less involved in Lady Ranelagh's work than he was in that of Bedlam or Montagu House, which were almost full-time commitments. His job was to find a reliable builder, negotiate a price, arrange a contract, draw up a plan, and make sure the contractor built to plan and schedule. Hooke was also regarded as a skilful and well-informed negotiator, who knew the quality of a house, understood the stratagems of builders and developers, and would drive a hard bargain for a reasonable fee.

Sometimes, as with Bethlem, Montagu House and the Physicians' College, Hooke saw his designs through to completion, but on occasions he provided a draft and had little more to do with the building. In March 1677 he prepared a plan of 'Maudlin Colledge' for Dr Burton, a Fellow of Magdalene College, Cambridge. This may be the building now known as the Pepysian Library, for which no other architect is known. Hooke does not seem to have made any contribution to it beyond providing a design. The same may be true of a house he designed for a Devonshire landowner, Sir Walter Yonge, in February 1677. He introduced Yonge to two trusted craftsmen, Roger Bates and Roger Davis, who apparently went down to Devon on his behalf in September. Since building continued into the 1680s, when his diary is patchy or non-existent, it is quite possible that he visited it and left no record of his journey. The house they built, Escot House, near Ottery St Mary, burnt down in 1808, but its gardens still remain. A plan and a drawing of the house, showing the Dutch influences familiar in most of Hooke's buildings, was published in Colen Campbell's *Vitruvius Britannicus* in 1715. It was, says Sir Nicholas Pevsner, 'an early and important example' of the compact country houses that began to appear in the later seventeenth century. Sir Walter Yonge and Hooke had a mutual friend in the City merchant John Pollexfen, who also built a house in Devon, at Wembury, in the same style as Escot.[5]

Although the demand for Hooke's services in measuring City plots and settling building disputes had declined after 1674, he still had plenty of work to do for the City authorities. In April 1677 he redesigned the water conduit at the foot of Snow Hill, which is now in the valley under Holborn Viaduct. He arranged the construction contract with Knight, the City stonemason, and supervised his work later in the year. Hooke was also involved that year in the rebuilding of Holborn Bridge across the Fleet Canal. The intention of both these improvements was to widen the route from Holborn into Smithfield and the City. He also continued to share with Wren the task of building the City churches before the coal-tax revenue that funded the work ran out, and visited about fourteen of them, either alone or with Wren, in 1677. He was most often at St Anne and St Agnes Gresham Street, St Peter Cornhill and St James Garlickhithe, where work was just beginning, and at St Lawrence Jewry and St Stephen Walbrook, which were almost finished by 1677.

*

IN HIS PERSONAL and domestic affairs, Hooke's life in 1677 reflected the range of pleasures and dangers to which Londoners of his time were exposed. He came close to losing his home when a fire broke out in the Gresham College stable on 26 January, but divine intervention (he thought) prevented it spreading, and on 20 December he thanked God for an even closer escape, when he was 'saved from the fire by a miracle. *Gloria Deo Solo*.' The profusion of wooden sheds and stables in Gresham College quadrangle was a serious fire hazard, as the Gresham committee noted in May 1681. Crime was also a threat, and on 30 May Hooke 'scuffled with a Rogue that kickd me in the shins and brak them'. He recovered from the 'great fit of melancholy' induced by this attack, and limped along to Exchange Alley for his usual nightcap in Garraway's and the nearby Fleece tavern. London crime and its punishment brought him pleasure, too. On 23 January, on his way to the Grecian coffee house, Hooke saw the heretic and visionary Lodowicke Muggleton pilloried in the Temple for blasphemy, getting the usual pelting with rotten eggs. No doubt Muggleton's followers regarded the fire that broke out in the Temple a week later as God's comment on their master's ordeal. On 24 January, at the court of the Bridewell to collect a £200 fee for his work on Bedlam, Hooke watched another regular London spectator sport, the whipping of prostitutes. There was better entertainment at Smithfield every year at the end of August, when the ancient Bartholomew Fair was held. Hooke found time to visit the fair at least three times in 1677. On Monday 27 August he paid 2d to see a tiger, and watched a child do 'strange tricks', on Thursday he saw a Dutch woman of prodigious height, and the following Tuesday he went with

Sir John Hoskins to see the famous boneless child. If this child was the one that had been exhibited in 1667, it was a tiny girl in her twenties, with no perfect bones except for her skull, who could sing, read and whistle.[6] Hooke's greatest pleasure was still the company of friends in taverns and coffee houses. Many of his days started and ended with visits to Garraway's or Jonathan's (which opened in July), where his closest friends usually met, but he also found time to call in at many other coffee houses, particularly Man's, Toothe's, Child's and the Grecian, and at several taverns, including the Crown (a Royal Society favourite), Hercules' Pillars, and the George and Vulture.

In his relationship with Grace, Hooke continued to combine the qualities of a guardian and a lover, allowing her to go to balls without him, but shielding her from disreputable company. When a 'jade', Kerry, came to see Grace in July she was refused entry, and paid Hooke back with scolding and abuse when they met later on Moorfields. Hooke made love to Grace on 9 August, and the next day she set off for 'the Countrey', probably to stay with her parents in Newport, Isle of Wight. As far as we can tell from Hooke's diary, she did not return to London until 7 June 1678. At the end of October Nell Young, his old lover and housekeeper, who had sources of information on the Isle of Wight, gave Hooke the worrying news that his niece was being courted by Sir Robert Holmes. Holmes was in his mid-fifties, but in other respects he would have been an extremely desirable husband – rich, heroic, healthy and popular at Court. The courtship was apparently taking place on the Isle of Wight, not in London, and Hooke wrote to his brother for news of it on 3 November. We do not have John Hooke's letters, and we do not know how the courtship developed, but there is some evidence that things went a little too far.[7]

This unlikely match, if it had come off, might have rescued Grace's father, John Hooke, from his growing financial difficulties. Hooke and his brother had always written regularly and sent gifts to each other, Robert sending news or books in return for John's turkeys, geese and honey, and John had stayed at Gresham College for a few days in April 1672. In 1675 John almost persuaded his brother to spend £4,000 – a very substantial proportion of his fortune – on a farm in Avington, near Carisbrooke, but Hooke backed out at the last minute. John's grocery business was clearly in trouble and in 1677 his letters took on a more desperate tone. He asked his brother for a loan of £50 in March and another £20 in June, substantial sums which Robert sent without much hesitation. His generosity was not enough, though, to drag his brother from the financial mire that was steadily sucking him under.

On 8 July Hooke took on a new assistant, Thomas Crawley, who came with a recommendation from Dr Scarborough and the advantage that he wanted no payment beyond his board and lodging. Hooke fitted out a workshop for him, and he turned out to be a fast learner, a skilful craftsman who could mend watches and grind lenses, and a pleasant coffee-house companion. By September 1678 Hooke had decided that his work was worth £5 a quarter. He stayed with Hooke until January 1680, and they remained friends for many years afterwards. Crawley's arrival was probably unwelcome to Hooke's young cousin Tom Gyles, who had to make room in his bed for the newcomer. Tom was a useful errand boy when there were quadrants or telescopes to be collected from Tompion or Cock, but he was lazy by Hooke's high standards. Several times he was threatened with being sent home to the Isle of Wight, and once, in March 1677, he only saved himself from this humiliation by bursting into tears. It would have been far better for him if he had got on the cart and taken the road to Portsmouth. On Saturday 8 September, just after Bartholomew Fair, he was taken ill with smallpox, one of the deadliest and most common of London's infections. First he felt a painful stiffness or crick in his back (an early symptom of the disease), and Hooke wrote to his brother John to have him taken home. The next day he 'rored with creck', and by Tuesday he was gravely ill, bleeding from the nose, mouth and bladder. Hooke went in search of five of London's best doctors, and found one, Dr Diodati, who was prepared to draw blood from Tom's arm and mouth. The next day, 12 September, Drs King, Mapletoft and Diodati met and agreed that 'pissing blood in the small pox' was fatal, and that Tom's position was 'irrecoverable'. Thus freed from the murderous attentions of London's finest physicians, poor Tom Gyles was able to die in peace just after noon that day. Hooke described his last moments:

> Tom spake very piously, began to grow cold, to want covering, to have little convulsive motions, and after falling into a slumber

> seemd a little refreshed and spake very sensibly, and heartly, but composing himself againe for a slumber he ratled in the throat and presently Dyed. It was about 14′ after 12 at noon, he seemd to goe away in a Slumber without convulsions.

Tom was buried in the churchyard of St Helen Bishopsgate, Hooke's parish church, on 13 September, where the registers list him only as 'a servant of Mr Hooke's in the Colledge'.[8]

17. Tiny Creatures and Springy Bodies

(1677–1678)

Another death in September 1677, seven days before Tom's, changed the course of Hooke's life. On 3 September he heard that Oldenburg had taken to his bed with an ague, or malarial fever, at his house in Kent. Malaria was endemic in marshy districts of southern England, and Oldenburg's estate at Charlton, between Greenwich and Woolwich, was in the Thames estuary marshlands. Oldenburg's death two days later (after being 'stricken speechless and senslesse') raised the prospect of radical change in the Royal Society, and Hooke acted at once to improve his own position. The secretaryship would give him increased status and power, control over the Society's minutes and correspondence, a place on the Council, and a return to the central position in the Royal Society's affairs that he had lost in the mid-1670s. On 7 September he talked long and seriously with Boyle about the secretaryship and the running of the Society, and bought a long run of back numbers of *Philosophical Transactions*, the periodical edited and published by Oldenburg, from John Martyn, the printer. At this point Tom's fatal illness intervened, but on 13 September, the day of Tom's funeral, Hooke took over Oldenburg's duties until a permanent secretary was appointed. During the next week he was busy with his usual duties as a builder and natural philosopher – supervising Lady Ranelagh's building, designing seven houses for Mr Hervey, delivering the Lord Mayor's draft for £1,500 to Wren's office, falling asleep over mathematical books in the Temple Library, discussing the mechanics of the orbital motion of planets with Wren, calculating the necessary dimensions of rainwater pipes for large public buildings, designing a monument for Garraway's dead daughter, directing a 'dung hole' for Dr Whistler of Gresham College. On 22 September there was something more out of the ordinary. He went to Boyle's house, where Krafft, who had the secret of phosphorus, had promised to demonstrate some of the amazing qualities of his new substance. By applying a particle of his white powder on a quill Krafft ignited a pile of warmed gunpowder, and a little smeared on paper ignited as soon as it was warmed. Hooke found the two experiments 'exceeding strange and much more than I had ever seen', and persuaded Boyle to provide an account of them for inclusion in *Cometa* the next year.

On 24 September the Royal Society's Council met and decided to postpone the choice of Oldenburg's successor as secretary until a full ballot could be held on the Society's regular St Andrew's Day elections on 30 November. Hooke, along with others who were worried by the decline of the Royal Society in recent years, saw this as an opportunity to revive the Society by replacing Lord Brouncker, its President since 1662, with a younger and more vigorous leader. Hooke met two of his closest allies among the Council members, Abraham Hill and Sir John Hoskins, on Monday 8 October to plan their coup, and the next day, in a coffee house, they 'desird Mr Aubery to spread the Designe of choosing new president to Mr Ent, Dr Millington, &c.'. On Thursday Wren and Nehemiah Grew, Hooke's fellow curator, were drawn into the plan, and 'things seemd to goe well for new president'. Brouncker's strongest defender seemed to be Daniel Colwall, the Society's Treasurer since 1665, who was 'canvassing for the other side'. The coffee-house plotters revealed their intentions at the Council meeting of 18 October, where Hoskins insisted on a ballot for the Society's officers, and 'Lord Brouncker in a Great Passion, raved and went out'. Hooke, who was not on the Council and had spent the day on business at the City's Court of Aldermen, met Hoskins several times after this meeting and found him resolute in his challenge to Brouncker.

Oldenburg's death was a grievous loss to the Royal Society. He was a man of enormous energy, diligence and open-mindedness, and it took many years for the Society to find another secretary with his qualities to act as an intermediary between the Society in London and the wider world of natural philosophy. Hooke performed Oldenburg's secretarial duties at the meeting of 25 October, but he was not certain of getting the permanent position, or of filling it as well as Oldenburg had done. In modern terms, he was a player, not a referee. He had the necessary international reputation and breadth of interests for the secretaryship, but his neutrality and honesty were not trusted as Oldenburg's had been. He was not prepared to devote himself so wholeheartedly to the burdensome duty of initiating and answering foreign correspondence, or to subordinate his own scientific career to the common good. It was unfortunate, too, that he was on bad terms with some of the Society's most important overseas members, especially Hevelius and Huygens. But from Hooke's point of view the removal of Oldenburg made the Royal Society a congenial forum once more, and drew him back again into the public demonstration of his many scientific interests.

In the meetings of autumn 1677 Hooke combined the roles of curator and acting secretary, performing his own experiments and demonstrations (often of quite old work) and reading accounts of work done elsewhere. Hooke's control of Royal Society minutes ensured that his weekly contributions were fully recorded, perhaps at the expense of other participants, giving the impression that his role in these autumn meetings was even more central than it had been ten years earlier. The meeting of 25 October, the first he minuted, was dominated by him from start to finish. First he demonstrated a waterpoise or hydrostatic balance, a device for establishing the comparative density or specific gravity of liquids with great precision. A large pear-shaped bottle, weighted with lead shot so that it was very slightly heavier than the water it displaced, was hung from a balance by a hair, so that it rested submerged in a container of water, neither rising nor sinking. If the composition of the liquid was changed by even a tiny amount the bottle would rise or fall, indicating a change in the density of the liquid, which could be measured by adding weight to the other side of the balance to restore the bottle's former position. When the audience was convinced of the accuracy of his waterpoise, Hooke moved on to tell them about a new type of waterproof leather from Paris, which he had first seen only a few days before, which might be used for riding coats, bottles, wading boots, tents, coach covers or inflatable swimming aids. Taking on Oldenburg's role as a reporter from the wider scientific world, he then entertained the meeting with a display of a self-propagating Portuguese onion, and reading letters on cider-making, Saturn, and Leeuwenhoek's microscopic discoveries.

This meeting gave Hooke an agenda for the next four weeks. In the early November meetings he showed that his waterpoise could detect changes in the density of liquids 'even to the hundred thousandth part of their bulk', and demonstrated his way of treating leather with wax and turpentine, making a material that was less waterproof than the French version, but smelt better. Throughout November he tried to repeat the work reported by Leeuwenhoek, who had seen 'multitudes of exceedingly small insects or animals wriggling amongst each other' in water, using a small single-lens microscope. Leeuwenhoek was very keen to publish his results, but he kept his methods secret. This made it difficult to replicate his observations, and Grew had already failed to repeat them several months earlier. Hooke's idea, which had failed to impress the Royal Society back in February, was that it might be easier to see the creatures if the water was held in very thin glass pipes, which would themselves have a magnifying effect. Using a good compound microscope, he tried pump water on 1 November and saw nothing, and strong pepper water on 8 November, and saw pepper. He worked on the problem at home, and on 10 November, using rainwater which had been standing a week with black pepper in it, he at last saw 'great numbers of exceedingly small animals swimming to and fro', so tiny that there were, he thought, many millions of them in a single drop of water. 'I was very much surprised at this so wonderful a spectacle, having never seen any living creature comparable to these for smallness', he wrote. 'Nor could I indeed imagine that nature had afforded instances of so exceedingly minute animal productions.' At the meeting on 15 November the creatures were seen by Wren, Henshaw, Hoskins, Moore, Croone, Grew, Aubrey and others, confirming (just as Royal Society demonstrations were meant to do) Leeuwenhoek's discovery of micro-organisms. In his minutes Hooke described the animalcules they had seen: 'Their shape was to appearance like a very small clear bubble of an oval or egg form; and the biggest end of this egg-like bubble moved foremost. They were observed to have all manner of motions to and fro in the water; and by all who saw them they were verily believed to be animals.' By the next Thursday Hooke was able to show the meeting micro-organisms in twelve different types of water, using a double microscope. Over the next two weeks, while he intrigued for the removal of Lord Brouncker and his own election as secretary, he was also trying to make a clearer and more powerful double microscope and a tiny single-lens microscope like Leeuwenhoek's. By 6 December he had made both microscopes, and Royal Society members at the meeting that day saw the little creatures far more distinctly than before. They were especially impressed by the single-lens microscope, but Hooke, in confident mood, promised that he would soon show them a single lens that would magnify a thousand times more than this.[1]

To make his single-lens microscope, Hooke held a small rod of pure glass in the flame of an alcohol or salad oil lamp and drew it out into very fine threads. By heating the end of one of these threads he made a tiny glass ball, perhaps as small as a tenth of an inch in diameter, and allowed it to cool. Then he took a thin brass or silver plate with a small hole pricked in it, and he fixed this tiny lens into the hole, making a sort of magnifying glass. He fixed a bent needle or tiny jointed arm to the metal framing his lens, so that the substance to be viewed could be held very close to the lens. For some observations he put the liquid in fine glass pipes, and for others he put the liquid between two fine glass plates fixed together in a screwed frame, so that it could be squeezed thin enough for individual creatures or globules to be seen. Hooke also noticed, and explained for the first time, that if the liquor being viewed was placed on a single glass plate, and was allowed to touch and adhere to the lens to form a liquid bond between the lens and the object, the magnification would be even greater and clearer, and the distortion caused by refraction reduced. Unlike Leeuwenhoek, Hooke was keen to pass on his methods, or (as those hostile to him would see it) to show how clever he had been, and he described all his techniques in a short essay, *Microscopium*, which was published with *Cometa* in a collection called *Lectures and Collections* in April 1678. Whatever his motives, *Microscopium* was the last manual of microtechniques to be published in England until the middle of the next century. In it, Hooke emphasized the scientist's duty to explain his methods and propagate his work:

> The manner how the said Mr. Leeuwenhoek doth make these discoveries, he doth as yet not think fit to impart, for reasons best known to himself; and therefore I am not able to acquaint you with what it is: but as to the ways I have made use of, I here

> freely discover that all such persons as have a desire to make any enquiries into Nature this way, may be the better inabled to do so.

For those who did not want to make their own microscopes, Hooke had shown Christopher Cock, of Long Acre, how to make all the equipment an enthusiast would need to continue his work. These tiny single lenses were difficult to handle, and using them harmed Hooke's eyesight, but the clarity and magnification – perhaps up to 1000x – they offered could produce spectacular results. Scientists could now watch bacteria 'swimming and playing up and down' in their drinking water, and it was unfortunate that it took almost 200 years – and countless deaths from typhoid, dysentery and cholera – for them to grasp the significance of what Leeuwenhoek had seen and Hooke had taught them to see.[2]

Meanwhile, the struggle for control of the Royal Society went on. Hooke's chances of replacing Oldenburg as secretary were improved by the refusal of John Ray, the great botanist, to take the post, but another candidate, Nehemiah Grew, Hooke's competent and respected fellow curator, was promoting his own claim, with the support of Abraham Hill, who had been secretary with Oldenburg in 1673–5. At the end of October Hooke met Grew and Hill in Jonathan's to discuss the imminent elections, perhaps with the intention of proposing a joint secretaryship, but Hill kept up his campaign for Grew. On 14 November Hooke was part of a group of eight Royal Society fellows, including Wren, Evelyn and Henshaw, that went to the Secretary of State Sir Joseph Williamson's house to urge him to stand for the presidency. The change could only be good for Hooke. He already knew that Williamson had a fair opinion of him, except for his 'minding business and profit too much'. He hoped that a friendly word from Wren would strengthen this good opinion, but Wren had doubts about the wisdom of Hooke's becoming secretary, and would not help. On 21 November Hooke wrote his proposal for the secretaryship, no doubt repeating the plans for a more exclusive and serious-minded research institution that he had propounded in papers and coffee-house debates for several years. Lord Brouncker's weaknesses were a regular topic of conversation in Jonathan's and the Crown that month, and by election day, 30

November, his position seemed hopeless. With Henshaw in the chair and Grew and Hooke reading and marking the votes, Williamson and Hill were chosen as President and Treasurer, Wren as Vice-President, and Hooke and Grew were both elected as Council members and joint secretaries.

This was a sweet moment for Robert Hooke. He was freed from a Royal Society leadership that had in recent years seemed hostile and oppressive, and he was finally a man of status and influence in the Society that he had served for so long as a humble curator. His new duties, if they were performed to Oldenburg's very high standards, would be very onerous, but they brought power and high company, and they would be shared with Grew. Hooke had always regarded Oldenburg's control over the minuting of Royal Society meetings as an important power, and one which had often been used to distort the record to his disadvantage. If he kept the minutes he would make sure that in future his contributions received their due credit. It was essential to establish a good working relationship with Grew, and three days after the election Hooke met him in Child's coffee house to discuss their duties. Hooke had proposed Grew for membership of the Royal Society in 1671, and had generally taken a friendly interest in his career, but the shared secretaryship proved to be a source of conflict between them. On 4 December, when the new officers were being sworn in, Hill, Grew's supporter and the new Treasurer, 'slandered' Hooke, and a few days later Hooke discovered that Grew had been 'going to Hill slyly' in pursuit of his ambition to sit at the secretary's table. Hooke sensed a conspiracy, and on 13 December he learned from Hoskins that Hill and Sir John Lowther had met secretly at Williamson's and 'plotted for Grew againste me . . . *Vindica me Deus*'. At the Society meeting that day Hooke showed experiments with a single-lens microscope and a waterpoise, and was furious to find Grew 'placed at table to take notes. It seemed as if they would have me still curator, Grew Secretary. I stayed not at the Crown. I huffed at Hill at Jonathans, Sir J. Hoskins being present.' Since Hooke had become a very active Curator of Experiments since Oldenburg's death, it is quite possible that Hill, Williamson and even Hoskins thought it would be impossible for Hooke to take notes on meetings at which he was the star performer. Perhaps, knowing Hooke, they valued his scientific talents above his secretarial skills, and knew that he could never be another Oldenburg. But Hooke had to be placated, and at a very full Council meeting on 19 December 1677 it was agreed that both secretaries should take notes at the meetings, but that Hooke should draw them up afterwards.

The Council of 19 December also agreed with Hooke's suggestion that in future those proposing experiments for Royal Society meetings should describe them and explain their purpose two weeks before the meeting, so that 'objections, answers, and confirmations may be timely thought of'.[3] This was Hooke's attempt to make life more difficult for dabblers and dilettantes, and to impose on the Royal Society a more structured and rigorous programme of research. The transformation had already begun. In all the meetings since 25 October Hooke had pursued three major subjects: the observation of microscopic organisms in water, the measurement of the density of liquids with a waterpoise, and (on 6 and 13 December) the use of a barometer or baroscope to detect variations in atmospheric pressure. No doubt sensing that some members were impatient with his apparently repetitive and didactic programme, Hooke explained his long-term purpose to the meeting on 13 December. Just as he had argued twelve years before in *Micrographia*, the microscope, barometer and waterpoise were all part of his grand plan to increase the power of the five senses over regions that were once 'inaccessible, impenetrable, and imperceptible'. 'Because this, as it enlarges the empire of the senses, so it besieges and straightens the recesses of nature: and the use of these, well plied, though but by the hands of the common soldier, will in a short time force nature to yield even the most inaccessible fortress.'[4]

Under the new rules agreed on 19 December it would be harder for men with a veneer of scholarship to impress an unprepared audience. The need for such rules was illustrated on 20 December, when Oliver Hill, a member recently proposed by Boyle, rose to deliver a paper asserting that worms in pepper water had been created by a process of spontaneous generation, without the need for seeds or eggs. This was a subject of which he, unlike Hooke, knew very little. Nevertheless 'King, an asse, admired him'. Two weeks later there was a long and fascinating debate between Hooke and Wren over the nature and existence of the aether, the mysterious medium which most natural philosophers (including Descartes, Hooke,

Huygens and Newton) believed filled all the apparently empty space in the universe, and the gaps between all particles of matter. The discussion moved on to the causes of mistiness in the atmosphere, and the effects of moisture and cold on the weight of the air. Hooke explained that the aether, 'the grand or universal menstruum' that encompassed the Earth, absorbed and dissolved most vapours in the atmosphere, and that when certain vapours, because of their 'incongruity' with aether, were not absorbed they gave the appearance of opacity because of the way they refracted the light. Wren, who knew a weak argument when he heard one, objected that all this was hypothetical, 'that it was not very evident that there was any such thing as an ether, much less was it understood what it was, and what properties it had; or that the air consisted of such parts, as was alleged.' In his usual style, Hooke answered that he could establish the existence and properties of the aether 'by multitudes of experiments', and that the Royal Society should be guided by 'the great schoolmistress of reason, experience; and not to be ruled by groundless fancies and conceits'. At this point Oliver Hill joined in the discussion and had the effrontery to declare in the presence of one of Europe's greatest authorities on the weight of air that 'there was no such thing as gravity in the air', and that he had a discourse and experiments to prove it. Hooke did not record his reaction in the minutes, but his irritation is clear from his diary: 'O. Hill an Ignorant coxcombly fool or an enthusiastick [fanatical] quaker'.[5]

Such frustrations aside, Hooke entered the new year with a sense of optimism. The new President treated him kindly, and seemed to share his desire for a more serious approach to research. He had welcomed the new rules on experiments, and on 31 December he had invited Hooke into his chamber and encouraged him 'to be diligent for this year to study things of use'. His unkindness at supper two weeks later, when he joked that Hooke needed a higher chair, could easily be forgiven. A much more important patron had also shown Hooke favour. On Christmas Eve he had been summoned to the King's bedchamber to show him moss seeds and pepper mites through the microscope, they had talked of the Tangier Mole, and the King had asked Hooke to make him a barometer. Hooke's optimistic mood is captured in a letter he wrote on the Royal Society's behalf to the York botanist Martin Lister in January 1678. The

Society's change of leadership, he said, had 'very much revived us and put a new spirit in all our proceeding which I perswade myself will not only be beneficial and delightful to the members of the Society, but to the whole learned world.'[6]

As part of his drive to expunge the relics of Oldenburg's influence, Hooke hoped to prevent the revival of *Philosophical Transactions*, the collection of papers and reviews that Oldenburg had published. When the Council discussed its future on 2 January 1678, Hooke spoke against resuming publication, and thought he had won the argument. But Grew and Henshaw disagreed, and Grew brought out issue 139 of the *Philosophical Transactions* in February, and three more issues over the following year. Though Hooke did not have Oldenburg's great interest in publicizing the work of others, he was happy enough to print some of the more interesting letters he received alongside his own essays and lectures, and eventually prepared seven editions of his own *Philosophical Collections*, mostly in 1681–2. But his reluctance to produce a regular replacement for *Philosophical Transactions* was a frequent cause of conflict between him and the Royal Society during the five years of his secretaryship.[7]

One of Hooke's New Year duties was to go through Oldenburg's books and papers with Grew, Hoskins and Abraham Hill, cataloguing them and weeding them out. Mrs Oldenburg had died a few days after her husband, and those in possession of Oldenburg's books had struggled to retain them, but at last, on Christmas Eve 1677, they were handed over. On Boxing Day the work of sorting and cataloguing began, and on 29 and 31 December Hooke found and transcribed the letters about his spring watch that Moray, Oldenburg and Huygens had exchanged in 1665. He saw intrigue and betrayal everywhere, so it is possible that reading these innocuous letters confirmed his belief that there had been a plot to rob him of his precious invention. On 5 January he found Hevelius's letter of 30 August 1675, in which he made fun of Hooke's inability to make or use the quadrants that he boasted about, and Flamsteed's 'paultry letter' confirming Hevelius' doubts about Hooke's practical abilities. Perhaps these hurtful jibes prompted him to resolve, as he did that day, to press on with some unfinished astronomical tasks: to perfect his refracting and reflecting telescopes, to measure the circumference of the Earth, to observe the celestial parallax (probably to improve upon his work on the movement of the Earth in 1669, which had been cut short by his illness and the breaking of his best object lens) and to test atmospheric pressure at the top of the Monument.

The debate with Wren on the aether and atmospheric pressure provided the main theme for the January meetings. On 10 January 1678 Wren repeated his doubts as to the reality of the aether, and challenged Hooke to prove its existence. For Hooke the existence of the aether was an important principle, because his wave theory of light seemed to depend on it, and his forthcoming publication, *Cometa*, explained the shape and appearance of comets by their reaction with the surrounding aether. Once again, Hooke claimed that he knew 'hundreds of experiments' to demonstrate the existence and properties of the aether, and that he would shortly perform some of them. What he meant by this, it became clear at the next meeting, was that he would go over the work on barometers and the springiness of the air that he had done with Boyle in the late 1650s. To judge from the minutes (which perhaps we should not) Hooke held a magisterial question-and-answer session for fellows who had forgotten the principles of Boyle's Law. Hooke was in his early forties now, and perhaps his most creative years were past. Repeating old experiments was an easy way to amuse and educate Royal Society fellows who had not seen his demonstrations in the 1660s, or had not understood them at the time, but more critical minds were not impressed. Flamsteed wrote to the astronomer and Oxford professor Edward Bernard in February: 'Mr Hooke fills all with discourse and mighty projects of inventions and discoverys but they are seldome seene and when they are *parturient Montes*'.[8]

Over the next three months the discussion moved on from atmospheric pressure to the measurement of water pressure at sea, the law of falling bodies and the process of respiration, all topics that had interested Hooke and the Society in the 1660s. In May 1678 two meetings were cancelled so that Hooke could organize a repeat of the experiments in barometric pressure at height which had been conducted on old St Paul's before the Great Fire. The Monument was now complete except for its inscription, and Hooke had always intended to use it for experiments on atmospheric pressure and the velocity of falling bodies. On 23 May he went with Hunt and Crawley to the top of the column, and took barometric readings at different heights. He repeated the experiment the following Thursday morning, using more accurate instruments, but a week later he 'brake the mercury glass in the Parlour', and the series of experiments ended. Only the novel location obscured the fact that it was very familiar work, interesting only to those who had not been paying attention for the previous twenty years.

The most important new work on display in the early months of 1678, the examination of micro-organisms in water, blood, milk and phlegm, was inspired by Leeuwenhoek's letters, though without Hooke's skill in making single-lens microscopes of the necessary power and clarity these observations could not have been repeated. A subject of special interest was the composition of semen, with its implications for the origins of human life. Hooke and Wren talked about 'tadpoles in Seed' in March, and Hooke had certainly repeated Leeuwenhoek's observations by 23 June, when he showed the work to two friends. Hooke also used his improved microscope to study the structure of lobster muscles, which he now discovered to be composed of fibres that resembled strings of tiny bladders whose inflation and deflation caused the muscle to shorten and lengthen. These tiny fibres, which he reckoned at four million to the round inch, would have been invisible to him in the 1660s. His revived interest in muscles and respiration was encouraged by coffee-house discussions with Christopher Wren, who had a long-standing interest in the subject, and with the brilliant physiologist Dr John Mayow, who shared Hooke's interest in the role of air in sustaining fire and animal life. Mayow had an Oxford fellowship and a Bath medical practice, but Hooke met him in London several times between 1674 and 1678, and proposed him for a Royal Society fellowship in February 1678. Mayow moved from Oxford to Covent Garden in 1679, but died there in September, aged only thirty-seven, before he could take up his fellowship.[9]

At the beginning of March 1678 Hooke received the appalling news from the Isle of Wight that his brother John, Grace's father, had hanged himself. The diary gives no explanation for John Hooke's suicide, but his grocery business was in serious difficulties, and this must have been hard for a man of his social pretensions, an ex-mayor of Newport, to bear. To add to his humiliation, it is possible that Grace was in a late stage of pregnancy in February 1678. Grace had probably been on the island since August 1677, and she did not return to London until 7 June, ten months later. In April and May 1678 Hooke exchanged several letters with Grace and a Newport doctor, Dr Harrison. Hooke's diary, which is frank about almost everything, does not mention the nature of Grace's trouble. If she was pregnant, as some have suggested, the likely father was Sir Robert Holmes, who had been 'courting' Grace since October (when Hooke heard about it), and probably longer. Holmes' efforts to preserve John Hooke's property for Grace and her mother certainly suggest a stronger interest than that of an ex-suitor. And there is a more intriguing piece of evidence: when Sir Robert Holmes died in 1692 he left a will which made provision for an illegitimate daughter, Mary Holmes, who was born to an unknown mother in 1678. She married Holmes' nephew, had sixteen children and died in Yarmouth on 7 March 1760. The man who thought he was her father is buried in Yarmouth Church, in an impressive tomb which is surmounted by a captured statue of Louis XIV, with Holmes' head substituted for the King's.[10] If Grace *did* give birth in May 1678, it is possible that she made a fool of Sir Robert, and that the true father was Hooke. Hooke and Grace made love on 9 August 1677, the day before she left London, and forty weeks before mid-May 1678. We do not know whether Hooke was sterile, or whether he used a form of contraception in his various affairs. He certainly knew that a sponge could be used as a pessary, and even mentioned this 'vile and sordid' practice in *Micrographia*. Did the sponge fail, leaving Hooke with a daughter and eventually a crowd of grandchildren on the Isle of Wight? On balance, it is more likely that Mary Holmes was the Governor's baby, as he believed she was, and that the mother was either Grace or another island girl.

As well as being a personal tragedy for Hooke, Grace and her mother, John Hooke's suicide had serious financial implications for his family. Self-murder was homicide, and was ranked as one of the highest of felonies against God and the King. Since the culprit was beyond punishment, the law took its revenge on his body and his family. As a suicide, John Hooke would be buried on a public highway with a stake through his heart, and all his goods and chattels would be forfeit to the King. This was a severe punishment, and like most sentences, as Hooke knew, it could be mitigated by the mercy of the sovereign. On Saturday 2 March he went with Crawley to see Sir John Hoskins, perhaps for legal advice, and then spoke with Wren. When he saw the King a little later to beg him to allow John Hooke's family to keep their estate, he found that Holmes had already been there on the same mission. After this Hooke sent Crawley, his reliable assistant, to the Isle of Wight on a borrowed horse, to sort out John's affairs and no doubt to help Grace and her mother. Hooke spent a wretched night sleeping in his clothes, and the next day Haak and Hoskins found it hard to console him. On Monday he spoke to Sir Joseph Williamson, who was still the King's Secretary of State, on his brother's family's behalf, and by the evening, which he spent in Garraway's and Jonathan's with Hill and Hoskins, he had regained his composure. Crawley returned on Wednesday, and on the following Saturday there was news that the Hooke estate would not be forfeited to the Crown, and Grace and her mother were spared this final humiliation. His body was buried privately, 'not in Comon burying place', but not at a crossroads, either. Grace stayed in Newport for another three months, and arrived back in London on the Portsmouth coach in early June. The longer term effects of the suicide on Robert Hooke are hard to establish, but it is notable that his recorded attendances at Sunday church services are far more frequent after March 1678 than they were before. His favourite clergyman seems to have been Gilbert (later Bishop) Burnet, chaplain of the Rolls Chapel, who was a popular preacher in many London churches in the 1670s.

In the weeks following his brother's death Hooke became involved in an enterprise that would occupy much of his time later in the year. On 20 March he went with his friend Theodore Haak to meet a London bookseller and publisher, Moses Pitt, who had a proposal for a new universal atlas. Hooke was impressed by the plan, and arranged for Pitt to discuss it with Wren. On 28 March the Royal Society decided to support the atlas, and chose a committee, including Hooke, Wren, Haak and Grew, to supervise its preparation. Pitt's plan was enormously ambitious, envisaging 600 pictures or maps and 900 printed pages in eleven volumes, covering the Earth, sea and heavens. Hooke had a long-standing interest in cartography, and he had produced several devices for easier and more accurate surveying, including the small sextant with telescopic sights and a variety of horse-drawn and self-propelled waywisers. He met Pitt more than a dozen times in April and early May 1678, and did everything he could to introduce him to men whose influence or money might help his grand project on its way. He was not a man to give his time for nothing, and it was soon clear that his interest in Pitt's atlas was not entirely altruistic. Hooke believed that his new way of printing place names onto a map cheaply with movable type, rather than engraving them onto the map itself, had commercial value. On 31 July he offered to give Pitt this secret and to help him in the production and inspection of the atlas in return for two delayed payments of £200. A week later the deal was done, and Hooke told Pitt his 'contrivance for the making mapps without Graving names but only printing them on sheets by [sic]'. Once the contract was signed Hooke threw himself into work on the atlas. On 10 August he began writing an introduction to the first volume, covering northern Europe and the north polar regions, and spent many hours directing the work of Richard Lamb, Pitt's engraver. On 27 September he drew a 'polar projection', which is probably the 'Map of the North Pole and the Parts Adjoining' that was printed at the beginning of the atlas. Hooke now felt free to describe his map-printing technique to all his coffee-house and Royal Society friends, but felt let down when Thomas Henshaw, 'contrary to promise', passed the secret on to others. As it turned out, Hooke's contribution was the most original part of the atlas, because Pitt decided to save time and money by simply reissuing the maps already published in Jansson and Waesburg's Dutch atlas, with corrections and new lettering, and with no new maps except Hooke's.[11]

In April 1678 *Lectures and Collections*, containing Hooke's Cutler lecturers *Cometa* and *Microscopium*, appeared, and as usual he gave copies of it to his friends and Royal Society colleagues. Influenced perhaps by the responsibilities of his new position, he did not use *Cometa* to attack the methods or morals of his rivals. Instead he printed short contributions from several fellow scientists, including Wren's method for calculating the parallax of a comet, Boyle's account of Krafft's demonstration of phosphorus, Cassini's observations of Mercury and Jupiter, and a letter from young Edmond Halley, who had gone to St Helena to make a catalogue of the southern stars, but found the island covered in almost perpetual cloud.

Hooke's own contribution to *Cometa* consisted of his brief observation of the comet of April 1677, long extracts from his 1665 lectures on the two comets he had observed with Wren in 1664–5, and a discussion of the shape, appearance, origins and paths of comets. He wrote interestingly of the various conceivable sources of a comet's light, with the new phenomenon of phosphorus adding another possibility to the eleven he had already considered. Since Hooke's favoured explanation of the comet's light involved its interaction with the aether he felt bound to reject the idea that this was an imaginary substance, and to assert that he could show, 'had I now time', many experiments to prove its existence. As to a comet's trajectory, he held to the view he had arrived at in 1665, rejecting Kepler's straight line theory in favour of a curved or circular path that would bring the same comet back into view at intervals which might in time be predictable. It is striking that Hooke's thinking on celestial motion had not progressed since his pioneering work on the deflection of a straight line into a curved orbit by gravitational force in 1666. Later, when Newton was defending himself against Hooke's charges of plagiarism at the time of the publication of *Principia* in 1686–7, the weakness of the analysis of motion and gravitation in *Cometa* would be a strong argument in his favour. In particular, Hooke's claim that he had been the first to develop and understand the inverse square law was weakened by the fact that he never mentioned it – even in code – in *Cometa*.

In May Hooke was drawn into dealing with the problem of Chelsea College, the King's gift to the Society, which had now fallen into disrepair. He surveyed the College with Harry Hunt and Abraham Hill on the 10th, dining there on cakes and ale, and spent some of the next week persuading Williamson and several Royal Society fellows that the college could be refurbished for the Society's use. But Sir Christopher Wren, whose opinion of Hooke's plan was bound to be influential, 'liked it not', and on 30 May the Council ordered Hooke and others to find a way of disposing of the tumbledown buildings to the highest bidder. Eventually, in 1682, the King decided to build a hospital for veteran soldiers on the site, and paid the Society £1,300 for the old college.

The Royal Society's members met through July and August in 1678, risking London's summer diseases to discuss the effects of animal poisons and the growth of plants in a vacuum. Hooke enjoyed scientific anecdotes, and cheerfully contributed his own accounts of children who ate henbane in mistake for parsnips, and of the awful effects of a rattlesnake bite. To give the meetings something more serious to think about, he decided that it was time to reveal the secret of springs, which he had published in code in *Helioscopes* in October 1675. His weekdays in late July 1678 were occupied by his work on Montagu House and Pitt's atlas, but his Sundays were free. On Sunday 21 July, three days after his forty-third birthday, and with his head 'mightily refresht by Tobacco', Hooke sat down to write his theory of springs. The next Sunday, restored to excellent health by his return to smoking ('Strangely Recoverd by taking tobacco'), he completed a discourse to be read at the 1 August meeting of the Society.

The lecture went well ('all were well pleased'), and during the rest of August Hooke spent his weekends trying to refine and amplify his ideas. Sometimes he did this by thinking alone (25 August – 'Puzzle about Velocity and time of Spring'), but at least five times that month he discussed his 'equation of springs' with his most important sounding board, Sir Christopher Wren. Hooke and Wren met at Child's, Cheapside, Jonathan's and Man's coffee houses and exchanged ideas on springs and spring balances, but at dinner with Wren on 28 August Hooke 'could not procure his judgement of springs'. Two weeks later Wren at last approved his spring theory, and in early October Hooke delivered his manuscript to the printer. By late November the book was ready, and Hooke could start distributing marbled or gilt copies to his friends and Royal Society colleagues – Wren, Boyle, Williamson, Hoskins, Hill, Ent, Papin and Dr Busby.

Hooke's discourse on springs, which was published as *Lectures De Potentia Restitutiva; or of Spring*, was much more than a description of the way that springs and other elastic bodies behave under tension. What he was trying to do in his lectures was to explore the fundamental causes of elasticity, and in doing so he put forward a theory of matter that was different to anything that had been proposed before, and similar in many respects to the kinetic theory of gases that was not generally accepted until the 1860s. Hooke began with a brief history of his theory, claiming to have discovered it eighteen years earlier, when he first proposed his spring-regulated watch, 'but designing to apply it to some particular use, I omitted the publishing thereof'. Then he taught his readers how to make a simple spring balance by coiling a length of steel, brass or iron wire into a helix around a cylinder, suspending the resulting coil on a nail, hanging weights of different sizes from its lower end and measuring the resulting extension of the spring. They would find that if a one-pound weight stretched the spring by one inch, then a two-pound weight would stretch it two inches, and so on. The same experiment tried with a watch spring or with wire, wood, or any 'springy body' would point towards the same rule of nature, 'that the force or power thereof to restore itself to its natural position is always proportionate to the Distance or space it is removed therefrom'. Knowing nothing about the electrical and chemical forces that bind atoms together, Hooke had nevertheless devised a law of springiness which is fundamental to modern engineering, and thereby established the science of the strength of materials and the behaviour of solids under stress. About 130 years later, Thomas Young used Hooke's Law to develop a way of defining the stiffness of different materials, 'Young's modulus', 'perhaps the most famous and the most useful of all concepts in engineering'.[12]

To explain the behaviour of springs Hooke devised a new theory of the nature of matter. He began with the basic proposition of mechanical philosophy, that the universe consisted of 'body and motion', and that 'the whole Universe and all the particles thereof [were] in a continued motion'. The relationship between body and motion was so close, he argued, that 'it is not impossible but that they may be one and the same'. A body could be defined simply as something that was 'receptive and communicative of motion or progression'. Then he took up and developed one of his favourite concepts, the principle of congruity, which he developed more fully in *Of Spring* than anywhere else. Congruous bodies consisted of vibrating particles of the same or a 'harmonious' magnitude and velocity, and because of this they clung together, defending their space and (if they were solid) their shape from the invasion of other particles, and resisting the tendency to disperse into their surroundings. The shape and size of any body, and therefore the space it occupied, was determined by the vibrating movements of its particles and by the constraining movement and pressure of the 'ambient fluid' – the air or aether – that surrounded it. Within a particular solid or liquid body, these vibrating particles repeatedly collided with each other, and these collisions sustained the speed of each particle. In normal circumstances, the outward pressure exerted by those vibrating particles that were on the surface of the body was balanced by the inward pressure of the surrounding 'fluid', and the body retained its natural shape. If an elastic body was compressed to (say) half its former size, the distance each particle vibrated would be halved and the number of collisions between them would be doubled. The increase in the number of collisions gives the compressed body a tendency to expand to its 'natural' size and shape, an outward spring which is directly proportionate to the degree of its compression. The dilation or stretching of a springy body would have precisely the opposite effect, creating longer vibrations, fewer collisions, and a tendency to retreat to its previous shape because of the pressure of the surrounding air or aether. So the longer the spring is extended, the greater the weight that is needed to prevent it from returning to its original shape, and the more a gas is compressed, the greater the pressure it exerts on the vessel surrounding it. Hooke's explanation covered bending as well as compression and expansion. In a bent rod the outer side was stretched and the inner side was compressed, creating exactly the same decrease and increase in collisions between particles, and the same tendency in the rod to regain its previous shape. He then went on to show that the laws governing the relationship between force and velocity explained why the vibrations of any spring must be of equal duration, regardless of how far it had been stretched or bent initially, and to explain how the distance travelled by an object propelled by a recoiling spring could be calculated.

To some extent Hooke's kinetic theory of matter was drawn from the work of earlier thinkers. The atomic or molecular theory of matter had its origins in the work of Democritus and the ancient Greeks, and was widely accepted in the seventeenth century. The idea that particles moved in particular circumstances, especially when they were hot, was put forward by Galileo, and the theory that they were in constant motion had been mentioned but not developed by Descartes in *Principles of Philosophy* in 1644. But Descartes's particles were soft and feathery, their motion was rotary, like a tiny vortex, and they did not collide with one another. The theory that hard particles were in random motion over relatively large distances, colliding with each other without loss of energy, and exerting a greater outward pressure the more they were compressed together, was Hooke's, and was drawn from his work with Boyle on the compression of air. Heat was not his concern in his work on springs, but he was sure in 1664, when he wrote *Micrographia*, that heat was 'nothing else but a very brisk and vehement agitation of the parts of a body'.[13] Hooke's kinetic theory of heat and matter was then forgotten for many years, until the Swiss scientist Daniel Bernoulli rediscovered the kinetic theory of gases in 1738. Bernoulli's work was in turn ignored until the idea was revived by two uninfluential English amateurs, John Herapath and John Waterston, in 1820 and 1845. Only with the work of James Joule in the 1840s, and Rudolf Clausius and James Clerk Maxwell in the 1860s, did the kinetic theory of heat and matter achieve general acceptance, and then its origins were traced back 120 years to Bernoulli, not 180 years to Hooke's pioneering work on elasticity.[14]

18. 'A Man of a Strange Unsociable Temper'

(1678–1680)

In September 1678 Hooke had to organize the removal of the Duke of Norfolk's library from Arundel House, which was about to be demolished to make way for four streets of houses between the Strand and the Thames. It was agreed that the library, which contained several valuable English and European collections, should be divided between the Royal Society and the College of Arms. Hooke and Henry St George, the Garter King of Arms, organized the division, arguing over works of common interest. On 16 September Hooke and Harry Hunt trundled four cartloads of books, probably over 3,000 volumes, from Arundel House to Gresham College, about a mile and a half along the rough and crowded streets of the City. Most of these books were sold off by the Royal Society in later centuries, but a substantial number, perhaps 10 per cent, remain in the Society's collection today. Ten days later Hooke met his old protector, Dr Busby, who was still the headmaster of Westminster School, and agreed to design a church for the parish of Willen in Buckinghamshire, which was to be built at Busby's expense. Hooke prepared a draft of the church in November, and showed Busby a second proposal in December. After Christmas he introduced Dr Busby to Roger Bates, who was engaged to build the church. Hooke met his old headmaster often in 1679 to discuss the work or arrange terms with craftsmen. In January 1679 he designed another country church, for Sir John Lowther, but he does not seem to have been involved in the building to the same extent.

During the autumn and winter of 1678–9 there was a national panic over the so-called Popish Plot, an imaginary conspiracy concocted by two rogues, in which Jesuits would set London ablaze (again!), kill the King, and use French and Irish troops to put his

Catholic brother James on the throne. Hooke noted some of the main events in the crisis in his diary, but seems not to have been caught up in the frenzy, and had no particular ill-will towards Catholics. On 31 October he was given a paper found in the Custom House, apparently threatening his life, but he burnt it, 'supposing it sent by some rogue to frighten me'. His rich client Ralph Montagu got involved in the political crisis surrounding the 'Plot' when he accused the King's chief minister, the Earl of Danby, of secret negotiations with the papal nuncio. But Montagu's temporary disgrace did not seem to interrupt work on Montagu House, to which Hooke gave the finishing touches in 1679–80. In January 1679 he went with the carpenter, John Hayward, to supervise work on the main stairs and the ornamental railings on the top of the roof, and at the end of February he had to widen the stairs. In May and June 1679 he directed work on the stairs, rails, arch, pillars and baths, and in August he dealt with the stairs, hall and doors. This was almost the end of Hooke's work on the great house, though there were some faults and repairs to deal with in 1680. He had installed sash windows in the house in the later 1670s, but they were a new device and joiners were still learning how to make them. On 2 February 1680 Montagu's agent showed him a 'sash window blown down', and Hooke had to arrange the repair. In July, when Hooke was trying to get Montagu to pay a £50 bill, he had to deal with some cracks in one of the turrets.

Montagu House was one of London's finest aristocratic mansions, one of the very few secular buildings in the city to compare in size and splendour with Bethlem. Its main building was in the French style, with two main storeys, and a third and fourth in the attic and basement. It was about 215 feet long, and stood at the back of a large square courtyard enclosed by two lower wings and a front wall and gatehouse on Great Russell Street. In October 1683 John Evelyn visited the completed palace, which was now decked out with old masters, and frescoes by the fashionable Italian artist Antonio Verrio. As for Hooke's work, Evelyn's judgement was mixed:

> The garden is large, and in good air, but the fronts of the house not answerable to the inside. The court at entry, the wings for offices seem too near the street, and that so very narrow and

> meanly built, that the corridor is not in proportion to the rest, to hide the court from being overlooked, by neighbours; all which might have been prevented, had they placed the house further into the ground, of which there was enough to spare. But on the whole it is a fine palace, built after the French pavilion-way, by Mr Hooke, the Curator of the Royal Society.[1]

Evelyn's description is valuable, because just over two years later, in January 1686, the fabulous interior of Montagu House was gutted by fire. Evelyn said the house was 'burned to the ground', but it is clear from Hooke's description of some of the exterior details of the building, and a pen and wash drawing among his papers, that when the burned-out house was rebuilt much of Hooke's exterior was retained, along with his gatehouse, with a cupola in the style of the Royal College of Physicians. Montagu, who seems to have been a self-seeking knave in private as well as public life, paid for the work by marrying the very wealthy but exceedingly insane dowager Duchess of Albemarle, who thought he was the Emperor of China. So two of Hooke's greatest buildings were used for the confinement of lunatics. Many pictures of the restored Montagu House survive, because in the 1750s it was bought and converted to become the first home of the British Museum. As the Museum's collection grew Montagu House became more crowded and inadequate, and in the 1830s and 1840s it was demolished piecemeal to make way for the present building. Coincidentally, one of the additions which made this expansion necessary was the Royal Society's own repository of curiosities, which Hooke had assembled, labelled, catalogued and cared for between 1663 and 1676.[2]

In the early months of 1679 Hooke finished a project that he had been working on intermittently since the early 1660s, the construction of a self-recording weather clock or meteorograph. Like so many of his interests, the weather clock had originated in the mind of Christopher Wren. Wren had discussed the idea at Oxford in the 1650s, and presented a design for a clock that would measure weather conditions, including wind direction, temperature, rainfall and atmospheric pressure, and record them on revolving discs or drums, to the Royal Society in January 1662. Wren's interest in the weather clock apparently faded after 1663, and it was left to Hooke, who had suggested improvements to Wren's scheme, to turn the paper project into a working machine. The Royal Society reminded him of this commitment in May 1670, and again in February and March 1673, when he was instructed to 'make such an engine with speed'. The first sign that the weather engine was actually being made came on 2 January 1675, when Hooke gave directions to Tompion, who was apparently making some sort of weather clock for William Petty. At a time when many ingenious men were interested in the same problems and had access to the same information and the same craftsmen, delay was dangerous. Hooke had learned his lesson over Huygens and the spring watch, and now he tried to keep his cards close to his chest. In 1678 he was worried about Sir Samuel Morland, a mathematician, inventor and royal favourite who shared many of his interests. Morland had made an ear trumpet, a calculating machine (which Hooke thought 'very silly'), and a man-powered pump that could raise water more than sixty feet, for domestic supplies or fire-fighting. He was also interested in designing new barometers. Hooke was anxious in the summer of 1678 that Morland was picking up information about his work on watches and barometers through Tompion, who worked with them both. In August Tompion 'confessed that Dr Samuell Moreland had seen my way [of watchmaking] in his shop', and a month later he heard that Morland had 'bought of Tompion my very Instrument', a weather glass, and taken it home.

By this time Hooke was working seriously to finish the weather clock. In September 1678 his assistant Thomas Crawley made wheels for the clock to Hooke's pattern, and three months later, on 5 December 1678, he demonstrated part of a clock that would keep a record of wind strength and direction, temperature, air pressure, humidity, rainfall and hours of sunshine to the Royal Society. A month later, on 6 January 1679, several leading members of the Society went into his turret to see the unfinished weather clock in action. In the early meetings of 1679 Hooke explained the construction of the self-emptying rain bucket and demonstrated two hygroscopes, one made with gut strings and the other with pieces of elm board stretching to at least twenty-four feet. (Elm was chosen because of its tendency to swell in moist conditions and to contract when the air was dry.) By May 1679 the complete weather clock was ready. The presence of strangers in the meeting of 22 May delayed its appearance, but a week later the Society went to Harry Hunt's lodgings to see the magnificent machine at work.[3]

Hooke's weather-wiser consisted of a large pendulum clock that would run for a week between windings, slowly unrolling a cylinder of paper. Every fifteen minutes a hammer operated by the clock activated a set of punches linked to a collection of meteorological devices, creating a punched paper record of the week's temperature, pressure, humidity, rainfall and wind speed and direction. The sunshine measurer was not mentioned again. Using a punch instead of a pencil to eliminate friction was an inspired idea, antedating its next 'invention' (by a French meteorologist, Changeux) by a hundred years. Most of the other instruments in the weather clock were also Hooke's own inventions, refinements of ideas he had been working on since the 1650s or 1660s. He adapted his wheel barometer, as described in *Micrographia*, so that the weighted cord on a pulley-wheel which moved as the float on the mercury rose and fell would alter the position of a punch, rather than turning a needle on a dial. His thermometer, with its scale calibrated from the freezing point of water (another Hooke invention), probably worked in the same way. He had devised a poised and pivoting rain bucket that would overbalance and empty itself when the water in it reached a set level, and then return at once to its former position. The bucket operated two punches, one to show the present level of the water, the other to show how many times it had tipped over. His anemometer or wind gauge was a weather vane which operated four punches to indicate each quarter of the compass, and three to indicate the number of revolutions the vane had made (in hundreds, thousands and ten thousands), and thus show the speed of the wind. His description of the weather clock does not reveal whether he used the oat-beard hygroscope he had described in *Micrographia*, or one of the newer ones, based on the swelling of elm or the twisting of gut, that he had recently demonstrated to the Royal Society. Hygrometers that measured the change in weight of absorbent or hygroscopic materials like sponge or sal-ammoniac were also available, and these would have been more practical than twenty-four feet of elm.[4]

By making a working weather clock Hooke met the challenge set for him fifteen years earlier, and demonstrated once again his outstanding mechanical ingenuity. Like his wonderful clockwork equatorial quadrant, though, it was devised before the technology and materials needed for its manufacture were available, and remained an interesting curiosity rather than a practical machine. The record it produced was difficult to understand, and in April 1684, when it was in need of repair, the Society had to ask Hooke to help Harry Hunt, no amateur, to interpret the punched paper rolls. Since the request referred to 'the first papers marked by the weather clock' it seems that the instrument had not been heavily used in the previous five years. Apparently it remained unused for the next five years, too. In December 1689 Hooke once again reminded the Society of the 'considerable uses that might be made of the exact observation of the weather and Desired that the weather Clock in the Colledge might be repaired', but nothing came of his request.[5] Automatic weather recording was convenient but there was no great advantage in having five observations in one machine, and it was undesirable to leave a barometer outside in the rain. Later weather clocks were sometimes constructed, including the 'atmospheric recorder' which George Dollond contributed to the 1851 Great Exhibition, but they never became popular.[6]

After the publication of *Of Spring* in 1678, Hooke's flow of new ideas rather dried up, and he turned instead to the re-examination of projects he had left unfinished during the previous twenty years. The idea of a universal language was still one of his regular coffee-house talking points in 1679. In January he spent night after night in Jonathan's or at Gresham College, talking the subject over with Sir John Hoskins and Francis Lodwick, an old Royal Society friend who was also a skilful linguist. He loyally regarded John Wilkins' *Real Character*, which was published in 1668, as the best starting point for a universal language, and in a letter of 10 March 1679 he encouraged one of Wilkins' admirers, Andrew Paschall, to bring the work to perfection. 'The whole difficulty lyes in finding out the person that hath will, ability and leisure, for those 3 must be joined. The first of these I have much of but want the latter two. All our Hopes are in yourself . . . Mr Aubrey, Mr Lodowick & I doe often discourse of it, but we can doe little more than talk . . .'[7] In fact Hooke spent two days in February 1679 writing about the universal language, but there was no time in his crowded life for the long and tedious task of converting the whole language of scientific discourse into a pattern of symbols.

He was busy, among other things, with the production of Pitt's Atlas, which seemed to be making fair progress towards publication in March and April. In the autumn of 1678 he had spent days working on maps and plates for it, and advising Lamb, Pitt's engraver, on mapmaking techniques. At some time, he also seems to have gone through the maps in the Dutch atlas that Pitt was trying to emulate, distinguishing between the maps that were accurate and in good condition, and those that could not be re-used, because they were worn, outdated or simply 'intollerable'. One of the early plates, a universal planisphere, was 'old, worn out, full of errors. Projection to a false meridian. Hath few of the late Discoverys and amendments. And is noe ways tollerable.'[8] On 24 February 1679 the first sheet was at the printer's, and within a month nine sheets were ready for printing. On 13 March Hooke told Pitt of his 'new contrivance for printing books'. Exactly what this involved is not clear, but the next day he told Hoskins of his 'contrivance for tinplates for Rolling presse'. If Hooke had in mind a hand-operated rotary press, as his words suggest, he was looking very far ahead of contemporary practice. All printing machines in use at this time, and for the next hundred years, were flat wooden presses, and the first recorded proposal for machines with revolving cylinders was not made until the 1780s.

Unfortunately, Hooke never explained his ideas on printing with a rolling press to the Royal Society. Instead most of his Thursday demonstrations went over old ground, reintroducing its members to ideas that he had first proposed ten or fifteen years earlier. In the first seven meetings of 1679 he repeated his experiments of 1665–6 on the nature of combustion and respiration, establishing once again that a 'nitrous' component in air was necessary to respiration and combustion. Although the work was not new Hooke found that he had to prove his theory all over again. Dr Croone, an experienced physician and anatomist, believed that creatures in a sealed container died because the steams from their breath and body killed them, and others suggested that coals in an enclosed space stopped burning because the pores of the air were filled up, leaving no more space for smoke to leave the coal. Hooke countered both arguments and confirmed his own theory by showing that life and fire were extinguished more quickly in rarefied air, where there would have been more 'room' for steam or fumes, and lasted longer in compressed air, where the reverse was the case.[9]

Hooke's interest in dissection and anatomy seems to have been revived around this time by his growing friendship with Edward Tyson, a young physician who had first visited him in May 1677 after coming down from Oxford. Hooke had treated Tyson to a five-shilling dinner in Filpot Lane, and encouraged his interest in anatomy and natural history from that time onwards. In February 1679 they often met and talked about anatomy, generally at Jonathan's, and on 20 February Hooke proposed Tyson for membership of the Royal Society, as 'a person very curious in Anatomy and one that would be very usefull to the Society in producing Observations of that kind'. On 21 March they went to Tower Dock to buy a flounder, which was dissected at Gresham College later that day.

Another of Hooke's old interests, the possibility of human flight, was also reawakened in 1679. On the last day of 1678 he heard the news from Paris that a Frenchman had mastered the art of flying. This was probably M. Besnier, a smith, who had devised a set of four wings, one for each limb, that could be moved diagonally, counterpoising each other, so that a wearer starting from high ground could fly 'a pretty distance' across a river. Hooke knew of a second proposal, the work of a Jesuit mathematician, Father Francisco Lana, which had been published in Italy in 1670. Father Lana's plan was to suspend a sailing basket from four light but strong copper balls, each twenty-four feet in diameter, from which all the air had been pumped. This was a distinct advance on earlier ideas that swans' eggs filled with mercury or flasks filled with morning dew might lift a basket, but it was impractical, nevertheless. It was Hooke's job to test and demonstrate such devices for the Royal Society, and in May he did so. From two experiments with cylinders he concluded that Lana's assumption that the thickness of the copper used in making a small ball need not be increased if the ball was larger was far from correct. On the contrary, he told the Royal Society on 5 June, the thickness of the copper had to be increased disproportionately to prevent the evacuated ball from collapsing, and therefore its weight would be too great for flight. As for the 'French Wings', Hooke made a pasteboard model to demonstrate Besnier's technique to the Royal Society on 8 May, and assured them that something better could be devised. Henshaw's alternative idea of a flying chariot with wings powered by springs did not sound especially airworthy. Hooke pursued the subject later in the month, when he made a new set of wings and discussed his 'flying module' with Wren and Papin, Boyle's assistant.[10] Perhaps this subject did not interest the wider public as much as it fascinated Hooke. In November he wrote lectures on the work of Father Lana to deliver in Gresham College, but nobody came to hear them, and he had to read them to his friends instead.

In the early summer months Hooke entertained the Royal Society with accounts of the effects of moonlight on the luminescence of phosphorus, and of the qualities of the milk from a Barbadian coconut (still a rare delicacy), and displayed, but did not demonstrate, a supposedly bullet-proof suit of armour made from thick quilted silk. In May he introduced the Society to Denis Papin, the young French scientist who had helped Robert Boyle with his air pump. Working on compressed air, Papin had invented an air gun and a digester, or pressure cooker, in which bones and horns could be softened at high temperatures. Hooke brought him along on 22 May to show what his digester could do to bone and hartshorn. Papin became a firm Royal Society favourite, and in subsequent weeks he cooked calves' feet, mackerel and oranges. Members seemed more interested in the culinary uses of the digester than its industrial ones, and eventually, on 12 April 1682, after Papin had left for a post in Venice, they enjoyed a sumptuous 'philosophical supper' of beef, mutton, pike and pigeon. John Evelyn recorded this mouthwatering marriage of science and gluttony: there was 'an incredible quantity of gravy; and for close of all, a jelly made of the bones of beef, the best for clearness and good relish, and the most delicious that I had ever seen, or tasted. We eat pike and other fish bones, and all without impediment; but nothing exceeded the pigeons, which tasted just as if baked in a pie, all these being stewed in their own juice . . . This philosophical supper caused much mirth amongst us, and exceedingly pleased all the company'. Perhaps Hooke, the custodian of Papin's digester, was the chef at this first pressure-cooked banquet.[11]

As if flight, a universal language, the weather clock, Pitt's Atlas and the secrets of combustion were not enough, Hooke was also busy in May and June 1679 supervising building work in Bloomsbury and the Strand for Montagu and Hervey, collaborating with Tompion in making a watch with a second hand for Wren, and looking after a raccoon. The raccoon's experimental purpose remains unknown, because on 10 May, having bitten Hooke on the finger, it had the good sense to escape from Gresham College. It is probable that he intended to dissect it, perhaps with Edward Tyson's help. Shortly after this Hooke accepted an important architectural commission from the Earl of Conway, whose wife, the theologian and philosopher Anne Conway, had just died. On 5 July he gave Conway a preliminary design for a great country mansion for a fee of ten guineas, and later in the year he became more deeply involved in the project.

Hooke's work as a City Surveyor was by now much reduced, and the £20 he was given in February 1679 was his last substantial payment. His miscellaneous duties included work on the Fleet River and the City water conduits, and occasional views of particular properties, for which he was paid a fee. With Sir Christopher Wren, he was still involved in the rebuilding of the City churches. It is probable that St Mary-le-Bow got its huge weather vane, Robert Bird's copper dragon, in 1679 and it is possible that Hooke's negotiations with Bird in February for a 'Coper Obelisk at 20d per pound', and his visit to the top of the church in March, were connected with this. As for St Paul's, Hooke frequently visited the construction site with Wren, and discussed the work with him in parks and coffee houses. Everyday management of the work was entrusted from 1676 to Wren's assistant surveyor, John Oliver, but Hooke often dealt with the craftsmen, and was so free and expert with his advice (4 September: 'Walkd with Sir Chr Wren in the Park, told him of double vaulting Paules, with cramps between') that the extent of his influence on the building of the Cathedral is likely to have been large.

In the summer of 1679 Hooke endured a serious domestic crisis. In late July Grace was struck down with smallpox. She fell ill on Thursday 24 July, apparently from the effects of eating fruit. On Friday she was 'desperately sick', and Hooke called in Dr Mapletoft, one of the doctors who had attended Tom Giles two years earlier. Mapletoft gave Grace burnt (hot) wine and an infusion of 'crocus metal' (probably crocus or oxysulphide of antimony, a traditional fever and smallpox remedy), but the spots had not appeared and he did not offer a diagnosis. After a bad night, Grace was 'exceeding dangerously ill', and Hooke gave her 'Dr Mapletoft's Clyster and Julep' – a rectal medicine and a sweet soothing drink. Then he went to Garraway's to consult Mr Whitchurch, an apothecary. When Whitchurch visited Grace later that day he diagnosed smallpox, and prescribed a medicine known as Gascoyne's powder, which 'made her sweat and brought out the small pox thicker'. Hooke knew from the death of Tom Giles how quickly smallpox could overwhelm its victims, and he sent for Grace's mother. Whitchurch returned on the two following days, and prescribed a herbal cordial, and Dr Whistler came on Tuesday, with more cordial. These prescriptions were all useless (as all smallpox medicines were until the twentieth century) but Grace was luckier than Tom Giles, and survived the attack. When Grace's mother arrived on Thursday 31 July the worst was over, though Grace was still sick, and seems to have stayed indoors for over a month after her illness. Thus she missed Hooke's visit to Bartholomew Fair on 1 September, where he saw a performing elephant that could 'wave colours, shoot a gun, bend and kneel, carry a castle and a man, etc.'. Grace was probably left with some permanent scarring, but on the positive side, she was now immune from further smallpox attacks.

During the worst week of Grace's illness Hooke continued to carry out his duties to the Royal Society. In late July he wrote and delivered a paper on juvenile myopia, in which he proposed that short-sighted people should be fitted with convex spectacles which enlarged and clarified distant objects. The disadvantage of these lenses, that they produced an inverted image, would be corrected over time as the wearer learned to correct the image in his own mind. In fact, Hooke argued (following Kepler), this was exactly what people with normal vision did, receiving an inverted image and correcting it to suit their natural or instinctive concept of the true appearance of things.[12] The mind's ability to interpret or misinterpret the messages received by the eye, and especially the refracted rays sent to Earth from celestial bodies, always fascinated Hooke.

It is not surprising with all these commitments that Hooke should have neglected the work that interested him least, and paid the least. When he had taken a half share in the secretaryship after Oldenburg's death, he had undertaken to maintain a full correspondence with the Society's overseas and provincial members, keep the minutes in the Royal Society's Journal Book, enter fuller accounts of experiments and papers in the Register, and publish *Philosophical Transactions* or a similar regular journal of essays and reviews. There were many signs by the middle of 1679 that Hooke was not managing to find the time in his overcrowded life to perform these duties as Oldenburg had done. In June 1678 Huygens complained that he was no longer being kept informed of Society proceedings, and a month later Grew and Hooke were reminded to reply to letters and keep summaries of correspondence in the Journal. In March 1679 Hooke was authorized to employ an assistant to write his letters, and in July Papin was given this job, at the rate of 18d for a short letter and 2s for a long one. In return Hooke undertook to enter all correspondence in the Society's letter book, and produce a fortnightly sheet of material drawn from the letters of his correspondents. A month later the Council was still unhappy with Hooke's work. On 7 August 1679 he was ordered to publish *Philosophical Transactions*, which had not appeared since February, at least once a month, and to 'proceed with the correspondence, and send away such letters as are already written', not forgetting to pay the postage both ways. At the same time, he was ordered to compile and print a complete record of all the experiments and demonstrations he had ever performed for the Royal Society. Leaving this massive task aside, he did his best to perform his tiresome secretarial duties, spending several days in August reading through and sorting his collection of letters. In November 1679 he distributed the first edition of *Philosophical Collections*, his replacement for Oldenburg's *Philosophical Transactions*. Publication thereafter was hardly regular – the second edition (dated 1681) appeared in October 1680, the third in December 1681, and numbers four to seven in the first four months of 1682.

The secretarial problem was exacerbated by the virtual withdrawal of Nehemiah Grew, Hooke's fellow secretary, from Royal Society business from the summer of 1679. There had been conflicts between Grew and Hooke since 1677, and Grew's preoccupation with the task of cataloguing the Society's Repository meant that he left most correspondence to Hooke. According to Hooke Grew had 'promised to resign his Secretary's place' in October 1678, and in 1679, probably because of his growing London medical practice, he gave up his secretarial duties and began withdrawing from scientific research. Things might have got even worse in September when Papin decided to return to Paris, but a salary of £20 a year and free lodging in Gresham College persuaded him to continue as Hooke's amanuensis until the end of the year.[13]

Sir Jonas Moore, John Flamsteed's wealthy and powerful patron, died on the road between Portsmouth and London in August 1679. Now that Flamsteed had no protector, the Royal Society decided that all its astronomical instruments should be returned from Greenwich to Gresham College before Moore's executors laid claim to them. On 26 September Hooke went with Hunt and Crawley to reclaim his quadrants, an excursion that Hooke recorded with obvious pleasure: 'To Flamsteads, I brought back by Hunt and Crawley, Iron Screw quadrant, little brasse screw quadrant, wooden quadrant and ring, 3 reflecting rules. Flamstead mad.' Flamsteed was furious at the loss of a quadrant which he had previously condemned as virtually useless. Later, he described Hooke's raid to Towneley: 'Mr Hooke . . . got an order to have that which I then used (which was that hee boasts so of in his Animadversions . . .) to be withdrawne this was done without my knowledge and so speedily and malitiously executed that I had not time to remedy it, nor could I get him to permit mee the use of it but for one forthnight though I urged it with great instance . . .'[14]

Over the next few weeks the quadrants were repaired, and Hooke turned his attention once again to the great questions of planetary motion and attraction. His continuing friendship and collaboration with Wren was an enormous advantage to him in such matters. Though Wren was no longer an active scientist he still shared many of Hooke's interests, and had the intellect to match or surpass his friend in philosophical discussions. Wren and Hooke talked about the problem of 'Elliptick motion' on 18 October, and discussed the 'Planetary catena [chain] and coyled cone for Celestiall theory' at Bruin's coffee house three days later. At Man's coffee house on 8 November Wren told Hooke of a way 'that he had found to make a circle equall in periphery to Ellipse, about central attraction'. Hooke's cryptic notes leave the precise content of these discussions unclear, but Hooke and Wren were plainly becoming interested again in the relationship between gravitational attraction and the elliptical orbits of planets, and now at last they knew the crucial inverse square law – gravitational attraction is inversely proportional to the square of the distance between two bodies. Hooke might have arrived at the inverse square law mathematically, using Kepler's third law and Huygens's work on centrifugal force, but instead he drew on an intuitive analogy with another natural phenomenon that weakened over distance as it spread – light. In lectures in 1681–2 Hooke explained gravitational attraction as a vibration in the aether emanating from a body, drawing other bodies towards it. The vibrations spread in a conical shape, like a torch's beam, and as they spread they become proportionately weaker, just as a light does. And since the area of the base of the cone increases with the square of the distance from the apex of the cone, the force of the attraction diminishes at the same rate.[15] This was probably enough for Hooke, and all he could manage. When he tried to devise a mathematical account of the inverse square law it was mistaken and inadequate. As Alexander Koyré says, 'The reason for Hooke's missing the great discoveries made by Newton is clear: Hooke lacked mathematical training which, in turn, can only be explained by a lack of understanding of the fundamental value of the mathematical approach to experience.'[16]

Edward Tyson the anatomist was a regular member of Hooke's coffee-house group by the autumn of 1679, and was starting to review books for the *Philosophical Collections*. In August and October Hooke recorded meeting him at least once a week, usually in Jonathan's on Thursday or Saturday, and he was probably one of those allowed to see the 'child stood 25 years in its mother's body' that had recently arrived in the Royal Society repository. Hooke obviously liked him, and did his best to launch his scientific career. On Friday 14 November Hooke saw a porpoise for sale, and after discussion with Grew and Tyson he bought it for 7s 6d. On Saturday afternoon he took it to Garraway's, the scene of several of his most interesting experiments, and cut it open and drew it. He spent all day Monday dissecting and drawing the porpoise, and on Tuesday and Wednesday, with Tyson's help, he completed his analysis and record of the creature. Tyson apparently took over the dissection, probably using Hooke's microscopes to inspect the tissues, and published the results, illustrated with Hooke's pictures, in May 1680. He was elected to the Royal Society on 1 December 1679. This was the start of Tyson's distinguished career as an anatomist, which founded the science of comparative anatomy in England and established the idea that there was an anatomical relationship between apes and man. Tyson became a physician at Bridewell and Bethlem hospitals, and gained his greatest success from the publication of his anatomy of a chimpanzee in 1699. Hooke's friendship with Tyson seems to have suffered none of the tiffs and jealousies that damaged his relationships with some other scientists, and they remained close in the 1690s.[17]

Hooke's renewed interest in planetary motion, together with the fact that he was under pressure to keep up the Royal Society's scientific correspondence as Oldenburg had done, led him to write to Isaac Newton on 24 November 1679. Since his letter began a chain of events that had important consequences for Hooke, for Newton, and even for the history of science, it is worth quoting its first part in full.

> Finding by our Registers that you were pleasd to correspond with Mr Oldenburg and having also had the happiness of Receiving some Letters from you my self, make me presume to trouble you with this present scribble. Dr Grews more urgent occasions having made him Decline the holding Correspondence. And the Society, hath devolved it on me. I hope therefore that you will please to continue your former favours to the Society by communicating what shall occur to you that is philosophicall, and in returne I shall be sure to acquaint you wth what we shall Receive considerable from other parts or find out new here. And you may rest assured that whatever shall be soe communicated shall be noe otherwise farther imparted or disposed of then you yourself shall prescribe. I am not ignorant that both heretofore and not long since also there have been some who have indeavourd to misrepresent me to you and possibly they or others have not been wanting to doe the like to me, but Difference in opinion if such there be (especially in Philosophicall matters where Interest hath little concerne) me thinks should not be the occasion of Enmity – tis not wth me I am sure. For my own part I shall take it as a great favour if you shall please to communicate by Letter your objections against any hypothesis or opinion of mine, And particularly if you will let me know your thoughts of that of compounding the celestiall motions of the planetts of a direct motion by the tangent & an attractive motion towards the centrall

> body, Or what objections you have against my hypothesis of the lawes or causes of Springinesse.[18]

The rest of the letter passed on news of scientific activity in Paris and London, just as Oldenburg might have done. Hooke's talk of friendly cooperation in spite of mischief-making and disagreements echoed the correspondence of January and February 1675, and once again he suggested that any exchange of ideas could remain private if Newton preferred. Whether Hooke was sincere in inviting Newton to comment on his work on springiness and planetary motion is not known, but it seems likely, knowing Hooke, that he was more interested in drawing Newton's attention to his best work. He had no way of knowing that Newton had been working on planetary motion intermittently since the early 1660s, or that Newton's understanding of these problems was in many respects far ahead of his own.

Newton's answer to Hooke, sent on 28 November, did nothing to alert Hooke to the advances that he had made in this field. Newton politely dismissed the idea of a 'philosophical correspondence' with the Royal Society, and told Hooke that he had given up natural philosophy except as an idle diversion, and no longer had the health or eyesight for astronomy. Having 'thus shook hands wth Philosophy', he was so out of touch with work in London, he said, that he had not heard of Hooke's ideas on springs or the hypothesis that a planet's orbit was a compound of a straight path and solar attraction, and was in no position to comment on them. Had Newton really never heard of Hooke's work on planetary motion, which had been registered by the Royal Society in May 1666 and published in *An Attempt to Prove the Motion of the Earth* in 1674 and 1679? Other remarks in Newton's correspondence suggest that he had already read of Hooke's theory, but it seems that he had not grasped its significance until he received Hooke's letter in November 1679. Richard Westfall, Newton's best biographer, points out that before Hooke's letter 'Newton's papers reveal no similar understanding of circular motion . . . Every time he had considered it, he had spoken of a tendency to recede from the center, what Huygens called centrifugal force.'[19] Elsewhere Westfall, who had no inclination to favour Hooke's claims over Newton's, writes:

> What was required at this point was exactly what Hooke had, not the power of careful mathematical analysis, but a leap of imagination to a superior conceptualization of the problem. The major step forward in the treatment of curvilinear motion from Huygens' *Horologium* to Newton's *Principia* was embodied in the substitution of the word 'centripetal' for the word 'centrifugal'. I do not know of any document in which Newton employed either the word or the concept before Hooke instructed him to do so. Without the concept of centripetal force, the theory of universal gravitation was inconceivable, and Hooke's contribution to gravitation was not then insignificant . . . [Hooke's] very willingness to speak of attractive forces was a break with the mores of mechanical philosophers. Hooke's suggestions exercised a profound influence on Newton's speculations about forces between particles, and these speculations in turn prepared his mind for the conception of universal gravitation.[20]

To sweeten his rejection of Hooke's invitation to correspond, Newton proposed an experiment to establish the Earth's diurnal rotation by dropping a pistol bullet from a high point and comparing its path with that of one suspended by a thread. Since ancient times it had been argued by those who doubted that the Earth rotated that if it did so a falling object would be 'left behind' as it turned, and thus fall to earth a little to the west of the perpendicular. Newton argued, conversely, that an object in a high place would be moving faster than the surface of the Earth at the moment it was dropped, and therefore when it fell it would outrun the Earth and land a little to the *east* of the perpendicular. He explained how the test might be done, and drew a diagram to show that in theory the body would fall in a spiral turning just one revolution before it reached the very centre of the Earth. This little diagram, dealing with a problem which neither Hooke nor Newton really understood, helped to bring about an unfortunate collision between them.[21]

When Hooke read Newton's letter to the Royal Society on 4 December 1679, omitting all the personal material, there was a hearty welcome for Newton's proposed experiment from the fellows, who liked nothing better than dropping things from high places. The interpretation of this meeting advanced in a 1997 biography of

Newton by Michael White, who describes Hooke 'gleefully' and treacherously reading Newton's letter to the Royal Society, 'revelling' in Newton's discomfiture, and crowing over the triumph of his 'elliptoid' over Newton's spiral, is based on guesswork and invention, and reflects the standard Newtonian view of Hooke's character. In fact there was no recorded discussion in the 4 December meeting of the shape of the path taken by a falling body, and Hooke specifically told Newton in his next letter to him that the question had not arisen.[22]

Over the next few days Hooke wrote again to Newton through an amanuensis (probably Papin) who could not spell, begging him not to desert philosophy, and doubting whether he really intended to do so: 'I know that you that have soe fully known those Dilights cannot chuse but sumetimes have a hankering after them'. Then Hooke went on, perhaps unwisely, to question Newton's assumption that a body would fall to the centre of the Earth in a spiral with a single revolution. Hooke applied his own theory of circular motions to the problem, and concluded that if the Earth were split along the equator, a falling body would describe a sort of 'elliptoid' or oval curve as it fell, revolving many times, and approaching the centre of the Earth a little more with each revolution. In practical terms, this would mean that a falling object in the London area would hit the ground to the south-east of the perpendicular, not the east. The letter ended in a friendly spirit, urging Newton to 'goe on and Prosper' in his work.

In 1686, when relations between Newton and Hooke had broken down completely, Newton still recalled Hooke's apparently innocuous letter with fury. He wrote to Edmond Halley that June:

> Should a man who thinks himself knowing, & loves to shew it in correcting & instructing others, come to you when you are busy, & notwithstanding your excuse, press discourses upon you & through his own mistakes correct you & multiply discourses & then make use of it, to boast that he taught you all he spake and oblige you to acknowledge it & cry out injury and injustice if you do not, I beleive you would think him a man of a strange unsociable temper.[23]

The fact that Hooke read his letter correcting Newton's spiral to the Royal Society on 11 December, turning Newton's simple mistake into public knowledge, would have infuriated Newton even more. Newton assumed that Hooke was enjoying a triumph over him in revenge for his defeat in the matter of light and colours, and this interpretation of Hooke's behaviour has often been repeated by later writers. Michael White presents Hooke as 'driven by a hatred born of humiliation', 'captivated by his own vengeful eloquence' and determined to damage Newton's reputation in the scientific world. To some extent this reflects Newton's later interpretation of Hooke's behaviour, but there is nothing in the sources to indicate that Hooke acted in a triumphalist or aggressive way in his dealings with Newton or the Royal Society in these months, or that he was doing any more than his duty as the Royal Society's Secretary and Curator of Experiments. Newton had been offered the option of privacy but he had not accepted the offer, and throughout the correspondence Hooke had made it clear that he was writing on the Society's behalf. It was his job to correspond with other scientists, to enter into informed and courteous debate, and to read the letters he received at Society meetings. In fact, it is widely accepted that Hooke's debate with Newton over the path of a falling object significantly advanced scientific understanding, just as the Royal Society intended such exchanges to do.

There was every reason for Hooke to perform his duties punctiliously at this time. Criticism of his failure to carry out his duties to Oldenburg's high standard were still strong in December 1679. In the annual Royal Society elections on 1 December Grew was replaced as joint secretary by Dr Thomas Gale, High Master of St Paul's, who had been a contemporary of Hooke's at Westminster School in the 1650s. On 8 December there was an important Council meeting which tried once again to bring order and purpose to Royal Society meetings by declaring that they should pursue one chosen subject until it had been 'brought to perfection'. Each week Hooke should prepare and read a 'distinct account and narrative' of the work done in the previous meeting. His failure in recent years to enter full accounts of Royal Society experiments and papers in the Register, to publish regular editions of *Philosophical Transactions*, and to enter copies of letters in the Letter Book, were all to be rectified. Hooke made no promises, but agreed to 'see what he could do'. Two days later the Council gave responsibility for all overseas correspondence to Gale, and on 17 December a committee was appointed to go to Hooke's rooms and look at the letters he had received and, it was implied, not dealt with properly. At a time when his competence was being challenged in this way it is not difficult to see why Hooke pursued his correspondence with Newton and read the letters out in Society meetings. He was not gloating, but just trying to show that he could do his job.[24]

Newton could barely bring himself to answer Hooke's second letter, he recalled later, but he did so on 13 December in order to undo the damage he felt he had done himself by his earlier careless error, and to show Hooke that he could cope with the most difficult gravitational problems. Unlike the first letter, which was sent to his 'ever Honoured Friend', this one was addressed to plain 'Mr Robert Hooke'. Omitting the customary pleasantries, Newton launched straight into a complex account of the path a falling object would take if it were imagined that the Earth did not resist its descent, correcting Hooke's elliptoid path. Unfortunately, perhaps intending to simplify a difficult problem for himself or Hooke, or because he wanted to keep the inverse square law to himself, Newton prefaced his calculation with the assumption that the falling body's 'gravity be supposed uniform'. Newton knew perfectly well that gravitational attraction did not remain uniform, but varied over distance in accordance with the inverse square rule. He had derived this rule from Kepler's laws more than ten years before Hooke, who had only discovered it in 1679. But because Newton worked in secrecy Hooke did not know this, and came to the mistaken but not unreasonable conclusion that he had arrived at the inverse square law before Newton. This mistake led him into his disastrous confrontation with Newton six years later.

Hooke read Newton's second letter to the Royal Society on 18 December 1679, and reported on his trial of Newton's experiment to prove the motion of the Earth by dropping iron balls from high places. Because of the movement of the air, he said, results had been inconclusive, but he had better hopes from indoor experiments. On Christmas Eve he discussed the letter with Boyle in Garraway's, and in the new year, he turned his mind to the great question of celestial motion. His diary for Sunday 4 January ends with this bold announcement: 'perfect Theory of Heavens'. Two days later he wrote again to Newton, apparently unaware of the anger and contempt his every letter provoked in his correspondent's mind. Hooke began, infuriatingly, by suggesting to Newton that gravitational attraction was not uniform, but 'in a duplicate proportion to the Distance from the Centre' – the inverse square law. He then added some confused arguments drawn from Kepler's mistaken law of velocity, and arrived at a conclusion which would (he thought) 'very Intelligbly and truly make out all the Appearances of the Heavens' and thus – an insight into Hooke's priorities – solve the problem of longitude. If this was Hooke's 'Theory of Heavens', it was nothing more than a demonstration of his inability to derive the velocity of planets or the elliptical shape of their orbits from the inverse square law. His letter went on to consider whether the inverse square law continued to operate beneath the Earth's surface (Hooke thought not), and to explain the failure of his experiments with an iron ball dropped in the open air. Finally, he told Newton of Halley's pendulum experiments in St Helena, which demonstrated a fact that Hooke had long ago tried to establish on Westminster Abbey and old St Paul's, that gravity diminished with height.[25]

On 16 January Hooke went to Garraway's with Harry Hunt, to try his falling ball experiment indoors. This involved dropping a ball from a height of twenty-seven feet into a box of tobacco pipe clay which had been marked with a cross to indicate the point at which the same ball touched it when it was suspended by a thread. In fact the southward deviation of an object dropped from twenty-seven feet would be less than two hundredths of an inch, and the forces deflecting the falling ball would also have deflected the hanging plumb line, making the experiment valueless.[26] But by chance the ball fell to the south-east of the mark, and Hooke regarded the result as conclusive. He wrote to Newton with this news on 17 January. Having described the experiment, Hooke urged Newton to put his mind to the question of the shape of planetary orbits in the light of the inverse square law. This passage, when it is not read with a jaundiced Newtonian eye, seems to be a friendly and respectful encouragement to work towards an important goal, and even an admission that Hooke could not reach it alone:

> It now remaines to know the proprietys of a curve Line (not circular nor concentricall) made by a centrall attractive power which makes the velocitys of Descent from the tangent Line or equall straight motion at all Distances in a Duplicate proportion to the Distances Reciprocally taken. I doubt not but that by your excellent method you will easily find out what that Curve must be, and its proprietys, and suggest a physicall Reason of this proportion. If you have had any time to consider of this matter, a word or two of your Thoughts of it will be very gratefull to the Society . . .

And that was the end of this fateful correspondence. Newton did not reply to Hooke's third letter for almost a year, when he had to write to him on another matter. He seemed to hate Hooke's letters, and to hate Hooke for sending them. But if Hooke's aim was, as he claimed, to encourage Newton to turn his mind again to the great scientific questions of the day, he succeeded beyond all expectations. Hooke's letters seemed to stimulate Newton into a short but terrifically productive burst of scientific enquiry, which transformed his natural philosophy in the winter of 1679–80. Hooke was the grit in Newton's oyster, the irritant that provoked him into producing his greatest work. As Newton wrote later to Halley, Hooke's infuriating criticisms and advice forced him to consider questions he might otherwise have ignored:

> his correcting my Spiral occasioned my finding the Theorem by wch I afterward examined the Ellipsis; yet am I not beholden to him for any light into y^t^ business but only for ye diversion he gave me from other studies to think on these things & for his dogmaticalnes in writing as if he had found ye motion in ye Ellipsis, wch inclined me to try it after I saw by what method it was to be done.[27]

It is hard to imagine that the 'other studies' in the fields of theology and alchemy Newton had to put aside because of Hooke were as significant as his work on gravity and motion. And although he was in no mood to admit it in 1686, it was not true that he learned nothing useful from Hooke. Hooke's mechanical explanation of orbital motion as a compound of direct motion and attraction towards the centre was probably new to him in November 1679, and became a cornerstone of his philosophy thereafter. It is possible that he accepted the idea of gravitational attraction operating between distant bodies so readily in 1679 because he had already thought of it for himself, but nothing in his writings shows this to be the case. Hooke's tragic mistake was to identify not this but the inverse square law as the vital ingredient which Newton had got from him in 1679–80, and thus to claim credit for the wrong idea. Newton, never one to concede credit to others where he could avoid doing so, cheerfully exploited Hooke's mistake, and maintained that all he had learned from Hooke was 'that bodies fall not only to the east, but also in our latitude to the south'.[28]

The Hooke–Newton correspondence is often read on the assumption that Hooke was a cantankerous and friendless troublemaker, writing with the intention of tripping up and humiliating his detested rival, while Newton was a reasonable and even-tempered genius, driven to distraction by Hooke's offensive intrusion into his scholarly isolation. In reality, Newton possessed many of the characteristics usually ascribed to Hooke, but in an intensified form. He was neurotic, self-centred, ambitious, intolerant, oversensitive, secretive, unforgiving and highly argumentative. It is hard to imagine Newton spending his evenings drinking coffee with a large group of congenial companions, or forming lifelong friendships with laundresses, sea captains, clerks and scientists. None of this should affect our opinion of Newton's matchless scientific genius one jot, but it is essential to our understanding of his relationship with Hooke.

19. All Trades

(1680)

The reactions of Hooke and Newton to the problems raised in the letters of November 1679 to January 1680 speak volumes about the differences between the two men. Newton embarked upon several years of intensive solitary mathematical study which eventually led to the publication of one of the greatest scientific works ever produced. Hooke took the task more lightly, and performed a few simple experiments to confirm the 'theory of heavens' he had perfected on 4 January. He reported the Garraway's experiment to the Royal Society on 22 January, and at the same meeting demonstrated that the weight in air or water of an alloy of two metals, copper and lead, was greater than that of either metal alone. With these two experiments, Hooke thought, he had invalidated Archimedes' specific gravity test for pure gold (the famous 'eureka' experiment), and demonstrated the rotation of the Earth. His diary captures the great satisfaction he felt on this day of double triumph: 'Archimedes refuted. Diurnall motion of Earth established'.

With the falling body problem solved – at least as far as the experimental method could solve it – Hooke moved on to other less important but more immediate concerns in the worlds of science, building and mapmaking. In February he was commissioned by the Lord Mayor to design a new conduit for the western end of Cheapside, near Foster Lane and St Paul's, to replace the medieval one lost in 1666. Hooke's first design for an obelisk was apparently rejected, but his second draft, shown to the Lord Mayor on 24 February and to the City Lands Committee on 5 March, seems to have been approved. The Cheapside conduit, which was never quite finished, became a lively meeting place for City chimney sweeps, until the spread of piped supplies led to its demolition in 1727.

Although Hooke later came to see the discovery of the laws of gravity and planetary motion as the greatest prize in natural philosophy, he does not seem to have grasped their unique significance in 1680. His natural instinct was to keep as many subjects on the boil as he could, rather than devote all his energies to a single project. He sometimes blamed the demands of the Royal Society for this, but since in December 1679 the Council of the Society had resolved on a more methodical approach, in which a single subject should be pursued over a long period until a satisfactory conclusion was reached, it was Hooke's choice to maintain the variety of his work.

Hooke's most consistent scientific interest in January and February 1680 was the old question of congruity, the tendency of some substances to penetrate into others when they mixed, becoming (as we would say) chemical compounds rather than simple mixtures. He had done experiments like this in 1668 and 1673, but his investigations had been left unfinished. From December 1679 to March 1680 he treated the Society to a long series of demonstrations in which various metals were mixed into different alloys, and weighed in air and water before and after mixing, using the highly accurate balances he had developed in the 1670s. By showing that some alloys had a higher or lower specific gravity than the average of their component metals he demonstrated that the interaction of metals involved a process more complicated than simple mixing. Others, including Francis Bacon, had suggested that this might be so, but no one before Hooke had proved it by systematic measurement. In a lecture he gave on 29 January 1680 Hooke tried to explain why 'some two metals compounded made a heavier, and some other two a much lighter, than they really ought'. It was all a question of congruity. A mixture of tin and lead, for instance, was lighter than the average of the two separate metals because 'there is an aversion in the joyning of those two bodys & a kind of recesse and the parts acquire a greater rarefaction of texture than they had before'. When two metals had an affinity or congruity, as copper and tin did, their particles penetrated into one another, making an alloy that was denser and heavier than either of the original metals.[1] The discovery of congruity, Hooke argued in an undated paper on the mixture of metals, was a good example of the value of making careful observations before leaping to theoretical conclusions: 'Nature it selfe then is to be our Guide and we are to spend some time in her school wth attention & silence before we venture to speak and teach.'[2]

The atmosphere in the 29 January meeting, and at their usual tavern supper afterwards, seems to have been strained. Hooke recorded in his diary: 'much cavilling, caballing. At Crown, supper – Croone, Gale, President, Colwall, Henshaw, against me'. Nevertheless, many Royal Society members appreciated the theoretical or practical value of Hooke's congruity investigations, and an informal committee was set up to help Hooke perform a series of experiments at Gresham College. Almost every Monday or Tuesday in February and March 1680 Hooke and a few Royal Society members mixed and weighed alloys of lead, tin, silver, copper, iron, antimony and gold, and Hooke brought weekly reports of their findings to the Thursday meetings. On 25 March 1680 Sir Joseph Williamson suggested that the subject of mixing metals 'might be brought to a conclusion, and another be pitched upon'. Hooke drew his alloy investigation to a close, but did not introduce another subject of similar weight to replace it, though he was reminded again in June to 'think of some other subject to be prosecuted for the future'. Hooke was more interested now in his building career than his scientific one, and devoted most of his time to it.[3]

Hooke was still helping Wren with the long job of rebuilding the City churches. In March the two visited St Peter Cornhill together, and in April Hooke 'advised' Wren about St Clement Eastcheap, where rebuilding had not yet begun, and went with him to St Benet Paul's Wharf and St Martin Ludgate. Rebuilding work on these two churches had begun in the late 1670s, and Hooke's involvement with both of them was very substantial. He recorded over thirty visits to St Martin Ludgate up to 1680, and probably made others before 1686, when the church was finished. In the case of St Benet surviving drawings in Hooke's hand strongly suggest that he was the chief architect of the church, working under Wren's overall authority but not (as in many other cases) as his assistant or building supervisor.

Some of Hooke's other building projects took him further afield. He dined with Dr Busby at Westminster twice in April, and early in May he made a rare excursion from London to check on the progress Bates was making on Busby's church in Willen. On Monday 3 May his coach left from the George Inn, Aldersgate Street, at four in the morning, picked up fresh horses at St Albans and Dunstable, and arrived at Newport Pagnell at five that evening. The next day Hooke measured the church, dined with Bates and the distinguished clergyman Dr Lewis Atterbury, who held a local living, and supped and slept at the Swan in Newport Pagnell. On Wednesday he rode to Dunstable to pick up the London coach, and was home in time to spend the evening in Jonathan's with Sir John Hoskins.

Perhaps the visit to Willen gave him a taste for travel, because within a few weeks he was planning another trip, to Ragley in Warwickshire. Hooke had submitted designs for a great country mansion to Lord Conway (a member of the famous Seymour family) in July 1679, and written to him at length in November with second thoughts about the design of the external stairs approaching the main portico of the house. Hooke had missed Lord Conway when he visited London in November 1679, but hoped to speak to him about the house before the foundations were laid. 'I conceive it will be much better for the work to begin the foundations somewhat later in the spring when the fear of frost is perfectly off, before which time I doubt not to be able (God willing) to be there to see everything put into a good order for the beginning and compleating thereof.' Hooke entertained Lord Conway at Gresham College on 30 April, and no doubt arranged to visit him in Ragley in June. Although there were regular coach services to Oxford by this time, Hooke chose to make this journey on horseback. He prepared for the long ride by buying himself a coat at the Fox and Goose, and also a whip and gloves. Leaving Grace £1 to cover household expenses for two weeks, he set out for Oxford at 5 a.m. on Saturday 19 June. He rode in the company of Roger Davies, the joiner who was to work on the house, and at Acton they were joined by Moses Pitt, who had business with printers in Oxford. The three men dined at Wickham, and were in Oxford by Saturday night. Hooke took lodgings for two nights at the Angell Inn, and spent Sunday meeting old friends, including Dr John Wallis, who had been Savilean Professor of Geometry in Oxford since 1649. On Monday, after a visit to the Sheldonian Theatre, one of Wren's earliest works, Hooke set off for Chipping Norton with Davies, who was by this time suffering an attack of malaria. On Tuesday, with Davies' malaria getting worse, the two riders reached Moreton-in-the-Marsh, and on Wednesday they arrived at Lord Conway's house. Lord Conway was away, and Hooke's reception was less than he had expected. He was 'roughly acosted' by one of Conway's employees, and given an inferior room. Hooke spent the next day looking round the neighbourhood, but on Friday, when Conway returned, Hooke was moved into the best room and started to enjoy himself. The house party now included Mr and Mrs Popham, Dr Johnson from Warwick, Dr Kast the chaplain, 'Leonard an ingenious German mechanick', and a Catholic carpenter called Holbert. The next day, Saturday 26 June, Hooke got down to business: 'Viewd module, shewd many faults, made a great many alterations, put the 2 great stairs into one and viewd the situation and ground round about. Dined and Supd with my Lord and Mrs Popham. Davys sick of Ague.' On the Sunday Hooke went with Conway to church – 'Mr Wilson, the Parson, I doubt a Sycophant or worse' – and on Monday 'spent most of my time in considering all matters'. On Wednesday, with the sickly Davies at his side and thirty of Lord Conway's golden guineas in his purse, Hooke rode to Islip, where he 'Lay very scurvily', and on Thursday they travelled through Beaconsfield to London, arriving in Gresham College by 6 p.m. Unlike poor Davies, Hooke was in fine shape after nearly fourteen hours in the saddle. 'I was not in the least weary. Went with Society to Jonathans stayd with them till 10 at night. Slept well. *Soli Deo Gloria*.'

Over the next two weeks Hooke and Conway exchanged letters, with Hooke doing his best to accommodate his client's ambitious ideas within a handsome and practical structure, without increasing the cost too much. On 20 July he sent new sketches and designs for each of Ragley's three main floors, to achieve, as he put it, £500 worth of improvements for £100. Learning lessons from Wren's mistakes with St Paul's and his own at Montagu House, he urged Lord Conway to have workmen and materials ready for an early start the following March, so that the walls would be dry and the house covered and safe from frost damage by the end of September. Apparently Lord Conway found Hooke's various designs confusing, and wrote asking him to return to Warwickshire to supervise the building. Hooke's reply, sent on 17 August 1680, gives some insight into how he managed his building projects:

My vocassions will not permitt my absence hence at this time . . . I humbly conceive it will be much better for Dispatch to send Leonard up with the old module and in a fortnight or thereabout he may Return with it back again compleated and Rectifyed, when it will be very easy for Mr Holbert or anyone else your Lordship shall imploy to proceed with the whole work without much if any further Direction. Here I can often be with him [Leonard] and he may have what help is needful for Expedition soe that he will [learn?] more in a week here than in a month in the country . . . My Lord when Leonard is come up your Lordship may be assured noe time shall be lost in the Doing of it, at least he will want noe help nor materiales Mr Davys having already upon my desire provided for him a very good workman and convenient place, nor shall he want any necessary Directions or overseeing that can be given him by

My Lord

yr Lordship's most humble and most obedient servant,

R. Hooke.[4]

Ragley was unfinished when Conway died in 1683, but an engraving by Johannes Kip in 1697 of Ragley, 'the seat of Popham Conway', shows an impressive country mansion in Hooke's style, with a basement, two main storeys, a balustraded roof, and two detached wings on either side of its great courtyard. The house is set in the middle of a huge formal garden, which was redesigned in 1758 by Capability Brown. Ragley was extensively altered by James Gibbs and James Wyatt in the eighteenth century, but beyond Wyatt's portico and Gibbs' great hall much of the present house, which is still owned by the head of the Seymour family, the Marquis of Hertford, is Hooke's.

Although the 'vocassions' that kept Hooke in town in the summer of 1680 were mostly connected with the building and surveying trades, he was also occupied with scientific and Royal Society work. His diary and letters to the Society's overseas correspondents indicate the range of his scientific interests that summer. He was interested in the news that a City apothecary, Warner, had developed a powder which enabled him to preserve a corpse without 'taking out the bowells, or mangling any part of the body', and that the astronomer Thomas Streete was publishing new astronomical tables that would make all others redundant. He was impressed by Johnson's new pocket herbal, with its 500 copper plates, by Dr Wood's new way of adjusting time by the Moon, 'soe as not to miss one day in 24000 years', and by Sir Samuel Morland's version of a German fire-fighting pump. On 17 August he went with Wren to the workshop of John Melling, who claimed to be able to grind a double convex lens that weighed a four-hundredth of a grain. Hooke apparently obtained some of Melling's minute lenses for the Royal Society Repository, and a few weeks later he made one of them for himself. Even on his days off, Hooke was at work. At the end of August he made his regular yearly trip to Bartholomew Fair, and took the opportunity to sketch a man walking on twelve-foot stilts. On the same day he spoke to Lord Ranelagh about Conway's house, saw Thomas Streete's astronomical instruments at Halley's, visited the new girls' ward at Christ Church, and found out about the technique of winding silk. And when he was at home with a cold three days later his reading was Herodotus, the Greek historian in whose works he still hoped to find evidence of the way earthquakes and volcanoes had changed the map of the world.

Hooke was still a City Surveyor, and for his diminishing work on their behalf the City Lands Committee generally gave him an annual gratuity of £5 in the early 1680s.[5] About once a fortnight, according to the patchy diary he kept over the summer of 1680, he carried out a view or survey of a plot of land in the City. Private work took up even more of his time. As soon as he got back from Warwickshire in early July he had to take charge of repair work on cracks in the turrets of Montagu House, after discovering that 'Bates had lagged and not followed Directions'. Bates, on whom Hooke relied so heavily, was plainly not at his best that summer. Dr Busby was unhappy with his work at Willen, and told Hooke so at dinner on 30 August. Negotiating contracts with reliable craftsmen and making sure they followed his instructions was one of Hooke's most important tasks. In July he tried to find a plasterer for Busby, without immediate success: on 23 July 'Doegood, plaisterer, would not agree.' A few weeks later, in St Paul's, he agreed a contract with Edward Strong senior, a master mason, for the building of St Augustine Watling Street, next to the Cathedral. The church was destroyed in 1941, and only its restored tower and spire remain. In July 1680

20. The Church of St Edmund the King, from Clement's Lane. One of the 'Wren' City churches probably designed by Hooke. Anonymous watercolour, 1820.

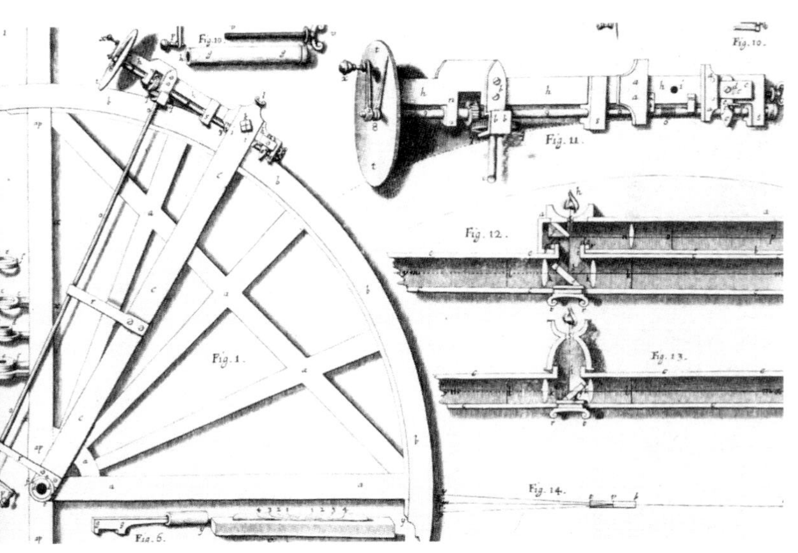

21. Hooke's large one-man quadrant (1) and its main components: a long adjusting rod (8), micrometer screw adjustment (11), telescopic and mirrored sights and eyepiece for seeing two objects at once, at any angle (12, 13). From *Animadversions on the Machina Coelestis* (1674).

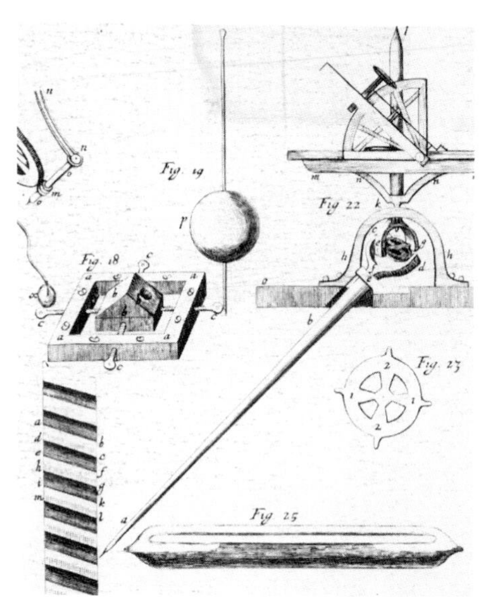

22. Hooke's universal joint, as part of the machinery for converting the rotation of the clockwork and axis (ab) into the horizontal rotation of the equatorial quadrant. With a pendulum bob (19), socket for the pointed shaft of the axis (18), inner plate of the joint (23) and bubble level (25). From *Animadversions* (1674).

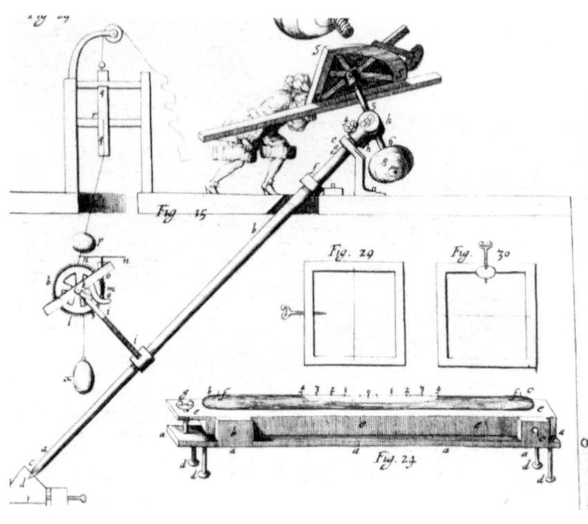

23. Hooke's equatorial quadrant, with its conical pendulum clock, rotating shaft and bubble level (24). From *Animadversions* (1674).

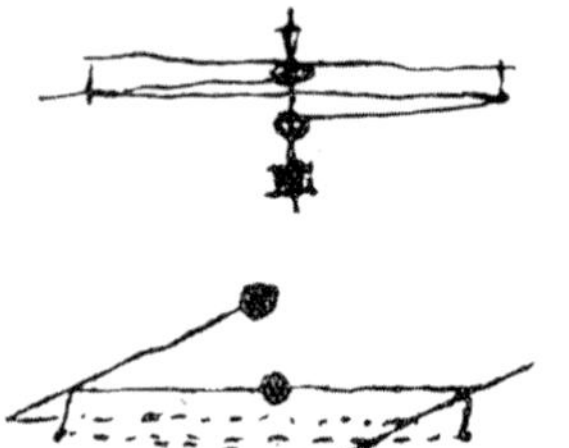

24. 'Invented the best way for a circular fly'. Hooke's design for a spring-driven watch that would be stable at sea, drawn in his diary, 12 October 1675.

25. Thomas Tompion (1639–1713), the 'father of English watchmaking', who constructed Hooke's quadrants and watches in the 1670s. Mezzotint by John Smith, after Sir Godfrey Kneller.

26. Sir Christopher Wren (1632–1723). 'Since the time of Archimedes, there scarce ever met in one man, in so great a perfection, such a mechanical hand, and so philosophical a mind'. Engraved from a portrait by Sir Godfrey Kneller.

IANVARY I. 1676

27. Hooke's diary for 1 January 1676, when the New Philosophical Club met (see p.222). The drawings represent light waves, crossed spurs for ice-walking, and an orgasm (on 2 January).

28. The Hooke–Wren Monument to the Great Fire, pictured from the west at about the time of its completion in 1677. At 202 feet, it dominated the new houses aroun it, and the street life of Fish Street Hill.

29. Montagu House, Bloomsbury, viewed from the nort around the time of its completion Damaged by fire in 1686, the rebui house became the first British Museum in 1755. Probably by Davi Loggan, 1680.

30. Bethlehem Royal Hospital (Bethlem, or Bedlam), Moorfields. 'So Brave, so Neat, So Sweet it does appear, Makes one Half Madd to be a Lodger there.' Demolished in 1814. Anon. engraving, *c.* 1676.

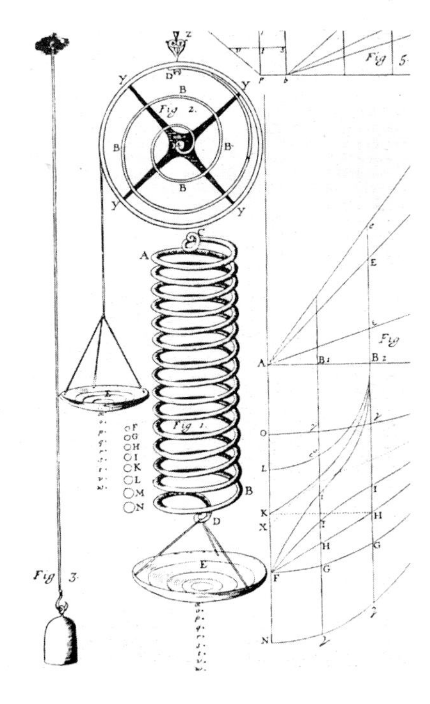

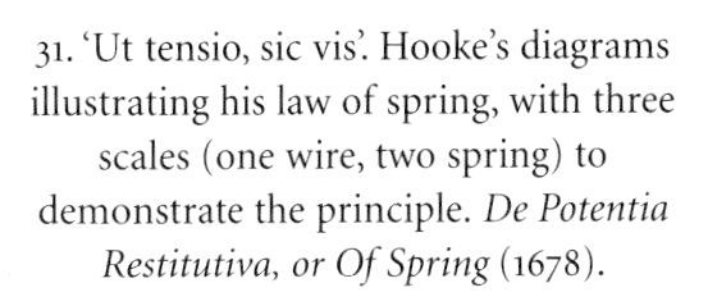

31. 'Ut tensio, sic vis'. Hooke's diagrams illustrating his law of spring, with three scales (one wire, two spring) to demonstrate the principle. *De Potentia Restitutiva, or Of Spring* (1678).

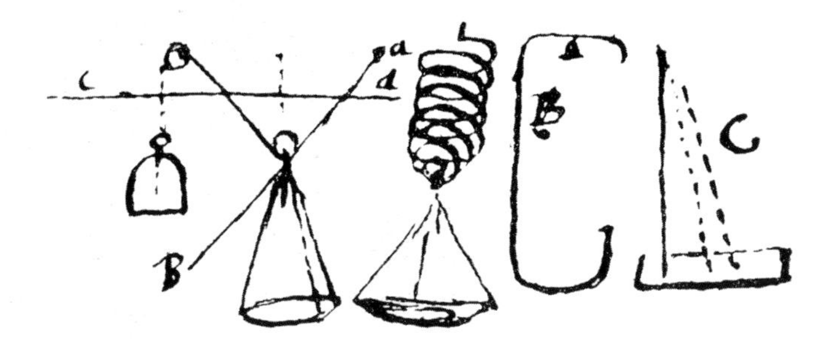

32. Sketches of Wren's rope scales, Hooke's new 'philosophicall spring scales', and two barometers (on the right), in Hooke's diary, 21 August 1678.

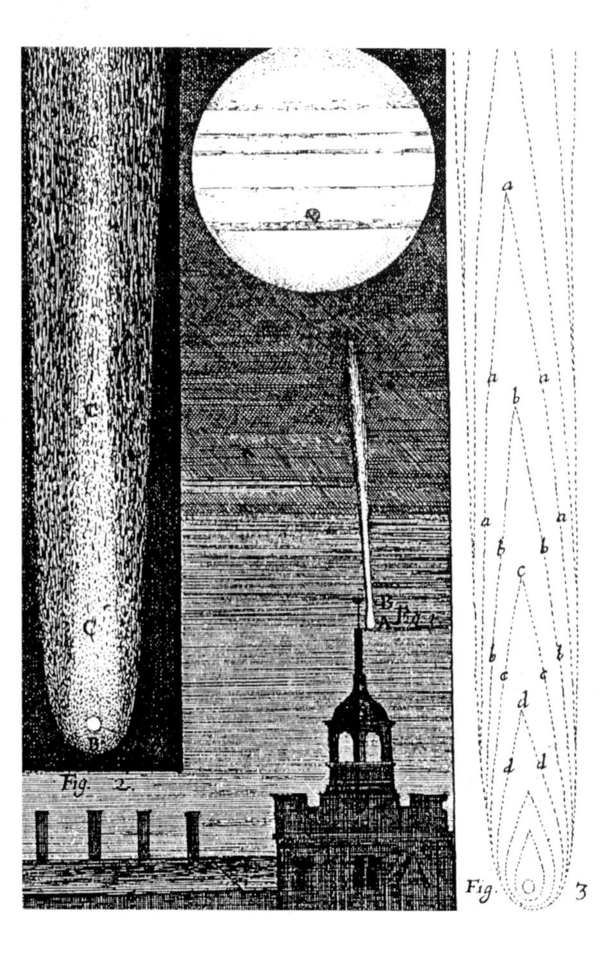

33. Hooke's drawing c a comet seen from Gresham College in April 1677, through a six-foot telescope an with the naked eye, with an analysis of th strength of its blaze. From *Cometa* (1678)

34. Robert Knox (?1641–1720), captive, East India Company captain and Hooke's main informant on the customs and wildlife of India and Ceylon. ngraved by Robert White (1691).

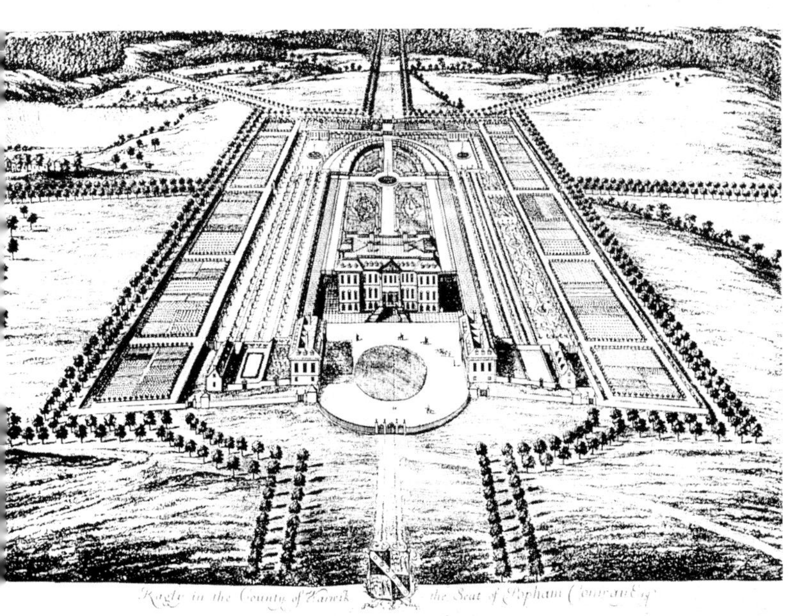

35. Ragley Hall, Warwickshire, designed by Hooke for the Earl of Conway in 1679–80 and built in the 1680s. Engraved by Johannes Kip (1697).

36. Haberdasher's school and almshouses (Aske's Hospital), Pitfield Street, Hoxton. Designed and built by Hooke, 1689–95, demolished in 1824. Anon. engraving, *c.* 1723.

37. Hooke's design for a portable drawing box, to enable seafarers to produce more accurate pictures of the lands they have seen. Illustration for a lecture to the Royal Society, 19 December 1694. *Philosophical Experiments and Observations* (1726).

Hooke started working on a house or houses in Bloomsbury belonging to Sir William Jones, a friend and neighbour of Montagu's who had lately resigned as Attorney-General. Hooke had known and advised Jones throughout the 1670s, and in August and September 1680 he often saw both his clients together, perhaps supervising some work on Montagu House, then looking at Jones' sash windows or chimneys. His companion on many of these visits was John Scarborough, who had worked with him as a surveyor or clerk of works on Montagu House, Bedlam and several City churches. Sir William Jones had another commission to offer Hooke. In 1676 he had bought an estate at Ramsbury in Wiltshire, on the River Kennet between Hungerford and Marlborough. He wanted a country house to match his status, and turned to Hooke as one of the leading architects of the day. Hooke met Jones at least five times in 1681, and probably provided him with designs for Ramsbury Manor at that time.[6]

The first volume of Pitt's Atlas was nearing completion in the summer of 1680, and though Pitt's repeated promises of payment sounded increasingly empty, Hooke continued to work on the project, and to meet him socially. He also engaged Pitt to print the second issue of his *Philosophical Collections* in October, since the Royal Society's own printer, John Martyn, had died in July. On 26 September Hooke wrote 'Finisht Atlas' in his diary, and ten days later, on the day on which Wren's second wife was buried, Hooke reminded Pitt of his original undertaking to pay him £400 for his work. Pitt asked for more time, and 'promised to pay at Christmas'. Hooke never got his money, and the atlas was a disappointment. Three more volumes were eventually published, but Hooke had nothing to do with them. Moses Pitt did not profit greatly from the enterprise, either. In 1685 he was arrested for debts of £1,000, and later spent two miserable years in the Fleet debtors' prison.[7]

Hooke was paid (or, as he often complained, not paid) to give lectures in Gresham College on most Thursdays in the year. In term time his theme was generally some aspect of geometry (the subject in which he held his Gresham professorship), and in vacations his Cutler lectures concentrated on the practical applications of science. His lectures were apparently not regarded as a high point in London's intellectual calendar, and he often recorded a disappointing or non-existent audience for them. On 17 June 1680 'none came. not one', and on 4 November there were 'none in hall but a Spy'. But at least Hooke, unlike most of the Gresham lecturers, made an effort to give his lectures, and lived in the rooms allocated to him. In August 1680, tired of years of absenteeism, the Gresham Joint Committee ordered all the lecturers to evict their tenants, reoccupy their rooms and 'personally read their lectures there at the usuall times appointed'. In the meantime, all salaries were to be suspended, and committee members ('spies', in Hooke's view) would check on whether lectures took place. Hooke appealed against the suspension of his salary on 26 October, and the committee, 'finding that he only of all the lecturers hath bin constantly resident, & for ought that appears hath bin ready to read when any auditory appeared, and besides hath printed many of his lectures for the common benefit', paid him his arrears. The laziness and absenteeism of the Gresham lecturers remained an issue through the 1680s, but Hooke's devotion to his duty was exemplary.[8]

For two weeks in February 1680 Hooke lectured to an unusually large audience of twenty, on the subject of light. He took the subject up again in November, and pursued it in 1681 and 1682. The lectures were not printed in his lifetime, but they were found among his papers after his death, and published in his *Posthumous Works* by his friend and biographer Richard Waller in 1705. He used these lectures to present some of his older ideas on light, especially the wave theory, along with his new discovery, the inverse square law, which he took to apply to light as well as gravity. He made no mention of his old *Micrographia* theory that colours were created by the distortion of white light on refraction, but neither did he admit the superiority of the optical ideas of Newton, whose name and work he ignored completely. In his February 1680 lectures Hooke assembled the arguments for his belief that light travelled in waves, rather than as particles of matter, as Newton had suggested. If light was composed of particles, he asked, how could it travel with such unimaginable speed, even through such very solid objects as diamonds? And would not the particles emitted by the Sun for so many years have gradually filled the universe, and left the Sun diminished in size? Wave motion was far more consistent with what was known about the behaviour of light, as long as one accepted that celestial space was filled with a dense but infinitely fluid matter, the aether, which could transmit vibrations instantly and without loss of power. Hooke doubted the ingenious argument put forward by Ole Roemer in 1675, that anomalies in the apparent movement of Jupiter's moons showed that light travelled with a finite, though unimaginably rapid, speed. Other explanations might be found for this anomaly, and on balance Hooke now believed what he had doubted in *Micrographia*, that light was transmitted across the universe instantaneously. He was sure, in any event, that light was a physical thing, subject to physical laws. Among these, he explained, were the well-established laws of refraction, reflection and his own discovery, inflection, where light was bent into a curve as it passed through a medium of gradually changing density, such as the Earth's atmosphere. Hooke used his February lectures to announce a newly discovered law which explained the diminishing power of light over distance, the inverse square law, and to hint that other actions of bodies at a distance from each other, gravity and magnetism, might be governed by the same rule. Thus, only a few weeks after his 6 January letter to Newton in which he told him about the law of 'duplicate proportion', Hooke demonstrated his knowledge of the new law in a public lecture.[9]

When Hooke resumed his lectures on light in November 1680 he began with an eloquent defence of the empirical method of scientific enquiry, in which 'what' should always come before 'why', and in which the development of explanations should not begin until the observable facts of nature had been established. He warned his listeners against looking for short-cuts or secret keys to understanding nature, as others had done: 'the *Peripateticks* with their four elements; the *Epicureans* with their Atoms; the *Chymists* with their three principles, Salt, Sulphur and Mercury; *Ptolomy* with his Orbs and Epicycles; *Kepler* with his Harmonicks; *Ghilbert* with his Magnetisms'. Understanding nature, he said, was like finding your way through a great labyrinth surrounded by impenetrable walls.

> He therefore that shall think immediately to fly and transport himself over these Walls, and set himself in the very middle and inmost Recess of it, and thence think himself able to show all the Meanders and Turnings, and Passages back again to get out, will find himself hugely mistaken and puzled in finding his way out again.
>
> Whereas he, that would march secure, must first find some

> open and visible Entry, and there enter with his Clew and his Instruments, and take notice of what Turnings and Passages he finds, and how far he can proceed in this and that way . . . then setting down and protracting all the ways he has there gone, and what he has there met withal . . . by comparing all which together, he will at last be able to give you the true Ground plat of the whole Labyrinth, and thereby to tell you which is the right and which the wrong way to find the middle . . .

This was advice Hooke found it easier to give than to follow. But trying hard to set a good example, he began by describing the behaviour of sunspots, and used these observations to draw conclusions about the nature of the Sun's atmosphere and the speed of its rotation. Sunspots puzzled scientists well into the twentieth century, but Hooke judged from observation of their drifting movement and changing shapes that they were clouds of gas or smoke rising in the solar atmosphere and blown by solar winds. As for the Sun itself, Hooke was convinced by its gravitational force and by the constancy of its rotation on a fixed axis that it was a solid body. He deferred his explanation of the Sun's gravity until another time, but his certainty that the Sun exerted gravitational force was based, he said, on its spherical shape and 'on the Motions of all the other primary Planets, whose Motions as I have many Years shewed in this Place, are all influenced and modulated by the attractive Power of this great Body'.

It was plain to him that the surface of the Sun was on fire, and that the Earth drew warmth as well as light from its flames. The coldness of high places was sometimes used as an argument against this, but the explanation for this was that lower air was also warmed by the heat of the Earth. Hooke accepted that a body on fire was likely to be diminishing in size, though there were no ancient astronomical observations to indicate that this was so, but even if the Sun's diameter were falling by a mile a year, this change would be invisible in so large a body. In 1572 Tycho Brahe had seen stars burst into bright flame and then burn themselves out in Cassiopeia, and the same might happen, Hooke implied, to our own Sun. As for the power of the light emitted by the Sun, it could be calculated by comparing sunlight with light from a lamp at a known distance, using the inverse square law (or law of duplicate proportion) to establish the strength of each source. Hooke explained this law in words and diagrams, using the mathematics of the cone to show that a beam of light became more widespread, and therefore (he argued) weaker, in proportion to the square of the distance from the apex to the base of the cone.[10]

Hooke ended 1680 in a fairly strong position. His authoritative and interesting lectures on light helped to reaffirm his stature as a natural philosopher, and his publication in October of the long-awaited second issue of his *Philosophical Collections* perhaps persuaded his critics that he was at last taking his secretaryship seriously. On 30 November 1680, a few weeks after the Royal Society reassembled after its long summer break, he was confirmed as a member of the Council and as secretary along with Gale. At the same meeting his long-standing friend and patron, Robert Boyle, was elected President of the Society. Boyle was plagued by ill-health and a strong sense of his physical frailty, and wrote to Hooke on 18 December refusing the presidency. But instead the Society offered the position to Wren, and his acceptance of the post in January 1681 must have given Hooke confidence that his position as curator and secretary was secure.

20. The Empire of the Senses

(1681–1682)

In the early 1680s Hooke gradually lost interest in keeping a diary. The daily record of his meetings and conversations got shorter, and long gaps started appearing in the autumn of 1680 and the spring of 1681. Whole months were left empty in 1681 and 1682, and in March 1683 he stopped writing the diary altogether. Unlike Samuel Pepys, who abandoned his diary at the end of May 1669 because he thought he was going blind, Hooke gave no particular reason for giving up his daily record, and simply stopped writing it. He began again, without explanation, in November 1688, and carried on until March 1690, and for a further eight months in 1693, when he was entering old age. It is possible that some volumes of his diary were lost when his papers were scattered after his death, but the many empty pages of the foolscap volume he had used since 1672 suggest that he simply found the task of keeping a journal too burdensome. This long break in his regular diary means that his life as it is known to us becomes more formal and scientific, the life recorded in his lectures, Royal Society records, and a few letters. No doubt his visits to Jonathan's and Garraway's went on as before, along with his games of chess with Theodore Haak, his visits to City churches and West End building sites, his meetings with Wren and Hoskins, his trips to London workshops in search of craft secrets, his meetings at the Guildhall and Christ's Hospital, his walks in Moorfields with John Godfrey, his dinners with Dr Busby, his tiffs and love-making with Grace, his dealings with Cock and Tompion, his contracts with Bates and Davies, his trips to the fair, and his experiments with new drinks and emetics. We can imagine all this going on unrecorded because we know the pattern his life had followed, just as Hooke could imagine the path taken by a comet when it was out of view. We can still glimpse Hooke's personal life from time to time, but we cannot trace it week by week, and we cannot know his private reactions to public events, as we could in the 1670s.

Just before dawn on 22 November 1680, a few days after he had given his 'Lecture of Spots', Hooke was distracted from his general study of light by the appearance of a large comet. He watched it through a hazy sky until it was lost in the morning light, and again the next morning, after which it disappeared altogether. Comets seemed to Hooke and other scientists of his time to have a particular significance for the study of light and celestial motion, and when a second comet 'with a very long blaze' appeared in mid-December Hooke, along with astronomers all over Europe, took an immediate interest in it. From 29 December to 10 February 1681, when it disappeared from his view, he observed the comet diligently, carefully noting down all its features for use in his lectures. He was particularly interested in the light and flame emitted by the comet, and thought he had spotted 'Mutations and strange Appearances, possibly never heeded by any before':

> That it waved, flared, or undulated to and fro: That it sometimes seemed to burn clearer and stronger, and sometimes fainter and more dim; sometimes on one side, and then on the other, and sometimes in the middle of the Blaze also, or in the Part opposite to the Sun: . . . Sometimes it appeared with a little or no Halo about it, but only the Nucleus or white Cloud with a little Stream or Blaze issuing from it . . . At other times, for the twinkling of an eye, or small moment, I could see a very small bright Point of Light in the middle of it . . . I think it is very plain, that the whole Blaze proceeded and issued from the Head in material streams, and that it is not at all produced by any manner of Refraction of the Rays of the Sun passing through the Head, as many are very apt to suppose and assert.[1]

Comets had fascinated astronomers for thousands of years, and Hooke set about reading what had been written about them from Aristotle and Seneca to Kepler, Galileo and Hevelius. Not surprisingly, when he resumed his series of lectures on light in January and February 1681 his audience were treated instead to a discourse on the changing understanding of comets since ancient times.

In 1680 Hooke had exchanged letters with the German natural philosopher Leibniz on the subject of a universal symbolic language, using his ageing but still active friend Theodore Haak to carry them to Hanover. Hooke's letter of 12 July 1680 told Leibniz that he had planned a simplified version of Wilkins' 'Universal Character' for use in 'the strict & philosophicall way of Reasoning', to do for natural philosophy what algebra did for mathematics.[2] Leibniz sent a reply, through Haak, which Hooke read out to the Royal Society on 12 January 1681. The letter is lost, but Leibniz told Hooke that through the universal algebra 'he had been able to perform very many considerable things, which the commonly known algebra would no way enable him to do'. Wren and Hoskins doubted whether the new algebra 'could be of so great use, as Monsieur Leibniz seemed to imagine', but Hooke was encouraged to keep up the correspondence, and he did so. He wrote to Leibniz in April 1681, and Haak delivered the letter in May:

> I have hitherto had the ill-fortune scarcely to meet with one man (except himself) whom I can perswade to concurr with me & though the greatest number look upon it only as a chymera and at best but a supplement for latine in the inner parts of Europe . . . yet my aymes have Always been much higher, vis to make it not only usefull for expressing and Remembring of things and notions but to Direct Regulate assist and even necessitate and compell the mind to find out and comprehend whatsoever is knowable.[3]

Hooke had been talking about this ambition for years, and sometimes spoke as though he was already halfway to achieving it. In *Micrographia* he claimed to have created a mechanical algebra, which might one day be extended to cover all physical enquiries, and in *Helioscopes* he said he had mastered 'an Art of Invention, or mechanical Algebra' seventeen years earlier. His letter to Leibniz in April 1681 shows that he had not yet produced his philosophical algebra, and the tone does not suggest that he saw himself as being in a race with Leibniz to complete it. No doubt in this, as in all his work, he was keen to establish or even exaggerate the extent of his own accomplishments, but this did not rank alongside the spring-regulated watch or the inverse square law as a matter on which he would assert his priority at almost any cost. The language that Leibniz eventually developed, the calculus, was more specific in its application than Hooke's universal algebra was intended to be. Leibniz's rival in the race to develop and publish the calculus was not Hooke but Newton, and the priority dispute that eventually broke out between these two in 1711 was as acrimonious and personal as any of those in which Hooke was involved.

In the first quarter of 1681 Hooke continued to combine the offices of curator and secretary at the Royal Society, but without the help of Denis Papin, who had taken a position with a scientific society in Venice. In his search for curiosities that would keep the Society entertained, Hooke was helped by his new friendship with Captain Robert Knox, who had returned to England in September 1680 after seventeen years as a captive of the Rajah of Ceylon. Hooke was very keen on the publication of travellers' accounts of unknown and distant countries, and he persuaded Captain Knox to write a description of Ceylon, with an account of his captivity and 'miraculous escape'. Knox was a modest man, but he had a fascinating story to tell, and he had studied the customs and language of Ceylon for nearly half his lifetime. Knox's *Historical Relation of the Island Ceylon*, the first account of the island in English, was published in August 1681, and it was soon translated into French and Dutch.[4] Hooke helped Knox to give the book the necessary organization and polish, and wrote a generous preface, in which he advocated the publication of similar memoirs and praised Knox's qualities as a writer:

> I conceive him to be no way prejudiced or byassed by Interest, affection, or hatred, fear or hopes, or the vain-glory of telling Strange Things. Read therefore the Book it self, and you will find your self taken Captive indeed, but used more kindly by the Author, than he himself was by the Natives.[5]

For the next twenty years Captain Knox provided Hooke with a useful supply of exotica and travellers' tales, and this strange friendship between a scholar who rarely travelled further than Westminster and an adventurer who had sailed the world's great oceans lasted until Hooke's death.

In April 1681 Hooke had a visitor, Frances Powell, from Newport, and Grace became 'very ill and discontented'. Although Hooke tried to revive her spirits by giving her £3 for a petticoat and 'other things', and getting himself a new suit, she was still 'melanchollyʼ in the middle of May. Hooke was not sure what to do. On Saturday 21 May he 'wrot to Sist. Hooke, but sent it not', and in the days afterwards Grace's condition grew much worse. She was 'most sadly afflicted' on Sunday, got worse on Monday and Tuesday, and was given an enema on Wednesday, which brought her a better night's sleep. By Friday she was much worse and vomiting, and on Saturday her mother – Hooke's sister-in-law – arrived from Newport. After this there is a six-week break in the diary, but Grace recovered, and by the autumn she was keeping house for Hooke as usual.

As for his own work that spring, Hooke showed an instrument for testing changes in the attractive power of a lodestone over distance, 'to reduce that power to a certain theory'. He had tried to measure magnetic attraction over distance in December 1673, but his equipment had not worked. He had improved the experiment by February 1674, but his tests then concentrated on the effect of various barriers on magnetic attraction. In 1681 his aim was probably to look for an inverse square law in magnetism, but the public experiments were not carried out. The Royal Society seems to have been more interested in seeing Hooke conduct more demonstrations of the 'digester', or pressure cooker, that Denis Papin had left in his care at Gresham College. To general satisfaction, Hooke showed that half an hour's boiling turned tortoise-shell into something like shoe-leather, and ivory into old Cheddar cheese.[6]

After his work with John Ogilby, William Morgan and Moses Pitt, Hooke was one of the most experienced map-makers in London, and a very desirable contact for those who wanted to try their hand in the trade. John Adams, an Inner Temple lawyer, had produced a large gazetteer map of England and Wales in 1677, and had given a copy to Hooke. In January 1681 he sought Hooke's advice on a much bigger project, a completely new atlas of England and Wales imitating the triangulation method used in the recent land survey of France, and with meridian lines based on astronomical observations. His proposal to the Society in May 1681 promised an 'actual survey of England by measuring the bounding line, the distances between places both in the road and the straight lines, by taking the latitudes and angles of position', all completed by 1685. Hooke joined a small

Royal Society committee to report on the plan, and a few months later Adams was on the road with his telescopic quadrant and pocket Bible. On Christmas Eve 1681 he wrote to Hooke from Shrewsbury asking for two or three seven-foot telescopes, and reporting on his progress in Wales and the West Country. 'I have travelled 7000 miles, and I will ply this undertaking Winter and Summer till I finish it.' This pioneering work, which involved the first systematic measurement of altitudes in England, was too enormous an undertaking for a single individual relying on private subscriptions, and by 1688, with no maps completed, Adams had run out of money and spirit.[7]

The problem faced by all map-makers was how to reproduce the spherical shape of the globe on a flat surface, retaining as far as possible the correct land shapes without misrepresenting distances and areas to too great a degree. The projection introduced by Gerhard Kremer, or Mercator (1512–94), and very widely used thereafter, retained shape at the expense of area, greatly exaggerating the size of territories further from the equator. Its great advantage to navigators, and the reason for its continued success, was that rhumb lines, the lines followed by a ship sailing a steady course, without change of compass direction, appear as straight lines on the Mercator projection, cutting all meridians at the same angle. On the globe itself, rhumbs are not straight lines, but curves. Hooke and his contemporaries, however, used the word 'rhumb' (or 'rhomb') in a now obsolete sense, to indicate any section of a voyage that followed a steady compass bearing, even when it was only part of a longer and more complex route. Seafarers who wished to take the shortest way between two ports had to follow a course known as a great circle, a straight line on the globe, but although this reduced the distance sailed it involved more complex navigation and very many changes of course. Navigators could compromise between these two methods, and simplify their great circle into a series of rhumbs, in Hooke's sense of the word. On Mercator's projection, great circle routes would be drawn as curves, unless they followed the equator or a meridian line.

On 16 March 1681 Hooke had a visit from a cartographer who had adopted the name of the great Mercator, and who wanted to tell him about his new system of projection for nautical maps. Since the announcement of Hooke's own projection, which he had already explained to Wren, was imminent, he refused to hear Mercator's system, in case his priority was jeopardized. So Mercator took his new projection to the King in search of a royal patent, and Wren explained it briefly to the Royal Society on 23 March. When Mercator met Hooke on 14 May, Hooke took the initiative by explaining his own 'planispherical projection', which allowed distances to be measured accurately. Mercator claimed that his projection was quite different from Hooke's, but seemed unable, at a second meeting on 16 May, to explain what it was, or how it differed from established systems. Hooke's diary records: 'Mercator & Aubrey here. Quarrelled for planisphere'. Hooke dissuaded Mercator from taking out a patent on 'so old, common, and practised a way of making maps', and reported the whole conversation (from his own point of view) to the Society. Mercator's pretensions were not heard of again.[8]

In May Hooke returned to the subject of light in his Gresham College lectures, after his diversion into the subject of comets. Much of this was old ground for him. He began with a survey of sources of light on Earth, especially those that involved combustion or friction, and took the opportunity to offer a definitive explanation of his original and important work on the role of air in combustion and respiration. Hooke had demonstrated experimentally that air itself, rather than the motion of air caused by breeze or bellows, was essential to combustion: 'tis the fresh Air that is the Life of the Fire, and without a Constant supply of that it will go out and Die'. The same was true in breathing – the fresh air itself, rather than the motion of the lungs, is essential to life. 'For whether the lungs move or not move, if fresh Air be supplied, the Animal lives, if it be wanting, it dies.' The fact that light on Earth was emitted by bodies that were on fire or had been rubbed or struck (diamonds, flints, 'Sugar, Black Silk, the Back of a Cat') led him to conclude that light was 'nothing else but a peculiar Motion of the Parts of the Luminous Body'. When the motion of the particles in heated, rubbed, hammered or decomposing matter reached the necessary speed light was emitted, as in the case of a piece of iron that was beaten until it was red hot. This motion was instantly transmitted through a perfectly fluid, perfectly dense, medium like the aether to the whole universe. The effect of a wave of light on the aether, he explained later, was like a blow on one end of a stick: one did not have to wait even an instant for the other end of the stick to move. Hooke then used the image of a cone of fluid transmitting light to explain the inverse square law more fully than before, and ended with the claim that 'this Power of Light'

> is the Power of Celestial Bodies by which they Act upon, and attract each other: and by which all the Primary Planets that move about the Sun are regulated in Velocities, Distances and Motions, whether circular or Oval. As also all the secundary Planets, as the Moon about the Earth, and the Satellites about *Saturn* and *Jupiter*, make their Periods. And from the true stating of this Power, and the Effects of it on Bodies at several Distances, all the Theory of Astronomy will be deduced *a Priori*, with Geometrical Certainty and Exactness; and consequently the Tables and Numbers will be easily adapted, which will tend to the Perfection of that Noble Science.[9]

It was Hooke's expectation that eventually, when he had the time to devote his mind to this great problem, he would be the one to reveal the mathematics of planetary motion to an admiring world.

He used his next lecture, in June 1681, to explain the structure and working of the human eye, drawing on the writings of others and his own anatomical work. For good measure, he told his listeners how to operate on a cataract, and how they could make their own artificial eye, or perspective box. This was an enclosed tube about five feet long, cylindrical for half its length, and then narrowing to a small aperture, in which was fixed a convex lens with a focal length equal to the length of the tube. At the wider end of the tube there was a circular base which could be pulled in and out, and with its concave inner surface painted white to form a screen. By looking through a large well-curtained hole cut in the side of the tube the viewer could see an image on the white base of the tube, and try the effects of moving the base in and out, and of putting pasteboard rings with different apertures onto the lens, mimicking the focusing and contraction of the natural eye. The use of a set of different rings was not a very elegant representation of the dilation and contraction of the pupil, and Hooke decided to devise a mechanically adjustable aperture, which might also be useful in telescopes. He demonstrated his new device to the Royal Society on 27 July, telling them that it 'would open and close just like the pupil of a man's eye, leaving a round hole ... of any size desired'. Details of this new aperture were not given in the journal, but it probably had a ring of overlapping plates like the iris diaphragm used in a modern camera shutter, and therefore represents yet another of Hooke's contributions to the technology of the modern world.[10]

Hooke was unusually busy on the Society's behalf in the early summer of 1681, because on 22 June he had promised the Council that in return for an extra £40 on that year's salary he would undertake 'a more sedulous prosecution of the experiments for the service of the society, and particularly the drawing up into treatises several excellent things, which he had formerly promised the world'. This deal led to a short burst of activity on his part. The late July meeting that saw his mechanical iris was also entertained by a demonstration of his new double-reflection helioscope which provided a perfect image of the Sun, and an experiment involving the production of musical and vocal sounds 'by the help of teeth of big brass wheels; which teeth are made of equal bigness for musical sounds, but of unequal for vocal sounds'. This was a machine begun by Hooke and Tompion in March 1676, when Hooke's long-running interest in musical vibrations had been revived by his discussions with Dr Holder. By turning a toothed wheel at different speeds, so that the teeth struck a piece of metal or card and created a sound, the machine made it possible to create a note of a known frequency, and to establish the rule that the pitch of a note (its position on the scale) is determined by its frequency, or the number per second of the vibrations producing it. So Hooke's boast to Pepys in August 1666, that he could tell the number of beats of a fly's wings from the sound they made, almost came true. A few notes in Hooke's diary suggest that he maintained his interest in his music machine over the summer of 1681 and into October, when he wrote: 'The wheel of unequal teeth: musick....' Once again, a public performance in the Royal Society did not ensure that an innovation would be remembered. Hooke's forgotten machine was reinvented by a French physicist, Félix Savart, in the early nineteenth century, and is now known as Savart's wheel.[11]

Although some of his biggest commissions – Bedlam, the Physicians' College, the Monument, Montagu House – were finished,

Hooke spent much of his time in 1681 in building and architectural work. His patchy diary records occasional visits to City churches, frequent views as City Surveyor, a few meetings with Lord Conway, many meetings with his clients Ralph Montagu and Sir William Jones, and frequent conversations with William, Lord Russell, who had just acquired Southampton House, a magnificent 1650s mansion in Bloomsbury Square. The work Hooke was discussing with Russell was probably cut short in 1683, as Russell himself was for his part in the Rye House Plot against King Charles and his brother.

About a dozen City churches were under construction in the early 1680s, but without a continuous and well-kept diary we can only guess at the visits Hooke made to the sites, and at the deals he negotiated with masons, bricklayers, carpenters, glaziers and plasterers. We know that the £1,355 he was paid for City church-building in the nine years 1680–88 was £40 more than he was paid in the previous nine years.[12] At least we do not have to guess at the finished work. St Benet Paul's Wharf, for which Hooke did the preliminary and final drawings, is one of the two City churches most surely associated with him, along with St Edmund the King, which was built ten years earlier. St Benet escaped Victorian restoration or demolition and Second World War damage, and remains as a perfect example of Hooke's Dutch domestic style. Looking at its charming red and blue chequerwork brick exterior, with garlands carved from white stone over the windows and a little lead dome and spire on a modest brick tower, and ignoring the noisy roads that now surround it, one can imagine oneself in the City of Hooke's day, freshly rebuilt after the Fire. Inside, the church is light and welcoming, with its ceiling, columns, galleries, pulpit, lectern, pews, reredos, doorway, font, communion rails and wall panels much as Hooke's craftsmen made them in the 1680s.[13]

It was clear by 1681 that Hooke's performance as secretary of the Royal Society fell far short of the high standard set by Oldenburg. Despite several warnings from the Council, Hooke was still not conducting the Society's correspondence or producing its publications with sufficient vigour. In March 1681 Hevelius, one of the Society's most important overseas members, complained again of the lack of news from London, and *Philosophical Collections*, which was meant to be published fortnightly or monthly, had now not appeared since

October 1680. During the summer and autumn Hooke began gathering essays and reviews for five editions of *Philosophical Collections* that appeared monthly between December 1681 and April 1682. The first of these, published on 10 December, contained articles by Hooke on 'a way for helping shortsighted or purblind eyes' and on the best way of arranging the sails of horizontal mills and sailing ships, a problem that had interested him since the autumn of 1674. Further editions, with contributions by Flamsteed, Tyson, Leeuwenhoek, Haak, Hevelius, Halley and from the journal of the great Dutch explorer Captain Abel Tasman, appeared in the first four months of 1682.

Hooke's position in the Royal Society was made more difficult by conflicts with newer members who had no personal attachment to him, and saw him as a spent force whose dominance over the Society's correspondence and experimental programme was stifling new developments. Chief among these opponents was John Flamsteed, the Astronomer Royal, whose experience with the mural quadrant in 1677–8 had convinced him that Hooke was a braggart whose scientific and mechanical achievements fell far short of his claims for them. Hooke's opinion of Flamsteed was hardly any higher. In June 1680, when Gresham College needed a temporary replacement for its Professor of Astronomy, Walter Pope, Hooke tried to get the job for Edmond Halley, and spoke to him about it in Jonathan's coffee house. 'He told me Flamstead Indeavoured to supplant Dr Pope, and to get an interest for himself.' The post went to Flamsteed, who was older and more experienced than Halley. He started lecturing at the College in April 1681, and continued to do so until 1684, when a dispute with other Gresham professors, no doubt including Hooke, forced him to resign. Like Hooke, he was sometimes disturbed by poor attendance, and he was inclined to blame Hooke, sitting there in the audience, for fixing this in some way.[14]

Flamsteed was a dangerous opponent, but Hooke was a tough and experienced controversialist, and he was happy to tackle Flamsteed in Royal Society meetings or coffee-house debates. On 2 November 1681 Hooke showed the Society a mechanical way of finding the focus of parallel rays falling on the spherical side of a plano-convex lens (a lens with one flat and one convex side), and promised to show the next meeting a way of finding the focus when the lens was turned round the other way. Flamsteed declared that Hooke's method was 'false and impossible', and Hooke appealed to the judgement of Sir Christopher Wren, who was not at the meeting, to support his case. At the next week's meeting Hooke announced that Flamsteed 'had now acknowledged that what he had formerly adjudged against the problem shown . . . was a mistake of his; and that upon considering it more seriously he had found out the demonstration, though he had not done it before the way . . . shown by Mr Hooke.' This dispute was a formal re-enactment of a coffee-house dispute that took place between Hooke and Flamsteed either before or after the 2 November meeting. Hooke's diary is blank for this period but we have Flamsteed's undated account of what happened. It seems that after a Royal Society meeting the members adjourned as usual to Garraway's, and Hooke had asked Flamsteed, 'somewhat captiously', whether, in making celestial observations, it was better to turn the plane or convex side of a plano-convex object lens towards the sky. Caught off guard, Flamsteed said that there was little difference between the two, but that it would be better to turn the plane side to the sky. Hooke mocked Flamsteed's ignorance of his own telescopes, and demonstrated, to the satisfaction of the others present, that the lens should be the other way round. Mortified by his defeat, and perhaps fearful for his reputation and career, Flamsteed left the scene and wrote an essay on what had happened, defending his own role and damning Hooke's. Hooke 'bore me downe with wordes enough & psuaded the company that I was ignorant in these things which that hee onely understood not I'. Hooke had a showy knowledge of optics, he said, and could impress the coffee-house crowd with a little pocket lens, but how deep was his knowledge of refraction? It was typical of Hooke, he continued, to devise machines to establish angles of refraction rather than calculate them, to boast about the performance of his inventions instead of putting them to the test, and to pick up knowledge in coffee houses which he then paraded before the Royal Society as his own.

> Whilest he pretendes to know better then others, hee makes questions to those he knows are Skilfull in them, & theire answers serve him for assertions on the next occasion. This unworthy dealing of his has a little exasperated mee I must confesse but I

> hope my Patrons and Readers will pardon mee & if they know but the impudence of this Mechanick Artist ... i endeavour to informe him better things then those on which hee grounds such monstrous boasts, & magnifies for such deepe knowledge as none besides himselfe is master of.[15]

Revitalized by his victory over Flamsteed, Hooke was on irrepressible form at the Royal Society meeting of 16 November. He told the Society of some new instruments he had devised for taking azimuths and altitudes at sea 'by a new way not before practised, which he designed shortly to publish', and of a new quadrant of unprecedented accuracy 'which he was now making, and would shortly produce', and demonstrated 'a new way of solving the phenomena of refraction'. In the same meeting he returned to another issue over which he had clashed with Flamsteed a month earlier. On 19 October Hooke had demonstrated a new device he had invented for drawing accurate lines indicating the path of a ship taking a steady course (without changes of direction) on a map drawn on a polar projection, and for converting a more complicated great circle course (the most direct route between two points on the globe, involving repeated changes of compass bearing) into a series of slightly longer 'rhumbs', each of which followed a single compass bearing. This was his 'method of finding the length of any part of such line; and of straightening the said line or any part thereof with ease and certainty, and thereby answering many questions in navigation without calculations by the help of a ruler and compasses without the use of tables.' Hooke's purpose was to make it easier for navigators to plan the shortest and most convenient routes, using the polar projection he had produced for Pitt's Atlas, without complex calculations. He 'straightened' the line on the map (but not on the globe) by converting the curved line of a great circle route into a series of straight 'rhumbs'. Apparently Flamsteed had expressed grave doubts about this method, because on 16 November, when Hooke showed the Society two new compasses he was making 'for describing all sorts of spiral lines for the rhombs', he announced that Flamsteed had now withdrawn his objections to his 'new way of measuring distances in great circles laid down on the planisphere projection of the globe by the help of a sector and compasses, without drawing lines or circles, or making any divisions, &c, which had been by Mr Flamsteed before the Society impugned as false; but now he acknowledged it to be true and real'.[16]

On 23 November 1681 Hooke produced the new compasses for drawing rhumbs and 'all manner of proportional spirals' for navigators and sea charts, and for the next three months he explored the possibilities of his new device. The fact that map projections could be formed into cones and cylinders led Hooke from cartographic to broader geometrical applications, and he crossed and recrossed the boundaries between practical and theoretical work in a way that typifies his whole career. On 14 December he came along with a sort of lathe that could cut helices on cones and cylinders, and also 'any fish shell; and all helices, screws, crenated, foliated, echinated, wreathed, &c. conchoids; and he observed, that this engine would be of great use for making divisions of mathematical and astronomical instruments, for turning wreathed work, and many other uses'. The machine was apparently a model or incomplete, and he was urged to discuss with workmen the cost of making the full-sized engine.[17]

The following week he brought in a one-foot globe fitted with an instrument which could draw all the rhumb lines on it 'most exactly', and on 4 January he had a refined version of his spiral engine, which could divide 'an inch into 100,000 equal parts, and that with the greatest ease and certainty imaginable; which he conceived to be the best way yet thought of in the world for perfecting all manner of astronomical and geographical instruments.' Over the next two weeks he demonstrated how his compasses could be adapted to draw a variety of conic sections (parabolas, hyperbolas and ellipses – shapes created when a plane intersects a cone at different angles to the cone's base), spirals and endless curves 'with as much ease and exactness, as a circle could be described'. Conic sections had fascinated mathematicians since the days of Archimedes and Euclid, and in the seventeenth century Kepler, Pascal, Descartes and John Wallis had taken a particular interest in them. Kepler's first law, that the orbit of a planet is an ellipse, gave them a vital role in understanding gravity and celestial motion, and the elliptical path of a falling object was the issue at the centre of Hooke's unfortunate correspondence with Newton in the winter of 1679–80. Hooke's demonstration provoked a discussion of the many interesting geometrical, algebraic and practical properties of ellipses and parabolas which must have stretched the mathematical abilities of some Royal Society fellows to breaking point.[18]

Flamsteed, as usual, was unimpressed by Hooke's 'idle and loose way of making a parabola', and asserted that the method he had demonstrated was false. The dispute was settled in Hooke's favour on 15 February by the enormous authority of the President, Sir Christopher Wren, who watched Hooke repeat his parabola demonstration and declared (according to Hooke's minutes) 'that it was true and certain, and the best way yet known of describing that curve, and never published before'. The absence of a detailed account of Hooke's device makes it impossible to say how it differed from the method described by Jan de Witt in 1650, or by Newton in *Principia Mathematica* in 1686, but Wren's endorsement suggests that Hooke had made a useful contribution to the subject. Flamsteed's opinion of Hooke could hardly fall any lower, but his hostility was particularly intense after this dispute. Writing to Samuel Molyneux of the Dublin Philosophical Society a few months later, he denounced Hooke's habit of promising wonderful inventions that turned out to be trivial or useless, keeping his ideas secret to magnify their importance, and responding to the discoveries of others by claiming to have proof in the minutes of the Royal Society that he had known about them for years.[19]

The range of Hooke's recorded activities in the winter and spring of 1681–2 was as wide and interesting as ever. He was still trying to think of a practical way to illustrate the inverse square law of gravitational attraction, and on 7 December asked for volunteers to help him to discover whether pendulum clocks at the top and bottom of the Monument would show a measurable divergence from each other. In February 1682 there was an unsuccessful experiment to examine the freezing of eggs and apples, and a lively demonstration of the effects of mixing oil of turpentine with alcohol and sulphuric acid. In March Hooke translated and read out a letter in Low Dutch from Leeuwenhoek on the structure of lobster muscles, a subject which both men had investigated. His knowledge of Low Dutch, which he had first learned from his lodger Richard Blackburne in the winter of 1672–3, was particularly useful to him that month. The sixth edition of *Philosophical Collections* appeared, with extracts from the journal of Captain Abel Tasman's voyage in 1643 to New Zealand

and Tasmania, translated for the first time by Hooke, probably helped by Lodwick. Hooke had a great interest in descriptions of distant lands and voyages of exploration, and got permission to seek out and buy other such accounts on the Society's behalf. His collection of papers in the Guildhall Library includes accounts of voyages to Java, the West Indies, the South Seas, Tartary, China and Hudson's Bay.

In March and April 1682 two letters to Hooke from Dr Martin Lister, a Royal Society member resident in York, sparked off a series of letters and debates about people who had vomited enormous worms (a favourite Royal Society topic), prompting him to argue that the seeds of all creatures were modified in the womb or body that carried them as they grew, and that the seeds of little worms took on the bodily size of their hosts. His most serious task in the spring was to continue, and effectively to conclude, his long series of lectures on light. In his last lectures, given to the Royal Society on 3 and 17 May 1682, he reaffirmed his conclusion that light was an actual wave motion, rather than a *tendency* to motion, as Descartes thought, and that its movement over any distance was instant. Using an image from a world he knew so well, he likened the passage of light from its source through a transparent medium to the eye to the action of a mason striking a steel chisel with an iron hammer and breaking a piece of marble. Hooke was at his best when he was explaining unfamiliar and difficult concepts such as gravity or the movement of light to sceptical audiences. His skill in constructing experiments or machines to demonstrate scientific principles was matched by his ability to devise persuasive illustrations drawn from the familiar world of craft or nature. He explained the ability of the aether to carry millions of rays of light in countless different directions at once by referring to the effects of dropping twenty drops of water into a pond. Amazingly, the rings of waves produced by each drop continue to spread across the pond just as if there had only been one drop. 'And though, I confess, after all this, it does seem not a little difficult to comprehend how one and the same Particle of Matter, or of the transparent *Medium*, should at the same Instant propagate through it a thousand different Motions, a thousand different Ways; yet since we are assured . . . that it is actually done in Nature, and that visible to Sense, though we cannot so clearly comprehend the Metaphysical Reason thereof: yet 'tis enough for a Principle to build upon it.'

To explain why there were so many things in nature that we could not sense and therefore struggled to understand, Hooke reverted to a favourite old theme, the limitations of human perception. Just as there were millions of particles in a visible point of matter, there were infinite instants in one perceptible moment. Countless separate motions were possible at one perceptible point and moment, transmitting millions of rays of light in millions of directions. He was sure that man's perception of time and space was crude compared to that of smaller creatures: 'I do not at all doubt but that the sensible Moments of Creatures are somewhat proportion'd to their Bulk, and that the less a Creature is, the shorter are its sensible Moments ... So that many of those Creatures that seem to be very short-lived in respect of Man, may yet rationally enough be supposed to have lived, and been sensible of and distinguished as many Moments of time as a Man ...' Our other senses also differed from those of smaller creatures. We could only make and hear sounds within a certain range, and see things of a certain size, but smaller animals could doubtless hear sounds produced by faster vibrations, and distinguish objects too small for us to see. Since our senses allowed us to experience such a small proportion of the matter and motion that actually existed, it was not strange that so many natural phenomena were hard for us to understand.[20]

Hooke was still the mainstay of Royal Society meetings. In those of May and June 1682 he demonstrated the creation of colours by the mixing of various chemicals, tried to explain how changes in the Earth's atmosphere might account for the changes in barometric pressure in different weathers, and discoursed on why variations in the time of a pendulum's swing in different parts of the world made pendulum clocks unsuitable for finding longitude at sea. At Wren's request, Hooke agreed to undertake the measurement of a degree of longitude in England – a task he had been set as long ago as 1669 – and to gather a set of accurate standard weights and measures for the Society's use. This was familiar territory, but on 21 June Hooke sprung a surprise on Society members by delivering a long lecture on memory, the soul and the perception of time. The relationship between the soul and the brain was not an entirely new subject, since Descartes, Hobbes, Boyle and Thomas Willis (Hooke's teacher and patron at Christ Church) had all written about it, but Hooke took the investigation further than anyone had heard it taken before. Hooke's discourse had such an impact that on 28 June several members who had missed it the previous week, including Sir John Hoskins, Sir William Petty, Sir Robert Southwell and Adrien Auzout, persuaded him to read it again.

Hooke's starting point was the discussion of the perception of time in his last lecture of light. Our five familiar senses were momentary, and could not in themselves record impressions beyond an instant. To explain the retention and chronological ordering of such impressions, Hooke argued, 'we shall find a Necessity of supposing some other Organ to apprehend the Impression that is made by Time. And this I conceive to be no other than that which we generally call Memory, which Memory I suppose to be as much an Organ, as the Eye, Ear or Nose, and to have its Situation somewhere near the Place where the Nerves from the other senses concur and meet.' The fact that the memory was an organ was indicated by its susceptibility to improvement by training and to damage by sickness, injury or drinking, and by the fact that it did not function when we are asleep. For the same reasons, the memory could not be identical with the soul, which was incorporeal, and not susceptible to physical damage or death. In Hooke's view the senses collected impressions (always in the form of motions) from the outside world, and carried them to the repository or storehouse known as memory.

The storage of these impressions by the memory was not an automatic process, but was governed by the soul, in a selection process he called 'attention'. Those impressions or motions that were retained were probably arranged in a spiral or coiled chain, with the ideas collected most recently at its centre, and those dating from childhood at its outer end: 'So that there is as it were a continued Chain of Ideas coyled up in the Repository of the Brain, the first end of which is farthest removed from the Center or Seat of the Soul where the Ideas are formed'. Hooke explained forgetfulness as the natural decay of motions over time, especially older motions that were 'closer and closer stuffed and crouded together' at the far end of the coil, distant from the seat of the soul. The memory's collection of impressions and ideas was the raw material from which the soul developed its understanding of past experiences and its sense of time, for 'Time, as understood by Man, is nothing else but the Length of the Chain of these Ideas . . . and the Notion of Time is the Apprehension of the Distance of Ideas from the Centre or present Moment'.

The memory had to be made of materials adapted to retain the impressions or motions delivered by the various senses, as phosphorus retains light and a warming stone stores heat. Hooke entertained his audience by working out that a man of fifty might have stored up nearly two million ideas, at the rate of a hundred a day. Each stored idea took up a distinct space in the memory, but it was quite conceivable, he said, that the memory could store a hundred million ideas. His work with the microscope made it possible for him to envisage the enormous storage capacity of the brain. After all, the microscope had shown that a hundred million functioning creatures could live and move around freely in a very small space, so 'we shall not need to fear any Impossibility to find out room in the Brain where this sphere may be placed'. The soul, he explained, was like the Sun, shining more brightly on new ideas than older ones, and sometimes leaving an old idea in shadow when a newer idea eclipsed it. Yet the soul could exert its will, and fix its attention on a particular idea, or on a hundred or a thousand ideas that harmonized with each other, to form a new and more complete idea. By vibrating in a particular way, and especially by vibrating in harmony with a new impression, older ideas might attract the soul's attention, inducing it to associate one idea with another, or 'call to mind' something it had almost forgotten. The soul's deliberate selection of impressions from the repository, in order to form a new idea from stored ones, was the process of thinking: 'Thinking is partly Memory, and partly an Operation of the Soul in forming new Ideas'. There was a higher process of the soul, called reasoning, which involved bringing together ideas or impressions from several parts of the repository, 'being sensible of the Harmony or Discord of them one with another', and uniting them into a more complete and compounded idea expressing a more general truth. Speaking as a man in his late forties facing competition in the Royal Society from a new generation of young scientists, Hooke emphasized that the soul of the older man could draw upon a richer and fuller repository of impressions to develop more complex, compounded and perfect ideas.[21]

Robert Hooke was a Christian, though not an especially devout or active one, and his friends, colleagues and clients were Christian believers, or appeared to be. Like Boyle, he was convinced that a mechanical philosopher could explain the world, even such mysteries as gravity, vacuums and the working of the human brain, in terms of matter in motion, and at the same time remain a Christian believer. He had no patience with the theologian and philosopher Henry More, who was too ready to give up the search for physical explanations for the most puzzling natural processes, and fell back on an incomprehensible 'Hylarchick Spirit'. Yet natural philosophers who set about explaining the hitherto hidden mysteries of God's creation as simply physical phenomena might be accused of heresy or atheism, as Thomas Hobbes had been. Hooke was careful in his paper to say that although the soul drew ideas from a material repository, the memory, and even occupied a specific location in the brain, it was itself incorporeal, a self-moving principle that irradiated and organized the physical body without being part of it. The process by which the immaterial soul and the material memory influenced each other was something that he did not try to explain. In the discussion that followed the second reading of his paper Hooke was accused of saying that the soul itself was mechanical, but he was able to defend himself against this dangerous charge by showing that he had made a clear distinction between the material memory and the spiritual soul. To the careful listener, the distinction was obvious. John Evelyn's diary record of the first meeting shows no confusion: 'To our *Society*, where Mr Hook read to us his ingenious *Hypothesis* of *Memorie*, which he made to be an Organ of sense, distinct from any of the five; placed somewhere in the braine, which tooke notice of all *Ideas* and reposited them; as the rest of the senses do their peculiar objects.'[22]

If Evelyn had read Descartes's *L'Homme* or Thomas Willis' *Anatomy of the Brain* (as Hooke certainly had) his admiration of Hooke's startling originality might have been a little weaker. What made Hooke's work new was not his discussion of the interaction between the physical brain and the spiritual soul, a concept familiar to ancient and medieval thinkers, but the clarity of his explanation of the supposed mechanism of the memory, and his ability to draw upon his experience as an astronomer and a microscopist to make his account credible. A few weeks later he explained to the Royal

Society that man stood in the middle of a material world that was infinite in its extent and unlimited in its divisibility. Further mechanical advances, he assured them, would one day reveal an astronomical and a microscopic world 'beyond the imaginations of most men hitherto known.'[23]

21. A Curator Again

(1682–1684)

A FEW BRIEF DIARY ENTRIES give us some clues to Hooke's work in the summer and autumn of 1682. He continued to supervise work on Montagu House, although he was finding it difficult to persuade Ralph Montagu to pay him what he was owed. Eventually Montagu's agent paid him £50, and on 6 August this went into his great trunk, along with £50 from Wren, £50 from the Gresham Committee, five golden guineas from Dr Busby and various other payments, £288 in all. The next day, 7 August, he set off by coach and four to view Ramsbury Manor, Wiltshire, which had been left unfinished when Sir William Jones died that May. Hooke had been down to Ramsbury for Jones' funeral, and now he returned to carry out his last request (added to his will on 30 April 1682), that Ramsbury should be completed. He travelled down with Jones' servant, Heblethwait, lodging at Reading, where he 'saw monastery & Drunken jugliers', and the next day visited Donnington Castle, which had been in ruins since the Civil War. He had supper that night with Thomas Pelham, Jones' son-in-law and executor, and three London builders with whom he had often worked before, Joseph Lem, Joseph Avis and Roger (or perhaps Thomas) Davies. It is extremely likely that this team of craftsmen, chosen by Hooke, were building Ramsbury Manor to his designs. The next day they all went to view the mansion, and Hooke then dined at the Swan and journeyed to Speenhamland in Berkshire for his lodging. After another day's travelling and a night at Maidenhead, Hooke was home again ('Deo Grat') on 11 August, in time for a drink of cider in Goodwin's tavern. The main structure of Ramsbury Manor was finished in 1683 (the date on the rainwater heads), but internal work continued until 1686. The house remains almost unchanged today, and stands as a perfect example of Hooke's country house style, dignified, comfortable, and built to the best modern standards.[1]

From 16 August to 10 September 1682 Hooke kept watch on a bright comet, the one later made famous by Halley's prediction that it would return in 1758. On 9 September, just before it disappeared, he saw it directly above St Mary-le-Bow church, where it seemed three times as long as the dragon on top of the steeple. He worked on a little quadrant and with Harry Hunt on a one-wheeled chariot, found a way of printing on plaster and cement, used oil of tartar (potassium carbonate) to clean tarnished tin and polish the mirror of his reflecting telescope, measured the amount a piece of wire stretched when different weights were suspended from it, and studied a book in Chinese. He discovered (so he said) a 'way to leap by spr[ung] shoes 12 foot perpendicular, and 20 ft. parabolical by Repetition rising bell-like to a great height', but if Londoners were treated to the spectacle of a wild scientist bouncing around the Gresham College quadrangle or along Bishopsgate in twenty-foot parabolas they left us no record of the event. Pursuing his interest in improving quadrants and telescopes, he erected a thirty-six-foot telescope through his ceiling, and tested ways of using mercury to keep optical devices level at sea, to make it possible to use them when the horizon was not visible. In September he perfected a 'Sea Waywiser', which was meant to chart a ship's progress by measuring wind and water flow and changes of direction, and over the next few weeks he investigated the problem of plotting a ship's course on a map based on a cylinder rolled flat – 'measuring distances difficult but not impossible'.[2]

When the Society reconvened on 25 October Hooke showed them his work on cylindrical projections, and a 'very easy instrument' for plotting a course by drawing rhumb lines, 'very fit for the use of navigators'. The mathematical properties of these and other lines drawn on a cylinder interested him all through the autumn, alongside problems of practical navigation. He told his diary, but not the Royal Society, of devices he had in mind to enable 'the steersman of a ship to sit above, and by an engine manage the rudder and care of the ship', and to add fins to a ship's keel to stop its bow dipping under the water.

Later that autumn he turned his mind to more fundamental questions. His lectures on the new comet developed into a more general account of his natural philosophy, and to an explanation of the nature and causes of gravity. His own philosophy of nature could hardly be simpler: 'I conceive then the Whole of Realities, that any way affect our Senses, to be Body and Motion ... These I conceive the two Powers or Principles of the World ... created by the Omnipotent to be what they are, and to operate as they do; which are unalterable in the whole, either by Addition or Subtraction, by any other Power but the same that at first made them'. God made the rules of nature that governed them, and it was the natural philosopher's task to discover and explain them. He was dismissive of those past scientists who had tried to explain nature by some favourite invention of their own – 'The 4 Elements, the 3 Chymical Principles, Magnetism, Sympathy, Fermentation, Alkaly and Acid, and divers other Chimeras too many to repeat'. He spoke at length on the compatibility of his interpretation of the universe with the early verses of the book of Genesis, which were an account of God's creation of body and motion, and of the two great laws of motion, light and gravity. He had already spoken about light, but his account of gravity in these lectures gives a good indication of how far he had gone on the road that Newton was also travelling, and how far he still had to go. In a remark that shows how much he was influenced still by his old idea of congruity between particular substances, he defined gravity as the power that caused 'Bodies of a similar or homogenous nature' to be drawn towards each other. He did not labour this point as he would have done a few years earlier, and in the rest of the lecture he spoke of gravity as a universal force operating over vast or even infinite distances and acting 'on all bodies promiscuously, whether fluid or solid'. One of the strongest arguments for the presence of gravity throughout the universe, he said, was the fact that without it the Sun, the Earth and all the rapidly revolving planets in the universe would have been dispersed long ago by centrifugal force. Diverting from this theme, he reminded his listeners that they should not take it for granted that the Earth was spherical – it might be shaped like an egg, or more probably like a turnip, with its equatorial regions pulled outwards by greater centrifugal force where it span faster. If this were so, as he suspected, then the force of gravity would be different in different parts of the world, and there could be no such thing as a standard one-second pendulum.

This new celestial system was far more plausible than the complicated magnetical theories of Kepler, or the vortices of Descartes. Its main advantages, Hooke said, was that it was simple, that it was consistent with observations made on Earth and in the heavens, and that it harmonized with the known laws of nature. The power of gravity, he argued, 'extended to a vast distance upward, even indefinitely', and diminished over distance according to a formula which also governed the power of light. His attempts to demonstrate this experimentally on St Paul's Cathedral or the Fish Street pillar had not been entirely successful, he admitted, because they were not high enough. But it was plain from the ease with which stones could be thrown into the air that gravity was a finite power, and that 'comparative to other Powers of Nature, tis weak'.

Finally, Hooke came to the greatest question of all – what was gravity? Spiritual and magnetical theories told us nothing, he said, but a simple experiment suggested by Newton in December 1676 and performed for the Royal Society by Hooke pointed towards a plausible explanation. When a plate of glass was rubbed pieces of paper rose and stuck to it, attracted by some sort of electricity. A similar attraction, which was not magnetic, could be created by rubbing amber or diamonds, and a ball barely floating in water could be drawn downwards by striking the bottom of the glass. Hooke demonstrated this to the Royal Society on 21 February 1683. The power of vibrations to attract could also be demonstrated by drawing a bow across the side of a glass of water, inducing the water to move towards the vibrating part of the glass, or by watching a workman tightening an axe-head onto its shaft by hitting the other end of the shaft. And though Hooke did not mention it here, his demonstration in March 1671 of the effects of striking and vibrating a shallow bowl of flour, causing the flour to climb to the top of the bowl, which he had thought 'might contribute to explain the cause of gravity', must also have been at the back of his mind, or at the end of the coil of his memory, perhaps eclipsed by more recent impressions.

'I conceive, then, that the Gravity of the Earth may be caused by some internal Motion in the internal or central Parts of the Earth'. This vibrative motion produced vibrations in the aether, and these drew other bodies towards the Earth, like paper to a piece of polished glass. There was nothing strange about forces working at a distance in nature, Hooke assured his listeners. Light and sound both worked in this way, and there was no reason why gravity should not do the same, with each force using a different medium – a different component of the multi-functional aether – to carry its vibrations. Like light and sound, gravity also obeyed the inverse square law: 'For this Power propagated, as I shall then shew, does continually diminish according as the Orb of Propagation does continually increase, as we find the Propagation of the Media of Light and Sound also to do; as also the Propagation of Undulation upon the superficies of Water.' Once again, Hooke's argument for the inverse square law was based on analogy and common sense, rather than mathematics.[3]

Although Hooke's lectures on light, comets and gravity showed that he was still a vigorous and original natural philosopher, it was equally clear that he would never be a good Royal Society secretary. Dissatisfaction with his handling of correspondence and Royal Society records, and with the spasmodic appearance of *Philosophical Collections*, was spreading among Royal Society members, and in the annual elections on 30 November 1682 Hooke lost his position as secretary and his place on the Council, the Royal Society's ruling body. Losing one's place on the Council was no disgrace, since ten of the twenty-one lost their places every year, and Flamsteed, Tyson and Pepys were also voted off in 1682. Hooke was re-elected to the Council in 1684 and again on many occasions after that. But for the moment he was back where he started nearly twenty years earlier, a mere curator of experiments. His five years of influence in the government of the Royal Society had not been a great success, and his weaknesses as an administrator and middleman had been exposed. He was too busy and too self-centred to be a second Oldenburg. On 6 December he handed the Council books and the keys to the Society's great iron chest and its press to the new secretaries, Francis Aston and Dr Robert Plot. A month later, the secretaries were still trying to get him to hand over Society books and papers that he kept in his rooms.[4] In March 1683, in another recognition of his secretarial inefficiency, a committee was set up to go through with him the journal books he had kept as secretary, filling in omissions and altering or striking out inappropriate passages in his records. A year later, in February 1684, Hooke's minutes were ordered to be copied into books that were uniform with those kept by Oldenburg, and as late as February 1686

Edmond Halley, the Society's new clerk, was trying to straighten out the confusion the Society's papers had fallen into in Hooke's custody.[5]

In the new year even Hooke's competence as a curator was questioned. On 10 January 1683, during 'a casual discourse concerning the entries of experiments brought in by the curator', two rules introduced under Williamson's presidency were revived, in order to reimpose a more planned and comprehensible pattern on Society meetings. In future, the curator should announce the following week's experiments at every meeting, and give the secretary a written account of the purpose of every experiment, to ensure an accurate entry in the journal. These were the rules Hooke had introduced in December 1677, in his moment of triumph after Oldenburg's death, but now they seemed to be directed against him rather than against irritating amateurs.[6] Hooke continued to bring experiments to the meetings in the usual way, and had to be reminded that the new rules applied to him. In February he was instructed to bring in written accounts of his 'way of measuring the rise and fall of quicksilver in the barometer', and his 'method of explaining the cause of gravity' by tapping the bottom of a vessel containing a barely floating ball. At a Council meeting at the end of February the lack of experiments at the weekly meetings was discussed, and it was agreed that two new curators, the anatomist Tyson and the chemist Dr Frederick Slare, should be appointed, undertaking to provide the Society with at least one experiment a week between them.[7] Thanks to the King's purchase of Chelsea College the Society was now richer than it had ever been, and Tyson and Slare were offered expenses and an annual payment of plate worth £20 for their work.

Hooke was not unhappy to work alongside other paid curators. He had worked well enough with Nehemiah Grew after 1672, and he was a good friend of Tyson's. Nor did he object to providing written accounts of his talks and experiments for the Royal Society. In 1683–5 he regularly produced such accounts, which were published over twenty years after his death by William Derham. On 14 March 1683 he gave an interesting demonstration of the effects of running a fiddle bow across the side of a glass of water, in which the vibrations drew the water towards the bow, 'giving a further explanation of gravity', and discoursed at length on the qualities of different clays and building stones. But there was a deeper problem, a divergence between the mechanical interpretation of nature favoured by Hooke and a chemical approach advocated by Slare. On 31 March, after a meeting in which papers from Slare and Tyson had left no time for his own planned experiment on magnetism, Hooke wrote to Sir John Hoskins, the new President, complaining that his work was being excluded from meetings. This was more than a matter of an overcrowded schedule. In Hooke's philosophy magnetism, like everything else, was a mechanical phenomenon – 'matter in motion' – not a chemical one, as Dr Slare and his friends maintained. His experiment was intended to show that a wooden model of a compass would move when its base was tapped, because of the attractive force created by vibrations. But it seemed to him that any experiment that contradicted the chemical explanation of magnetism was unwelcome now: 'I could not be admitted the last two meetings to shew it, and when I doe I expect to have it slighted.'[8] The next week he performed the postponed experiment, but the minutes rebuked him for failing to explain its purpose. For the next two months he contributed hardly anything to Royal Society meetings, and on 6 June the Council resolved to punish him for his failure to perform. In future, they declared, he should prepare two experiments each week, giving the Society a spoken explanation in advance of their purpose, and a written account afterwards for inclusion in the journal. 'And that at the end of every quarter there shall be a meeting of the council, where his performances shall be considered, and that a gratuity ordered him accordingly; and that from this time he have no other salary'. In effect, Hooke was dismissed from the salaried curatorship he had held for almost twenty years, and reduced to the status of a freelance curator. A copy of the order was left at his lodgings, and two weeks later he was called before the Council to agree to the new conditions.[9]

Hooke's failure to provide experiments for Royal Society meetings in April and May probably had more to do with his other commitments that spring than a sulk over a postponed experiment. During the winter of 1682–3 his long-running dispute with Sir John Cutler over the latter's failure to pay his £50 a year salary for over twelve years reached a climax. Hooke had delivered very many Cutler lectures at Gresham College during the 1670s, though not always in the numbers or at the times specified in his original agreement with

Cutler. Since Cutler had not paid him since June 1670 some shortcomings on Hooke's part were understandable. For at least ten years Hooke had been asking Cutler to pay the money he owed him, and Cutler had made repeated promises to do so, often before witnesses. In July 1680 Hooke had gone to Cutler's house with two Royal Society heavyweights, Wren and Hoskins, who was a lawyer, but Cutler threatened to have Hooke arrested. Two years later the Royal Society decided to give Hooke some official support, and sent a deputation of councillors, led by Sir William Petty, to speak to Cutler about Hooke's arrears. Nothing came of this, so on 23 October 1682 Hooke's lawyer, Charles Ballet, applied at the Court of King's Bench for an order forcing Cutler to pay the £1,000 specified in their original agreement as a penalty if he defaulted on the regular annuity. Cutler opted for a jury trial, and in the meantime Sir John Hoskins and Daniel Colwall wrote to him in December 1682 on the Society's behalf, urging him to settle with Hooke out of court. They assured Cutler that Hooke had done 'much more than he is ever sayd to have been obliged to . . . at least till stoppage of Payment began', and that he had given up 'other more profitable Employments' to meet his obligations to Cutler. If Cutler would not pay up willingly, he would find that his agreement with Hooke bound him as much as if it were 'the hardest Smithfeild Bargain'.[10] The surviving version of this letter is in Hooke's hand, either because he wrote it or because he made a copy of it afterwards. On 23 January 1683, the day set for the jury trial, Cutler met Hoskins, Petty and the distinguished physician Dr Daniel Whistler, and agreed to allow them to arbitrate in the dispute.[11] As a result, Hooke received £200 – four years' arrears – as an interim payment on 25 January, and the case went into arbitration. Cutler's defence was that Hooke had not lectured as often as he should have done, and had not stuck to the agreed subject matter, the practical applications of science. Apparently, the Royal Society was impressed enough by this argument to urge Hooke to compromise with Cutler. Feeling badly let down, Hooke wrote to Sir John Hoskins, the Royal Society President, on 31 March 1683:

> I am extremly Disatisfyed with the proceedings concerning the Sir J.C. businisse, and Doe exceedingly want the Comfort & councell of a true freind who is not to be known but at such a time for

> I find that many of those I thought I might have Relyed upon are nothing soe and I am like to Reap noe benefit by my delay but rather much detriment. Besides I find others very active in indeavouring to find my faults and defame me to make way for their own designes, and whatever I have done though to the best of my abilitys are – [minus] and at best but = 0. Soe that I find I must study other principles of Algebra then what I have Hitherto been acquainted with.

What Hooke could not stomach was a new agreement binding him as he had not been bound before to lecture on specific subjects. As he wrote in a long briefing paper for Petty, this would destroy the element of trust and liberality that had inspired the original agreement: 'For that would bee a diminution to Sir Johns generosity and make that seeme a bargaine which was a guift'. Adopting an uncharacteristically uncommercial tone, Hooke said 'hee would receive an honorarium but scornes hire', and warned that if he were not trusted his performances might 'degenerate into perfunctory wall lectures'. He went on to explain why he was no longer prepared to lecture on 'the Mechanick part of trades', as Cutler evidently wanted him to:

> 1. The reputation of Sir John Mr H., the RS. & England will bee most advanc't by teaching perpetuall & universall knowledge whereas the mechanicke is mostly pro hic & nunc [for here and now] & will not as the other, bee a part of naturall history.
> 2. All tradesmen know the operative part and that cannot perfectly bee taught by words onely but the speculative & rationall part may.
>
> . . . .
>
> 4. Jealousyes will bee put in tradesmens minds in that this may prejudice them, unless some philosophicall introduction vindicate it from contempt of gentlemen & contumely of the vulgar, and this suits with promoting naturall science, the work of the RS
> 5. Mr. H. is prepared with unheard of discoveryes & demonstrations in physicks, but must compile mechanicks though already hee know them.
> 6. The first and continued understanding of the designe was to

> explaine nature first, then art and soe he intends, and fancy can never be forc't & gratifying it is possibly, what most inclines Mr Hooke to these studyes, who could gaine much more by other imployments.

Whatever Hooke might have thought in the 1660s, he did not see himself now as a teacher of applied science to tradesmen. The best way to advance the public good, he now believed, was to lay solid scientific foundations, not to teach craftsmen their crafts. Petty did his best to persuade Cutler to pay Hooke's arrears and his future annuity without imposing conditions on subject matter that Hooke could not accept. By May 1683, with no settlement achieved, Petty was getting irritated with Hooke for his refusal to compromise with Cutler, and his letters to him were frosty: 'Mr Hooke, I am sorry you cannot trust the Society in such a matter as I propound . . .' But Hooke was determined to pursue his dispute with Cutler to a conclusion, and was in no mood to follow the Society's timid advice. The case was heard at the Court of King's Bench at Westminster Hall on 16 May 1683, and six days later the jury decided that Cutler was bound to pay Hooke the bond of £1,000 that had been agreed in 1665 as a penalty for non-payment of the annuity. Cutler answered by sending a Bill of Complaint to the Court of Chancery on 15 June, in which he claimed that he had only stopped his payments because Hooke had long ago abandoned his duty to read the Cutler lectures. He argued that the lectures Hooke had given were those he had been obliged to deliver as Gresham Professor of Geometry and as the employee of the Royal Society, and that Hooke should be asked to show 'what Lectures hee read and when in particular' in fulfilment of his obligation to Cutler. Moreover, Hooke had deliberately had the £1,000 bond worded so that he could get the money 'without doeing any thing for it', and he was in a conspiracy with 'severall persons unknowne unto your Orator [Cutler] how to gett from your Orator some grate summ of money and to share and Devide the same among themselves'. Cutler claimed to have paid Hooke £1,000 over the years – more than twice the true amount – and asked to be freed from his £1,000 bond.

Hooke had his answer ready by 26 June 1683. He went back to the original agreement between Cutler and the Royal Society, in which Cutler agreed to pay £50 of his £80 Royal Society salary, leaving the Society to pay the remaining £30. His recollection of the original agreement, he said, was that Cutler's payment was an unconditional gift which he had not solicited, and which was not dependent on his giving a certain number of lectures on particular subjects, except as directed by the Royal Society. He protested again that by devoting himself to providing lectures as Cutler and the Society required he had given up studies that would have been 'farr easier and much more profitable', and that he had published several lectures under Cutler's name, though he had no contractual obligation to do so. Much of his discontent was directed against the Royal Society, which had imposed the Cutler agreement on him in the first place and drawn all the financial benefit from it, leaving him with an inadequate salary of £30 a year during the 1670s. No doubt still angry at his loss of the secretaryship in November 1682 and of his regular curator's salary in June 1683, he recalled that in the 1660s the Society had caused him 'to sitt up and watch for the most part of the night for neere a whole yeare together for makeing of diverse observations . . . which was extreamely prejudiciall to his this Defendants health and had like to have cost the defendant his life'. He offered to forgo the £1,000 bond if Cutler paid his arrears and legal expenses and agreed to pay the annuity in future.

In November the Council of the Royal Society resolved to give Hooke its full support, but the case remained in limbo until 1 February 1684, when after further arguments between lawyers the matter was settled in Hooke's favour. Hooke received his arrears and expenses of £475 three weeks later, and if he had still been keeping a diary there is little doubt that it would have given us a flavour of his sense of triumph over Cutler, and also over the Royal Society appeasers who had left him to fight his battle virtually on his own. Although Cutler did not resume his annual payments in 1684 Hooke continued to deliver Cutler lectures, many of which survive in manuscript, in the knowledge that the Chancery judgement would enable him to collect any further arrears at a later date. In future, he kept a written list of the names of his listeners, probably so that he could prove if necessary that the lectures had really been delivered.[12]

The payment-by-results arrangement imposed on Hooke by the Royal Society in June 1683 stimulated him into a period of vigorous activity. On 27 June, after ten weeks in which meetings had been dominated by Dr Slare's dramatic chemical demonstrations and Tyson's dissections, Hooke reasserted himself with some interesting experiments. One of these was a demonstration of a clever device for testing the water pressure inside a pipe, an 'instrument of great use for water-works'. A foot-long glass tube, sealed at one end, was stuck into the side of a wooden water pipe, and the extent to which the air in the glass pipe was compressed would indicate the pressure in the main pipe. The following week he explained how he had used his pressure gauge to test the strength of earthenware pipes and connecting cements, and showed the meeting how to calculate water pressure from the height of the column of water that entered the glass pipe. On 11 July he brought along a model of a new windmill of his own invention, which was lighter, stronger and more efficient than those already in use. The mill's novel characteristics were that it could stand directly on the ground without the support of a millhouse, that it turned itself into all winds without human assistance, that it captured the wind equally on every part of its vanes, that its vanes could therefore be shorter and lighter, and that it transmitted a circular or beating motion without the use of troublesome and inefficient cogwheels. Hooke had been working on windmill design since at least 1674, and had published a paper on horizontal sails in *Philosophical Collections* in 1681. His latest version, he was sure, was 'the most plain, simple, cheap, and easy to be made and used, that has been yet made; and yet the most powerful in its effects, and the most universally applicable to all purposes; (as grinding, bruising, beating, sawing, pumping, placing, twisting, drawing, turning, lifting, &c.) that can be made of equal bigness.' As far as we can tell, nothing came of Hooke's invention. The fan-tail, a set of vanes on a pole projecting from the windmill to catch the wind and rotate the mill into the best position, was not patented until 1745, and the next scientist to take the study of sail design seriously was John Smeaton in the 1750s.[13]

Hooke had to defend his methods and achievements from the challenge of rising stars like Dr Slare, who had been impressing the Society with dramatic chemical experiments since his appointment in February. Watching phosphorus, turpentine and sulphuric acid burst into flames might be more exciting than observing vibrations in a glass of water, Hooke implied in his lecture on comets and gravity, but 'one plain but pertinent Experiment, apply'd with Judgement, may be more significant than thousands of such as are pompous, amusing, and excite Admiration. And I am satisfied that more Discoveries in Nature may be made by the most plain, obvious and trivial Experiments to be everywhere met with, than by the far-fetchd and dear bought Experiments which some seek after.'[14]

Hooke gave several lectures on navigation at Gresham College in June 1683. Among other matters, he stressed the importance of preserving navigators' journals so that what they had learned would not be forgotten, and repeated the possibility that the Earth might be turnip-shaped. If he wanted to discuss these questions with someone who had actually spent some time at sea, he could not have found a better companion than Captain Robert Knox. Since his return from Ceylon, Knox had sailed to Connecticut and been given command of an East India Company ship, the *Tonqueen Merchant*, but he was in London that summer. Hooke apparently gave Knox a quadrant or telescope and a longitude clock, probably hoping for a report on their effectiveness when Knox returned from his next voyage, a slaving trip to Madagascar and St Helena. In return, it seems, Knox gave Hooke a collection of Chinese curiosities, mostly seeds and vegetables, for the Royal Society's repository.[15]

Hooke drew on the work he had done over the summer to keep the Society entertained in the autumn. On 31 October he showed them a technique he had heard of for making transparent impressions of coins and medals using fish-glue (isinglass), and a week later he showed them how the same material could be used to make paper-thin transparent plates for copying maps or pictures. He also showed them some parts of the naval waywiser he had made the previous year. The purpose of the waywiser was to enable a navigator to keep a track of a ship's location by measuring and recording the speed and direction of its journey. One component was a wind-gauge, which Hooke and his colleagues tested by walking up and down the gallery of Gresham College with the windows closed, and the other element, a device for measuring the flow of running water, was based on work Hooke had done twenty years earlier. There was still more work to be done on it, he told the Society, but when it was finished the waywiser would 'keep a true account, not only of the length of the run of the ship through the water, but the true rumb or leeward way, together with all the jackings and workings of the ship'. How well the instrument would perform in conditions that were rather breezier than the Gresham College gallery was apparently not discussed.[16]

Hooke's relations with the Society seemed to be on the mend. On 24 November, the Council decided to support him in his dispute with Cutler, and voted him £15 as part of the fee he would be paid when he provided an account of the year's experiments. Hooke went one better, and promised to write an account of all his experiments before the Royal Society, along with 'an idea of the natural philosophy built upon them'. The Council agreed to cover the cost of redoing those experiments that had not been completed or properly recorded. Three weeks later, after elections which brought two of Hooke's critics, John Flamsteed and Martin Lister, onto the Council, the mood was frostier. Hooke was called in and asked to produce the account of his recent experiments, and he promised to deliver it to the Secretary by Christmas Day. The Council urged him to make a full record of his experimental work since Oldenburg's death in 1677, and insisted that it would only pay for the redoing of his experiments if they took place before appropriate witnesses. In the middle of January 1684, his six-monthly account had still not been produced.[17]

There had often been conflict between Hooke and the Society over unfinished paperwork, but now the tension seemed to run deeper. Hooke's mechanical interpretation of scientific problems, which had been central to the Society's general approach for so long, was beginning to seem rather out of date. Criticism of his approach seems to have come mostly from Slare, the chemist, Robert Plot, a well-known naturalist and antiquary, and Martin Lister, the naturalist and physician newly arrived from York. Arguments between the two sides, and especially between Hooke and Lister, ranged over several issues. On 5 December 1683 Nehemiah Grew brought some 'figured stones' along to the meeting, and Hooke identified them as petrified shells. Lister, who was a great expert on shells, disagreed. He claimed that they were gypsum crystals (selenites), and argued that 'there was no shell-fish known' that resembled the shape of the stone. The argument continued the next week, when Hooke mentioned petrified oysters. Again, Lister maintained that there were 'but two sorts of oisters in Europe, with either of which the rock oister-shells had no similitude'.

Long ago, in *Micrographia*, Hooke had suggested that petrified shells might represent extinct species, but he did not return to that argument in this meeting.[18]

The main focus of the dispute between Hooke and Lister was magnetism, the issue that had caused Hooke trouble in March. For Lister and Slare magnetism was the result of the presence of a sulphuric chemical component – pyrites – in iron and lodestones, not, as Hooke and other mechanical philosophers believed, some sort of molecular motion. For them, all Hooke's work on vibrating glasses and wooden compasses was misguided nonsense, not worth wasting the Society's time on. In June 1683 Lister produced what he saw as a proof of the chemical nature of magnetism, the story of a ship's compass that had been struck by lightning and reversed its polarity. This, in Lister's view, happened because lightning contained pyrites, which was transferred to the compass needle. Hooke could argue, just as persuasively, that the lightning had hit the compass like a hammer, and that the change was therefore mechanical, not chemical. Neither of them knew, of course, that the effect of lightning was not chemical or mechanical, but electrical.

In December Hooke proposed that the dispute should be settled by repeating a recent Oxford Philosophical Society experiment in which the polarity of pieces of iron and lodestone were changed by hammering and drilling. The Council had other ideas. On 23 January 1684 it resolved that no mechanical experiments should be allowed in meetings 'without a finished design upon paper, as used formerly to be done', and that the curator should be given instructions in writing on what experiments to prepare for the following week. Two weeks later he tried to show the Society an experiment that involved drilling brass to see its effects on the magnetism of the brass and the steel drill bit, instead of the planned investigation of the 'magneticalness of lightning'. When it became clear that Hooke wanted to undermine Lister's interpretation of the effects of lightning, and to show that all hard bodies could be induced to show 'a quality much resembling that of the magnetical', the experiment was cancelled. It is probably a sign of the Society's dissatisfaction with his work on magnetism that on 27 February the Council asked Edmond Halley, the astronomer, to become a temporary curator of experiments, and to start by bringing in experiments on magnets.[19]

The dispute ended inconclusively in March 1684. Lister and Hooke both found that drilling did not change the polarity of various metals or of a steel bit, so Hooke shifted his ground and showed the Society a series of experiments which demonstrated that bringing a magnetized steel rod to red heat destroyed its magnetism, and that hammering steel rods would change their polarity if they were placed in the right direction while they were hammered. The state of knowledge at the time meant that it was impossible for Hooke or Lister to draw useful conclusions about the nature of magnetism from these experiments, and they let the subject drop.[20]

Hooke had his difficulties with the Royal Society, but it would be wrong to depict his scientific life in the winter of 1683–4 as one of constant conflict. He was keen to earn his fees, and to do so he presented a different weighing machine at almost every Royal Society meeting from 5 December to 23 January. The first was a light wooden model of a steelyard, a balance suspended off-centre so that a weight in one pan would balance an object ten times its weight in the other. This was an ancient principle, but Hooke adapted it so that the user, moving weights or material from one pan to the other, could repeatedly divide by ten, and thus weigh the tiniest quantities. As always, he was an enthusiastic propagandist for his own ingenuity:

> And how obvious soever it be now known, yet I do not find it hath been taken Notice of by any Writer of *Mechanicks:* nor did I ever know any that has used it, or taken Notice of it, for this Purpose; and though it may be said to be a *Stilyard*, yet 'tis as differing from the common Use of the *Stilyard*, as that is from a *common Beam* ... This proportional balance, will be of general Use and to such, particularly, where Weights are troublesome to carry and remove; and, I suppose, the only Reason why it has not been used, is, because it has not been thought of; though it were altogether as obvious, as to set an Egg on End.[21]

Hooke's second offering was a brass spring balance for finding the exact weight and specific gravity of gold and silver coin, and his third and fourth were variants of his first steelyard. Finally, on 16 and 23 January, he showed the Society a rather complicated balance for discovering the comparative weights of two objects, and two balances from Japan. Perhaps the Society found five weeks of weighing machines tedious, but Hooke enjoyed showing that there was a multitude of ways of solving a problem, and that he could handle them all. He saw that he could contribute to the progress of science by teaching the importance of accurate measurement, and to him the wheel barometers, standardized thermometers, telescopic sights, quadrants, worm gears, micrometer eyepieces, spring watches, weather clocks, spirit levels, hydrostatic balances, universal units of length and accurate scales were all part of the same campaign. One of the most important lessons scientists learned in the seventeenth century was the need for quantification based on precise measurement, and nobody taught that lesson better than Robert Hooke.[22]

There were so many bitter winters in the later seventeenth century that the period is sometimes called the Little Ice Age, and the winter of 1683–4 was the coldest and longest of them all. John Evelyn kept a record of the terrible conditions, with 'men and cattle perishing in divers places, and the very seas so locked up with ice that no vessels could stir out or come in. The fowls, fish, and birds, and all our exotic plants and greens, universally perishing'. Food and fuel were exceptionally dear, and in London coal smoke hung so thickly in the air 'that one could hardly see across the streets, and . . . one could scarcely breathe. Here was no water to be had from the pipes and engines, nor could the brewers and divers other tradespeople work, and every moment was full of disastrous accidents.' One bright spot in this hungry and disease-ridden winter was that the Thames froze over for about seven weeks, from 23 December to mid-February. This gave Londoners a useful alternative to the blocked and slippery roads, and a huge open space for games and markets. London tradesmen were not slow to spot an opportunity for profit, and by early January the Thames was 'filled with people and tents, selling all sorts of wares as in the City'. This soon became a full-blown Frost Fair, the first for 120 years, with stalls, shops and booths set out in formal streets, and coaches and sleighs running regular services from Westminster to the City. Evelyn recorded 'sliding with skates, a bull-baiting, horse and coach-races, puppet-plays, and interludes, cooks, tippling, and other lewd places, so that it seemed to be a bacchanalian triumph, or carnival on the water'.[23]

Hooke always went to the Bartholomew Fair in Smithfield, and we can be almost sure that he enjoyed the Frost Fair too, although we do not have his diary to prove it. We do know, though, that the fair led him, as a practical man of science, to take an interest in the load-bearing strength of ice. Using a machine of his own construction, he supported a fifteen-inch bar of ice three and a half inches thick and four inches wide on beams a foot apart, and found that it would not break until a 350 lb weight was placed on it. Conditions were different on the Thames, he told the Royal Society on 6 February, but the safety of the ice on the river could not be discovered until tests such as his had been done. The experiment led him on to examine the other properties of ice. He disagreed with Dr Croone's contention that water expanded before it froze, and made some trials which seemed to show that what appeared to be expansion of water as it approached freezing point was really the contraction of its glass container as it cooled. In fact water expands by 0.01 per cent between 4 degrees and 0 degrees centigrade, but for practical purposes Hooke was right in his assertion that water expanded when it solidified, not before. At the next meeting, on 13 February 1684, Hooke used a rather complicated process to establish that a lump of ice weighed seven-eighths as much as a similar volume of water, and that therefore when it floated on water an eighth of its bulk would be above the surface. The common opinion that ice sank during a thaw was thus disproved, and the fact that icebergs showed only about an eighth of their true size, allowing for variations in water density, was established.

Since ice was plentiful in London for most of February, Hooke took the opportunity to probe its mysteries a little further. On 20 February he tried to show that the 'blebs' or bubbles in ice contained ordinary air, and demonstrated that ice would float even in boiling water. The fact that water, apparently uniquely, expanded when it froze was a fascinating puzzle to him. Nobody had yet given a satisfactory explanation of the phenomenon, but he was not prepared to reveal his own, 'which some may possibly have a prejudice against'. Instead, he suggested some simple, cheap but instructive experiments, to show the power of freezing water to break a sealed container. A week later, when the thaw was setting in, Hooke had an inconclusive argument with Dr Croone over whether water expanded just before it froze, and used the weight of an iron ball in cold and hot water to demonstrate that freezing water was about 3.5 per cent heavier than boiling water. Since the weight of a specific volume of water was later to become the basis of standardized weights in Europe these findings were significant, if not quite accurate.[24] For Dr Croone, a colleague of Hooke's since the 1660s, this was a final dispute. He died of fever in October 1684, in his early fifties.

22. Newton's Triumph

(1684–1686)

HOWEVER HARD HE TRIED, Hooke's position in the Royal Society was seriously weakened. He was now one of four curators (along with Halley, Slare and Tyson), and he had to earn his money like a tradesman, by giving an account of the work he had done. On 2 April 1684 Denis Papin, back in London after three years in Venice, was employed as a temporary curator at £30 a year, Hooke's old salary, with the task of providing weekly experiments for the Society. For the next three years Papin played the part that Hooke had once played, entertaining the Society almost every week with an air pump or his pressure cooker or digester. Two payments Hooke received in April 1684 for the work he had done before and after Christmas, amounting to £17 10s, had the air of a final settlement, an unspoken dismissal. But Hooke did not regard his employment as over, and continued to read papers to the Society, though less frequently than before.[1]

Even with such a plethora of curators, Royal Society meetings in the spring and early summer of 1684 were less well served than they had been in Hooke's inventive heyday. Papin weighed iron in a vacuum and stuffed a rabbit with plaster of Paris, Tyson dissected Thames snails, Lister examined the theory that blackamoors were black because their blood was black, and Hooke explained the working of his weather clock and gave a paper on long-distance communication. He had shown the Society how to send messages across the Thames in 1672, but the idea had been given a contemporary relevance by the great siege of Vienna in 1683. If Vienna had not been relieved by the Polish King, Jan Sobieski, the Habsburg Empire and perhaps the whole of Christendom would have been exposed to invasion by the Ottoman Turks. In case of another such emergency, Hooke reminded the Society, a chain of signallers standing on hills or high buildings, using telescopes and huge black letters or symbols representing basic phrases, could transmit vital messages over hundreds of miles 'with certainty, security, and expedition', on condition (though he ignored this problem) that the enemy did not shoot them.

The density of water, accurate weighing machines, and even long-distance signalling were all fascinating subjects, and typical of the wide range of Hooke's interests over the years. But none of these was the great question of the age, and answering them was not the way to achieve scientific greatness. This great question was celestial mechanics, the explanation of Kepler's laws of heavenly motion in mathematical terms. We do not know whether Hooke was working on this problem in the bitter winter of 1683–4, but we know that he was encouraged to do so. One Wednesday in January 1684 he was having a conversation with two of his closest friends, Sir Christopher Wren and Edmond Halley, and the discussion turned to the inverse square law and planetary motion. We only have Halley's account of the conversation, as he told it to Newton in June 1686, in an attempt to reassure and pacify him in the heat of the Hooke–Newton dispute of that year. Halley recollected that Hooke claimed that he had worked out the laws of celestial motion from the inverse square law, something that Halley had not managed to do. His letter continued:

> Sir Christopher to encourage the Inquiry sd, that he would give Mr Hook or me 2 months time to bring him a convincing demonstration thereof, and besides the honour, he of us that did it, should have from him a present of a book of 40s. Mr Hook then sd that he had it, but that he would conceale it for some time that others triing and failing, might know how to value it, when he should make it publick; however I remember that Sr Christopher was little satisfied that he could do it, and tho Mr Hook then promised to show it him, I do not yet find that in any particular he has been as good as his word.[2]

There is no evidence that Hooke took up Wren's challenge and went to work on celestial motion in early 1684. Even after a truce had been called in the magnetism dispute in March 1684 his contributions to Royal Society meetings generally referred to work he had done long ago on barometers, tubeless telescopes and the weather clock. In July he produced a series of three discourses on Roman tombs and the style of rowing used in ancient galleys, and in the autumn he lectured at Gresham College on astronomy and the art of navigation. These lectures dealt with modern mapping techniques, the magnetic poles and the shape of the Earth, and would have given him an ideal opportunity to reveal his latest thinking on planetary motion if he had wanted to do so.

Of course, Hooke was not inactive in these months. He was probably still involved in the building of Escot House, and he seems to have designed a house in Spring Gardens, off Whitehall, for his friend Sir Robert Southwell: Southwell paid him a fee of five guineas, probably for this design, on 17 June 1684. Later that year he designed some almshouses in Buntingford, Hertfordshire, for Seth Ward, the Bishop of Salisbury. Ward's Hospital, a two-storey brick building on three sides of an open courtyard, was built at the end of the decade, and still survives.[3]

About seven months after his conversation with Hooke and Wren, Halley went to see Newton in Cambridge, to ask him what he had asked Hooke and Wren: under the inverse square law, what shape would the orbits of planets around the Sun be? Like Hooke, Newton bluffed, and told Halley that he had mathematical proof that the answer was an ellipse, but that he had lost the calculation. But unlike Hooke he then dropped almost everything else and devoted himself for two years to discovering the answer to the problem. No social life, no coffee-house company, no professional commitments other than a few lectures, no competing scientific or technical interests diverted Newton from the task in hand. His income of about £300 a year, half from his Cambridge posts and the rest from a small landed estate, relieved him of the need to make a profit from his scientific work. Even if Hooke had possessed the intellect to do what Newton did, he certainly lacked the temperament to devote himself single-mindedly, even obsessively, to one task for two years.[4]

The first fruit of Newton's efforts was a short treatise called *De Motu Corporum in Gyrum* (*On the Motion of Bodies in Orbit*), which he sent to Halley in November 1684. This was a brief and incomplete sketch of the far more complex and thoroughly reasoned ideas he presented in his *Principia* in 1686. In *De motu* he introduced the term 'centripetal force' to describe an attraction towards the centre, and showed that orbital motion was an interaction between the centripetal force of the Sun and the inherent rectilinear motion of a planet – its tendency, along with every other moving body, to continue moving in a straight line unless impeded. These ideas had been present in Hooke's Royal Society paper of May 1666 and his letter to Newton on 24 November 1679, and there is no evidence that Newton had considered this way of explaining planetary orbits before he received this letter from Hooke. Imagining an orbit as a sequence of infinitely small triangles, Newton showed that the areas swept out by a planet's orbit in equal periods of time were equal (Kepler's first law) when centripetal force obeyed the inverse square law, and that such an orbit could be elliptical.

Almost as soon as he received *De motu* Halley returned to Cambridge to consult Newton, and then, on 10 December, he announced its existence to the Royal Society. It is possible that after this Hooke was given a chance to read the tract. This is the implication of a letter written to Newton by Flamsteed, who told Newton on 27 December that he looked forward to reading Newton's papers, 'tho I beleive I shall not get a sight of them till our common freind Mr Hooke & the rest of the towne have beene first satisfied'. But Flamsteed is not a very good source for Hooke's activities, and the fact that *De Motu* was not discussed in the Royal Society suggests that its contents were not generally known. *De Motu* was not a finished mathematical solution to the problem of orbital motion, but it was a clear signal that Newton was on the way to finding one. If Hooke had worked out his own solution to the problem, as he told Halley and Wren he had, it is almost certain that he would have rushed to announce it after hearing of *De Motu*, and realizing that Newton might beat him to the prize. As Halley told Newton in his letter of 29 June 1686, 'according to the philosophically ambitious temper he is of, he would, had he been master of a like demonstration, no longer have concealed it, the reason he told Sr Christopher and I now ceasing'.[5]

In 1685, while Newton worked tirelessly to turn the ideas introduced in *De Motu* into his great masterpiece, *Principia*, Hooke's mind was apparently on other things. Perhaps he recognized that the mathematics needed to produce the proof required by Wren and Halley were beyond him, and believed that his earlier work on orbital motion and the inverse square law would ensure him a fair share of the credit when Newton's work was complete. Hooke's most important work in the spring of 1685 was a long paper, read to the Royal Society in February and early March, on land and water vehicles and the problem of friction. He was responding to reports from Oxford of an experimental cart with five-inch wheels, but the main subject of his discourse was the very efficient wheeled sailing chariot made by Simon Stevin for Prince Maurice of Orange early in the seventeenth century. Stevin's chariot had carried a load of twenty-eight passengers forty-two miles in two hours, and seemed to be 'the swiftest Carriage yet known, for so great a Burthen, and for so long a way'. Hooke discussed the stability of the vehicle and its method of steering, and suggested that a similar sailing carriage could be developed for use on Salisbury Plain, Banstead Downs, and other flat and windy districts.

Hooke went on to consider ways of making a carriage that would far outrun Stevin's, and examined the problem of friction in wheeled vehicles. There was nobody in England better qualified to deal with this subject. He had practical experience of the problems of transporting heavy loads through London streets, as well as of constructing the streets and coaches themselves, a great knowledge of mechanics and motion, and an intuitive understanding of mechanical problems. In a highly sophisticated analysis, he identified the two main causes of friction between a revolving wheel and the surface on which it rolls: 'The first and chiefest, is the yielding, or opening of that Floor, by the Weight of the Wheel so rolling and pressing; and the second, is the sticking and adhering of the Parts of it to the Wheel'. A hard surface, he argued, created least friction, and a springy one only a little more. But soft and sinking surfaces which do not resist the weight of the wheel create the most serious loss of forward motion. A bumpy surface, though it causes discomfort, does not cause serious loss of motion, because whatever is lost in ascending a bump is regained in descending it. The second cause of friction, the adhesion of the wheel to the ground, is worse on sticky clay surfaces, and when the wheel is broad-rimmed. 'For in such Ways, the Wheel doth not only lose a Part of its Motion, by the yielding and pressing of the Clay against the fore Parts of the Wheel, but by the cleaving to, and holding of it to the hinder Parts, which makes all Carriages move very sluggishly and heavily in such Ways.' The second issue that he looked at was the friction between the wheel and its axle. He recommended wheels that were fixed to the axle, whose ends were fitted with metal pivots or gudgeons which revolved in oiled metal sockets:

> The less rubbing there be of the Axle, the better it is for this Effect; upon which account, Steel Axes, and Bell-Metal Sockets, are much better than Wood, clamped, or shod with Iron; and Gudgeons of hardened Steel, running in Bell-Metal Sockets, yet much better, if there be Provision made to keep out Dust and Dirt, and constantly to supply and feed them with Oil, to keep them from eating one another; but the best Way of all, is to make the Gudgeons run on large Truckles, which wholly prevents gnawing, rubbing, and fretting.

Bell metal is a copper–tin alloy with more tin than is usual in bronze, and truckles are rolling wheels or bearings, which were not used in road vehicles until the eighteenth century.

Finally, Hooke turned to the design of wheels, and argued that large narrow wheels, so long as they could be made light and strong, were best for fast travel, because they would sink into the ground less and ride over bumps and holes more smoothly than small ones, and create less friction and adhesion than wide ones. His essay on chariots, obvious as much of it seems today, placed the study of friction in vehicle and road design on a scientific footing for the first time. It appeared at an important time, when coach services were spreading all over England, but when roads had fallen into an appalling condition. Hooke's advice, that hard roads were more important than smooth ones, and that it was more efficient for wheels to have a large diameter than a wide rim, was sound, but generally ignored. Governments continued to insist on wide-rimmed wheels, and coachmen who wanted to drive on hard, well-drained roads instead of toiling through rutted clay had to wait for the work of John Metcalfe and John McAdam in the later eighteenth century and afterwards. But if the ideas presented to it did not spread beyond Gresham College this was the Royal Society's fault, not Hooke's.[6]

Perhaps this was not a good time to suggest new forms of transport based on scientific principles. In December 1684, only two months before Hooke's paper, a double-keeled (catamaran) sailing ship, built to a design Sir William Petty had been working on for years, was given a trial in Dublin harbour. William Molyneux sent a report of the ship's utter failure to the Royal Society: 'she performed so abominably, as if built on purpose to disappoint in the highest degree every particular, that was expected from her; . . . the seamen swear they would not venture over the bar in her for 1000 pounds a piece. Even right before the wind she does nothing. So that the whole design is blown up. What measures Sir William will take to redeem his credit, I know not'. Petty had designed several successful catamarans, including one that had sailed from Dublin to Holyhead, but such embarrassing failures were long remembered and laughed over, and made it all the more difficult for Hooke's more practical schemes to cross the gulf that divided scientific exposition from practical application.[7]

One of those who had gained the greatest amusement from the Royal Society's more impractical investigations, Charles II, died earlier in February, at the age of fifty-four. He had suffered an apoplectic fit – a stroke – on 2 February, and was immediately attended by Dr King and other eminent physicians. Being a man of robust constitution, he managed to survive their treatment, a combination of bleeding and purges, for four days before giving up the ghost.

Despite the doubts over his status as a curator of experiments, Hooke continued to play a full part in Royal Society meetings in 1685. From December 1684 to the end of November 1685 he was on the Council, and though the leading role in providing weekly experiments fell to Denis Papin, assisted by the operator, Henry Hunt, Hooke contributed comments or papers to many meetings, on his usual range of subjects. In May he explained his way of drawing a huge circle – big enough for the base of the dome of St Paul's – using a small truckle wheel fixed into a thin ruler or brass pipe on a hundred-foot wire. He reminded his audience that many inventions that had changed the world – the telescope, the compass, movable type, the pendulum – had seemed trivial or useless at first, but should not therefore be ridiculed. In June he lectured the Society about the causes of a mysterious 'glade of light' that sometimes appeared in the sky after sunset in February and March, and suggested that it might be light reflected from an atmospheric or vaporous trail, like the wake of a ship, left when the Earth was moving at its fastest because of its closeness to the Sun. As usual, he reminded his listeners of some of his 'big ideas', like the changing distribution of sea and dry land on the surface of the Earth, and the theory that planetary motion was a combination of direct linear motion and solar attraction, 'which I long since have explained and shewn to this Society'.[8] In July 1685 he spoke to the Society on Chinese, a language which attracted him because of its symbolic nature, and made models of Chinese and Roman abacuses to compare two ancient methods of calculation. This lecture was a sign of the growing interest in the ancient past that was to characterize the last years of his career.

One of the biggest ideas of all, for Hooke, was still the possibility of establishing a ship's position at sea by finding its longitude, and he gave a series of lectures on this subject in Gresham College in June and November 1685. Although he had produced excellent spring-regulated watches, he no longer seemed to believe that accurate timekeeping at sea was the best way to establish longitude. Observations of Jupiter's moons using good instruments and improved astronomical tables were the best way, and he announced that he intended 'shortly to publish' a new theory of celestial motions which would help to make these tables good enough for the purpose. In the meantime there was a second method, the careful measurement of a ship's route around a great circle along a series of rhumbs, for which Hooke had devised a nautical waywiser and an instrument for drawing accurate rhumbs on flat maps. The instrument was a variant of his device for drawing huge circles, an adjustable ruler with a truckle wheel set into its end, sharp enough to mark a line on a map when the ruler was pivoted from a centre pin.[9]

During the summer of 1685 Hooke worked on his own theory of celestial motion, probably hoping to produce something before Newton published *De Motu* or a fuller work on the subject. The result of his efforts was a handwritten paper entitled 'The Laws of Circular Motion', and dated 1 September 1685, in which he discussed the velocity of falling bodies, the relationship between the inverse square law and elliptical orbits, and his long-standing idea that circular orbits were a compound of linear motion and central attraction. But to the crucial questions of why the inverse square law produces elliptical orbits, and why planets sweep out equal areas in equal times, he could not supply the reasoned explanations that Newton provided. It is possible (as Flamsteed thought) that Hooke had read *De Motu* before writing this paper, but if so he had not learned very much from it.[10]

Hooke's new theory of celestial motions was never published. In February 1686, while Newton was working feverishly to complete the first book of *Philosophiæ Naturalis Principia Mathematica*, he was explaining to the Royal Society a better way of making barometers, advocating a tiny drop of mercury as the basis of a standard system of weights and measures, and going over his old dispute with Hevelius about telescopic sights. In March, he demonstrated a way of levelling a small instrument with a triangular pendulum, produced at the Society's request a full account of all the work he had done on navigation and the study of the sea, and continued his work on Chinese characters. His efforts were not unappreciated. In March the Council discussed a long-term programme of experiments with Hooke, and in April William Molyneux of the Dublin Society, the recipient of several of Flamsteed's diatribes against him, wrote to the Society in praise of his continuing ingenuity: 'Indeed I have always had a great esteem of his mechanical inventions, of which I look upon him to be as great a master as any in the world; and that is a most curious part of philosophy, and really useful in man's life.'[11]

On the day this letter was read out, 21 April 1686, Halley, the new clerk of the Royal Society and Newton's closest friend in the Society, announced that Newton's treatise on celestial motion was nearly ready for publication. The following week the manuscript of Book I of *Philosophiæ Naturalis Principia Mathematica*, which explained 'all the phænomena of the celestial motions by the only supposition of a gravitation towards the center of the sun decreasing as the squares of the distances therefrom reciprocating' – the inverse square law – was presented to the Society. Just as in 1672, Newton's wonderful work landed in the humdrum Royal Society world of barometers, quadrants, monstrous infants and pressure-cooked chocolate like a bombshell. We cannot know quite how Hooke reacted when he realized that the great theory of celestial motion which he had expected eventually to complete had been produced by his old rival Newton. No doubt he felt some anger and perhaps humiliation that he had been beaten to the greatest prize, but he also felt a sense of injustice. He was sure that Newton had got the inverse square law, the keystone of his new system, from him, in the letter he had sent him on

6 January 1680. If Newton had known it already (which in fact he did), why had he not mentioned the rule in his two letters of December 1679, both dealing with the question of falling bodies? Hooke has been mocked as the universal claimant, and no doubt he knew that others thought this of him, but he was sure of his ground, and he could not bear to see Newton so acclaimed without some notice being taken of whose shoulders he was standing on.

There is no record of Hooke's reaction in the Royal Society minutes, but we know roughly what happened from a later letter of Halley's to Newton. When Book I of Newton's *Principia* was presented to the Royal Society on 28 April 1686 the chairman of the meeting, Sir John Hoskins, joined the general praise of the book by declaring that 'it was so much more to be prized, for that it was both Invented and perfected at the same time'. Hooke was angry that his own earlier work was dismissed by Hoskins in this way, and 'they two, who till then were the most inseparable cronies, have since scarce seen one another, and are utterly fallen out'. After the meeting the members went to a coffee house, where Hooke tried to persuade them that he had given Newton 'the first hint of this invention'. But the universal view, Halley said, was that since there was nothing of Hooke's work on the matter in print or in the Society's journals Newton's claims could not be challenged, and that 'if in truth he knew it before [Newton], he ought not to blame any but himself, for having taken no more care to secure a discovery, which he puts so much Value on'.[12] In fact Hooke had summarized the main points of his theory, the inverse square law and planetary orbits as a compound of linear motion and central attraction, in his lectures on light, delivered to the Royal Society in the autumn of 1682, and referred to them frequently, in his usual style, since then. But he must have known, after thirty years, that his lectures to the Royal Society were too easily forgotten and too briefly minuted (even when he was secretary) to count as a form of publication.

Halley was a good friend of Hooke's, and had sat up with him a few weeks earlier watching the Moon's eclipse of Jupiter. But more than anyone else Halley recognized the genius of Newton's work, and regarded it as his duty to coax Newton into publishing his great work in full. To prevent the news of Hooke's claims reaching Newton from a less friendly source, Halley wrote to him on 22 May 1686:

> There is one thing more that I ought to informe you of, viz, that Mr Hook has some pretensions upon the invention of ye rule of the decrease of Gravity, being reciprocally as the squares of the distances from the Center. He sais you had the notion from him, though he owns the Demonstration of the Curves generated thereby to be wholly your own; how much of this is so, you know best, as likewise what you have to do in this matter, only Mr Hook seems to expect you should make some mention of him, in the preface, which, it is possible, you may see reason to præfix. I must beg your pardon that it is I, that send you this account, but I thought it my duty to let you know it, that you so may act accordingly; being in myself fully satisfied, that nothing but the greatest Candour imaginable, is to be expected from a person, who of all men has the least need to borrow reputation.[13]

Newton replied to Halley within a week, apparently in a conciliatory mood. He assured Halley that where Hooke had played a part in advancing understanding, as in the matter of planetary motion, his name would be mentioned, but that nothing Hooke had done, and nothing in his letters of 1679–80, had contributed to the work he had presented in Book I. If this was untrue, he said, 'I desire Mr Hooke would help my memory'. Newton recalled from a conversation in 1677 that Wren had known the inverse square law then, before Hooke, and asked Halley to find out when Wren had first learned of the law.[14] Three weeks later, he wrote to Halley again, at much greater length, and much angrier than he had been at first. He went through the 1679–80 correspondence in more detail, recalling (wrongly) that Hooke had erroneously claimed that the inverse square law operated right to the centre of the Earth, and he claimed that in a long paper on light sent to Oldenburg and the Royal Society in December 1675 he had made remarks about planetary and solar attraction which had implied the inverse square law, 'tho for brevities sake not there exprest'. In fact this obscure reference cannot bear the weight Newton asks it to carry. Newton's strongest argument was that Hooke had only guessed the right answer, just as Kepler had guessed at the ellipse, whereas Newton had demonstrated that both were true:

> without my Demonstrations, to wch Mr Hook is yet a stranger, it cannot be beleived by a judicious Philosopher to be any where

> accurate. And so in stating this business I do pretend to have done as much for ye proportion [the inverse square law] as for ye Ellipsis & have as much right to ye one from Mr Hook & all men as to ye other from Kepler. And therefore on this account also he must moderate his pretences.[15]

Before sending this letter to Halley, Newton received a much more vivid description of Hooke's reaction to his book from his friend Edward Paget, a fellow of Trinity (Newton's college) and of the Royal Society. Paget was probably hostile to Hooke. In 1682, when the incompetent Master of Christ's Hospital Mathematical School, Robert Wood, was removed from his post, Paget had been chosen on the recommendation of Newton and Flamsteed, in preference to Hooke's candidate. He had turned out to be a drunken absentee, and might well have clashed with Hooke, a governor of Christ's Hospital, over his conduct. According to Paget's account, Hooke had made a 'great stir pretending [Newton] had it all from him & desiring they would see that he had justice done him'. Newton's response to this fresh news of Hooke's complaints, in a postscript to his letter to Halley of 20 June 1686, was ferocious and powerful. First, he told Halley, Hooke had not originated the explanation of planetary motion that he claimed as his own. Giovanni Borelli's *Theory of the Planets*, published in 1666, had suggested that planetary orbits were produced by a balance between attraction and centrifugal force. Newton had used a similar argument, that Hooke was not the originator of his own ideas, in the dispute over light waves, but his point was not a fair one. Hooke's theory of planetary motion had certainly not been plucked from thin air, and drew openly from the work of earlier scientists, especially William Gilbert, and from his work with Wren on comets and falling bodies. His theory suggested a compound of straight motion and attraction, not centrifugal force and attraction, and he announced it in a lecture to the Royal Society in May 1666, before the publication of Borelli's work, and certainly before it was available in England.

Newton's second point was stronger, and perfectly identified the difference between Hooke the clever speculator and Newton the brilliant mathematician. The difficult task, and the one that deserved greatest credit, was not thinking of an idea, but proving that it was true. Hooke 'has done nothing & yet written in such a way as if he knew & had sufficiently hinted all but what remained to be determined by the drudgery of calculations & observations, excusing himself from that labour by reason of his other business: whereas he should rather have excused himself by reason of his inability. For tis plain by his words he knew not how to go about it.'

> Now is this not very fine? Mathematicians that find out, settle & do all the business must content themselves with being nothing but dry calculators & drudges & another that does nothing but pretend & grasp at all things must carry away all the invention as well of those that were to follow him as of those that went before. Much after this same manner were his letters writ to me . . . And upon this information I must now acknowledge in print I had all from him & so did nothing my self but drudge in calculating demonstrating & writing upon ye inventions of this great man.

Newton went on to recall with bitterness the correspondence of 1679–80, in which Hooke had dared to correct him on two or three points, and to tell him of the inverse square law as if he did not already know it. In fact, Newton said, the French mathematician Bullialdus had already revealed the rule in 1645, and a letter from Newton to Huygens, sent via Oldenburg in June 1673, had (Newton wrongly recalled) pointed this out. Hooke had possession of Oldenburg's letters when he was secretary, and if he had taken the inverse square law from this letter 'what he wrote to me afterwards about ye rate of gravity, might be nothing but ye fruit of my own Garden'. Hooke certainly went through Oldenburg's letters, but he did not find the inverse square law in them. Nor did he find it in Newton's long 'Hypothesis on Light', which he sent to Oldenburg on 7 December 1675. In this paper, Newton recalled, there was a discussion of the causes of gravity in which the inverse square law was almost explicit. Once again, Newton's memory led him astray: there is nothing in the 'Hypothesis on Light', which Hooke certainly read in the Royal Society Register Book, that can claim to be an implicit statement of the inverse square law. More justly, Newton argued that Huygens had revealed the inverse square law for centrifugal force in 1673, and that any competent mathematician could have worked out a similar law for gravity from that. But Hooke, Newton said, had taken five years to work it out, added a mistaken extension of the law to the centre of the Earth, and had the effrontery to claim the invention for himself. 'And why should I record a man for an Invention who founds his claims upon an error therein & on that score gives me trouble?'[16]

In his fury, Newton mixed injustice with justice. In his letter of 6 January 1680 Hooke had devoted several sentences to explaining that he did not believe the inverse square law operated to the very centre of the Earth – 'on the Contrary I rather Conceive that the more a body approaches the Center, the lesse will it be Urged by the attraction' – but only that mathematical calculations about 'attraction at a Considerable Distance may be computed according to the former proportion as from the very Center'. There was no error here for Newton to crow over. The simple fact was that Newton was as jealous of his priority over his inventions as Hooke was, and he was apt to deny his debts to others, especially when they laid claim to them. Hooke had been the first to publish the inverse square law (though certainly not the first to discover or prove it), and his theory of planetary motion as a compound of linear motion and solar attraction had helped to shape Newton's own thinking. Newton had done the lion's share of the work, and done so with a mathematical genius that was far beyond Hooke's reach, but Hooke had made a significant contribution, and a less tortured character than Newton would readily have acknowledged it.

Halley was the Royal Society's clerk, and although Newton's powerful attack on Hooke was not read out in a meeting it was lodged, either then or later, in the Society's records. Hooke certainly saw the letter at some time, because he made a copy of it which is now among Newton's papers. Perhaps Halley, who acted honestly and without malice throughout the correspondence, showed it to him. Halley also did as Newton requested, and asked Wren whether and if so when he had learned the inverse square law from Hooke. In a letter of 29 June he told Newton the story of the prize Wren had offered in 1684 if Hooke could prove the inverse square law, of Hooke's failure to claim it, and of the circumstances which had induced Hooke to complain that his work on gravity had been ignored. Halley told Newton that Hooke's claim had been 'represented in worse colours than it ought', and urged him not to withhold the third book of *Principia*, as he had threatened to do, just because of his anger at one discordant voice among the general acclaim for his work.[17]

Newton was mollified by what Halley told him about the meeting of 28 April, of Hoskins' remark and the causes of Hooke's complaint: 'Now I understand he was in some respects misrepresented to me I wish I had spared ye Postscript in my last', he wrote on 14 July. Newton went through Hooke's claims again, accepting only that he owed Hooke the idea that a falling body in England is deflected to the south-east, and that Hooke's letters had prompted him to begin his search for the mathematics of celestial motion. Conceding these minor points, Newton wanted the dispute at an end, and the burden of Hooke's letters to be lifted from his shoulders.[18]

The outcome of Hooke's protests was that his work received less acknowledgement than it would otherwise have done in *Principia*. In Book III, which Hooke had not yet seen, his name was mentioned in the manuscript, along with Borelli and 'others of our nation', as one of the originators of the principle of attraction in planetary motion, but before publication Newton struck the reference out. In other places, where Hooke's name was mentioned in relation to the observation of comets, his name was reduced from *Clarissimus Hookius*, 'the very distinguished Hooke', to plain *Hookius*. After Halley's reassuring letter of 29 June Newton added a revision to his account of the discovery of the inverse square law, accepting that 'our countrymen Wren, Halley & Hooke' had all arrived at it independently, and demonstrating that any good mathematician could have derived it from Huygens's centrifugal force formula.[19]

Hooke lost the battle for recognition as the true discoverer of the theory of universal gravitation and the laws of celestial motion, not because of the force of Newton's letters to Halley, but because nobody in the coffee house on 28 April 1686 believed him. He had claimed authorship of other people's work once too often, and it was plain even to members of the Royal Society that Newton's *Principia* was a work far beyond his capability. Even Wren, who knew that there was an element of justice in his complaints, did not (as far as we know) speak publicly on his behalf. But Hooke, understanding the importance of Newton's work more than almost anyone else, could not let the matter drop. For years after the publication of the whole of

Principia in the summer of 1687, he continued to press his charge of plagiarism against Newton, in letters, lectures and coffee-house conversations. His sense of injustice and betrayal may have clouded the last years of his life. The dispute between the two men was never resolved, and it was not until after Hooke's death in 1703 that Newton felt able to attend Royal Society meetings regularly and to accept the Society's presidency.

23. The World Turned Upside Down

(1687–1688)

ALTHOUGH HOOKE'S DISPUTE with Newton damaged his reputation in the long run, it does not appear to have done immediate harm to his position within the Royal Society. In June 1686 the Council agreed to give him £60 as a salary for the previous two years, a payment which he eventually received in 1688. At the end of November he was elected onto the Council after a year's absence, and for the rest of 1686 he continued to contribute comments and papers to Society meetings, on his usual eclectic range of subjects – magnetic variation, seamless shirts, Chinese lacquer, petrified shells, better barometers and Arctic tides. Hooke's friendship with Sir John Hoskins seems to have been repaired, too, since they were amiably discussing errors in the mapping of China and the East Indies in January 1687. It was not Hooke's position that was in question, but that of the Society's clerk, Edmond Halley, who had acted as Newton's midwife during the difficult labour that eventually produced *Principia*. At a Council meeting of 5 January 1687 it was ordered that Hooke and others inspect the Society's books, to make sure that Halley was keeping them as he should. At the same meeting Hooke was invited to submit a written proposal for a new agreement between himself and the Society, and a week later it was agreed, 'after much debate', that his salary should be made up to £100 a year, with £50 coming from the Society and the rest being extracted, with the Society's help, from Sir John Cutler. In return for this substantial increase it was agreed that Hooke would produce one or two experiments a week, with a written discourse, and 'that the said experiments should proceed in a natural method'.[1]

This agreement was especially generous in view of the fact that the Royal Society was once again in a difficult financial position. In 1686, partly because of the influence of Martin Lister, it had published a lavishly illustrated and costly edition of Francis Willughby's *History of Fishes*. Too many copies were printed, and the Society suffered such a severe financial loss that it was unable to underwrite the cost of publishing Newton's *Principia* later that year. The worst sufferer from this ludicrous situation was Halley, who was told to publish *Principia* at his own expense. In February 1687 the Council asked Hooke to use his contacts in the Amsterdam book trade to exchange 400 copies of the book for £200 in cash and £300 worth of books, but the Society still had far more copies than it could sell. In July 1687, when the question of paying Hooke's and Halley's salaries came up, both were told to take fifty copies of the *History of Fishes*, which were hardly worth a shilling each. Halley had to accept, but Hooke, with thirty years' experience of dealing with his miserly employer, played for time, and in November the Society agreed that he should get his £60 arrears for 1684 and 1685 in cash. The £50 promised in January 1687 was apparently never paid.[2] It is possible that Halley's mistreatment by the Society in 1687, at a time when he deserved so much better, was inspired by Hooke and his friends in the Society, to punish him for his championship of Newton. But there is no evidence in Halley's letters that this was so, and the Society had never needed special reasons for treating its employees meanly. Hooke remained on good terms with Halley, and probably did not have enough support to turn the Society against him, even if he had wanted to.[3]

As if the public humiliation of Newton's triumph was not enough for him to bear, Hooke suffered a terrible personal disaster early in 1687. His niece Grace, who had lived with him in Gresham College since 1672, and been his mistress since 1676, died in February, aged twenty-six. She was buried in St Helen Bishopsgate, 'in the South Ile over against the first Piller from the Quier', on 28 February. We have no idea of the cause of her death, but it is likely that she fell victim to one of the many infectious diseases that were prevalent in late seventeenth-century London, perhaps tuberculosis, typhus, diphtheria or influenza. In any event, it seems that Hooke, who had cared for Grace since she was eleven, was deeply affected by her death. Richard Waller, his friend and biographer, thought that Hooke's grief at her loss 'hardly ever wore off', and that from the time of her death he grew 'less active, more Melancholly and Cynical'. We know from his later diaries that Hooke did not lead a solitary existence after Grace's death, and that he continued to live the life of intellectual endeavour and coffee-house company that he had enjoyed when she was alive. But he could no longer go home to the domestic comforts that women in those days were expected to provide, or count on a partner's support as he slipped into old age. Still, one should not over-sentimentalize. Early death was common in the seventeenth century, and Hooke was far from alone in losing his loved ones before their time. Sir Christopher Wren lost two wives and a son in the 1670s, Lodwick lost three sons, John Oliver's son killed himself in 1689, and tragic John Evelyn endured the deaths of all six of his sons and two of his three daughters.

Hooke's favourable deal with the Royal Society in January 1687 was struck in the middle of a lively series of lectures which showed how valuable his fertile intellect was to the Society. He returned to his old argument that shell-shaped stones were truly petrified shells, rather than nature's tricks, and that they carried messages about the age and history of the planet, as surely as buried coins were clues to human history. His work as an architect and builder must have often brought such stones to his attention. Even in St James's Fields, the land on which he and others had built St James's Square, he had found a layer of sea sand containing shells and old bones at the bottom of brickmakers' wells. On 27 October 1686, and again the following week, he showed the Society the impression of a two-foot nautilus shell, bigger than any such creature then living, that had been found in a Portland stone quarry.[4] Perhaps encouraged by the interest in his nautilus, Hooke prepared a course of lectures on fossils and their origins, which he delivered to the Society between December 1686 and March 1687.

Hooke introduced his lectures as an answer to those people who asked 'what the Royal Society hath done for so many years as they have met'. He knew that the popular answer to this familiar question was 'just nothing', and he also knew that his own career had been the object of particular derision. Leaving his personal reputation to one side for the moment, he said, the Society's best response was to show how they had gathered information and drawn general axioms from it, just as they had promised to do. Before he took up his main subject, he discussed the two alternative ways of pursuing scientific truth. One, the 'synthetick' method, proceeding from effects to causes, or from facts to general theories, was the way usually favoured by the Royal Society. The other, the 'analytick' method, started with a theory, and tested it by observation and experiment, proceeding from causes to effects. This might not be the 'proper' way to do things, Hooke argued, but it had the great advantage of speed. An example of this method, he took the opportunity to mention, were his own lectures, given to the Royal Society 'some years since', on the motions of celestial bodies. 'I understand the same thing will now be shortly done by Mr Newton in a Treatise of his now in the Press'.[5]

A second example of the analytic method, he told the Society, was his own theory to explain the discovery of the petrified remains of sea creatures on dry land, including some that no longer existed as living animals. He repeated what he had said in 1667–8, that fossils should be seen as evidence for the history of the Earth, especially in the days before the invention of writing. As for the scientist (perhaps he meant Martin Lister or Robert Plot) who still denied the organic origins of fossils, Hooke wondered what would persuade him. 'I would willingly know what kind of Proof will satisfy such his doubt, and by what Indications or Characteristicks he will know a Shell of an unknown Species (for such may be shewn him) when it shall be presented to him'.[6] With minds open to the evidence and ready to embrace such ideas as the changeability of species, Hooke was sure that much that seemed mysterious about the origins of the Earth would soon be understood. His own hypothesis was that the Earth had once been covered in water, not in a Flood that lasted less than a year, but for much longer. It was also possible, he argued in January 1687, that the axis on which the Earth rotated had shifted over time, changing the climatic zones and explaining the discovery of apparently tropical species in England. To explain how this shift in the Earth's axis – an idea that was original to Hooke – could have caused great movements in land and water, he had to call upon another of his ideas, the probability that the Earth was an oval or flattened sphere which was at its widest at the equator. If the equator shifted, then this equatorial swelling would shift too, bringing about some of the great upheavals or earthquakes that had given the planet its present ragged shape. The fact that we had 'no Histories or Records that have preserved the Memory of them' did not matter, because the cataclysms predated written history, and the shape of the Earth itself, and the fossils that had been discovered all over it, provided the evidence that was needed.[7]

Hooke presented this only as a hypothesis, but he was obviously growing impatient with those who questioned his ideas without offering an alternative theory. It was easy to play the critic, 'but if he know better let him not hold his tongue but tell us'. His ideas about an oval Earth, he said, were not 'frivolous Suppositions taken up at random to solve one Phænomenon', but propositions based on a large body of evidence. In February 1687 he drew on the whole range of his scientific and technical experience to present his sceptical audience with all the evidence he could muster for his theory that the Earth was a flattened or 'prolate' sphere.[8] First, he offered a practical example from glassworking, where a hollow ball of soft glass would take on a flattened oval shape when spun, and a simple demonstration with a dish of water, in which the liquid sinks in the middle of the dish and rises at the edge when the dish is revolved. Secondly, careful measurement of the Sun and Jupiter, which were known to rotate, and of Mercury, Venus and Saturn, which probably did, might confirm that they were oval, and suggest that the Earth was the same. Hooke undertook to carry out the measurement of the Sun in June, when the problem of refraction (a constant worry in precise astronomical calculations) would be least. There was already evidence that pendulums swung more slowly nearer the equator, because they were further from the centre of the Earth, and gravity was therefore weaker.[9] Finally, the Earth's swollen equator could be explained by the accepted laws of motion. Equatorial swelling was simply a consequence of the law of inertia operating on the fluid and solid matter of the Earth where its rotation was fastest. The outer surface of the Earth received from the speed of the Earth's rotation a tendency to move forward in a straight line at a tangent to the circular motion of the Earth, and was held back by the Earth's gravitational attraction. Near the equator, where the speed of rotation was faster and gravity probably weaker, the tendency to move forward was more pronounced, creating an equatorial bulge.

Hooke had an exaggerated faith in the ability of the experimental method to establish whether or not these very long-term changes were still taking place. He urged the Royal Society to turn away from aimless experiments, however 'surprising and pleasant' they might be, and to embark on a series of investigations to test the truth of his theory that the Earth's axis had shifted, and the poles had moved.[10] This had been a very long process, but in his view it might be possible to make such accurate measurements of London's latitude that minute shifts might be found after a few years. In February and March he explained to the Society how these highly accurate observations could be made, using telescopes with the finest thread-sights, or tubeless telescopes hundreds of feet long, with their object lenses fixed to the spires of City churches. The point of reference might be a constellation invisible to the naked eye that Hooke had discovered ten years earlier, which he had named the English Rose. He thought ancient measurements of meridians of latitude were unlikely to be accurate, but he wondered whether old buildings that had been built on east–west lines, especially medieval cathedrals and Westminster Abbey and the great dial-stone in the Privy Garden of Whitehall, might be checked to see if they were still aligned with the compass points. On 9 March 1687, in his final lecture of the series, he argued that the atmosphere around the equator was swollen by centrifugal force, weaker gravity and greater heat, giving it a more oval shape than that of the Earth or sea. This imbalance, he said, would create a flow of the lower air from the poles towards the equator, and of the upper air in the opposite direction, a circulation which might account for the Earth's prevailing winds and other phenomena.

Hooke, more than any other English natural philosopher of his day, was proud of his strictly rational and scientific approach to his subject. Nobody was freer from the clutter of medieval beliefs and superstitions – alchemy, astrology, magic spirits – that still confused and preoccupied other intelligent men of his time. He was an occasional churchgoer, but his guiding beliefs were in the mechanical philosophy of nature, his own rightness and the endless possibilities of scientific advance. Even the history of the Earth and the creatures on it had to be pieced together from fossil evidence and established scientific laws, not biblical texts. But in a society in which belief in the authority of the Bible was almost universal his essentially secular approach was bound to arouse opposition, and the combative tone of his lectures shows that he was aware of his isolation. In the spring of 1687 there was a confrontation between the new science and the old religion which was in some ways as significant as the famous Royal Society debate on evolution between Thomas Huxley and Bishop Wilberforce in Oxford in 1860.

On 15 February 1687 Edmond Halley sent an account of Hooke's hypothesis to Professor John Wallis, the veteran Oxford mathematician. Wallis and Hooke had known each other since the 1650s, and they had met when Hooke went to Oxford in 1679. Wallis had been involved in the new science since the 1640s, and was one of the founders of the Royal Society. He had won fame and success during the Civil War because of his unusual ability to read coded letters, and though he had served the Parliamentary side he had managed to prosper under Charles II. In the 1660s he had caused a sensation by teaching the deaf and dumb to speak. He was a brilliant mathematician, and his work in the 1650s prepared the way for the development of the differential calculus. He was also the inventor of the modern symbol for infinity. In this age of polymaths Wallis was a theologian, an expert on classical texts, and a skilled controversialist who had beaten the great Thomas Hobbes in a mathematical dispute.

Wallis read Halley's letter to the Oxford Philosophical Society, which discussed Hooke's hypothesis twice. Wallis' reply on the Oxford Society's behalf, which Halley read out to the Royal Society on 9 March, was a stinging attack on Hooke's position. Hooke had provided no convincing evidence for his theory that the Earth's axis had shifted, he said, and the Oxford Society 'seemed not forward, to turn ye world upside down (for so 'twas phrased) to serve an hypothesis, without cogent reason for it; not only, that possibly it might be so, but that indeed it had been so'. As for Hooke's proposal on the shape of the Earth, this was 'but a conjecture, of what may be, without any Observations that so it is'. The Earth's circular shadow in a lunar eclipse and the fact that things fell at right angles to the Earth's surface suggested that Hooke was wrong. The central argument in Wallis' criticism of Hooke's account of earthquakes and polar shifts was that the theory opposed or ignored the whole of biblical and classical culture:

> Yet so vast a change as is now suggested, could not possibly have been (within the reach of Histories now extant) but that some foot steps thereof would certainly have been found in History;

> Since it is so many Hundred years (not to say thousands) since Astronomers have been curiously inquiring into such matters. And that . . . the whole face of the Earth should have been (as he speaks) many times all covered with water and dryed again; seems extravagant for us to admit. For, in what ages of the world (or, before the world) should this have happened? If we give credit to the story of Genesis (the most ancient certainly of any extant;) we shall find the world (soon after Noah's floud) so divided amongst its inhabitants & so planted (Gen 1.10) as agrees very well with the present Geography . . . So that (unless it were before the Creation of Adam) we cannot find a time wherein the Earth should (so often) have been tossed & turned upside down, for the Equator & Poles to change places, & the top of the Alps become a sea only to enable us to give an account of some Fish-shels found there.[11]

Hooke was clearly stung by the slightly mocking tone in Wallis' letter, and his reply, which he read to the Royal Society on 27 April, posing as the injured party but hitting back hard, shows him at his most aggressive. The letter, he said, was 'made up partly of misrepresentation, partly of designed Satyr, rising . . . cheifly from some prejudice ['and evill design' was deleted here] conceived againste me and my performances, which has formerly Discovered it self in print and has not it seems as yet all spent its ['poison' was deleted here] self, though there has never been by me the least cause given for such proceeding'. He made fun of the Oxford Society for 'being frighted with the Bugbear of turning the world upside down' since all Copernicans knew this happened every twelve hours, and reminded Wallis that he had not been 'soe cautious of twirling or swinging the world when he published An hypothesis of the moon'. This was an unkind reference to Wallis' erroneous theory of the tides, published in 1666. Hooke counted it 'noe ill sign of a good hypothesis when An Adversary is not able to produce soe much as one argument against it but such as a carman or porter would have made'. As for the shape of the Earth, Wallis seemed 'ignorant of the proofs [Hooke] brought, of the different Gravitation in different Latitudes and of the Expt of the Pendulum', while the argument about the Earth's shadow in an eclipse would only seem convincing to those 'that know not what an argument is'. Hooke was especially irritated by Wallis' assertion that he thought the Earth had been repeatedly flooded (though this was Halley's mistake rather than Wallis'), and that the world had been tumbled and tossed upside down: 'I think Jeering though never soe witty will never prove a math: or Phil: Argument. In short since the Dr could not find arguments to make it fals yet that he might shew his good will he has done what he could to make it Ridiculous though upon the whole I conceive he has prevailed but little.' Harvey, Copernicus and Galileo had been attacked in their time, and 'yet at last truth hath prevailed'. Wallis had also mentioned Steno, the Danish geologist who had written about fossils in much the same way as Hooke, but shortly after Hooke's 1667–8 lectures on the subject. Hooke used the opportunity to declare that Steno had been given information from his lectures by Oldenburg, and used them without acknowledgement. This is a familiar charge from Hooke's pen, but it is quite possibly true.[12]

Hooke believed that accurate measurements of latitude, showing a gradual change even over a few years, would be the best way to answer his opponents. He pressed the Royal Society to get on with them in the spring of 1687, using a new telescope he had devised, fitted with a compass and two plumb lines.[13] Meanwhile, his equatorial bulge theory had won an important supporter. Newton had previously doubted that the Earth was anything but round. In January 1681, in a letter to Thomas Burnet, who was writing a Bible-based *Sacred Theory of the Earth*, he had argued from analogy with the Sun and planets that the Earth was 'spherical or not much oval'. 'If it's diurnal motion would make it oval that of Jupiter would much more make Jupiter oval'. But in rewriting Book III of *Principia* in 1687 Newton added a new Proposition, number XIX, arguing that the Earth's diameter was greatest at the equator. After receiving the manuscript of Book III on 4 April Halley sent the news to Wallis that Newton 'now falls in with Mr Hook, and makes the Earth the shape of a compressed sphæroid, whose shortest diameter is the Axis'.[14] Here was another weapon for Hooke to use in his campaign to prove that he was the originator of Newton's most important ideas.

Wallis may have been disconcerted by this news, and he was certainly surprised by the venom of Hooke's response to his first letter. It is also possible that the death of his wife in March, after a forty-year marriage, had undermined his fighting spirit. In a reply to

Hooke on 26 April 1687 he protested that he had not written 'with any disrespect to Mr Hook: (whom I did not so much as name) but from the liberty we used to take (& do allow to others), to express our thoughts freely in matters of Philosophy'. He repeated the Oxford group's belief that the Earth was spherical 'or very near it', and that though 'some little variation' of the poles was possible, it was not likely, 'the latitude of places being still the same'. Once again, he asserted that Hooke's story of many great floods was 'contrary to the History of all Ages, which represent seas to have been allways where now we have seas, & Land where we now have Land'. Hooke's floods and shifting poles 'cannot be, without over throwing the credit of all History, sacred and profane'. This was the great argument, drawn not from science but from theology, that stood in Hooke's way, and which all geologists and paleontologists had to face for the next 200 years.[15]

Hooke read out his reply to Wallis in a Royal Society meeting on 11 May. He accepted Wallis' concessions on the possibility of polar movement and equatorial bulge with a bad grace, and dismissed arguments based on the Earth's shadow and falling objects. Wallis, he said, had used the old trick of 'fathering an assertion on me which was none of mine' and then arguing against it. Hooke had never said that there had been *many* floods, or that the *whole* world had been submerged, or that the shifting polar axis had been the *only* cause of the Earth's upheavals, and his ideas were hypotheses not assertions. As for the biblical and historical texts, they did not tell the whole story of the Earth's history. If anyone showed an unwarranted certainty about things that were not yet known, it was Wallis: 'I observe that he seems to have a better account of the Geographicall description of the whole earth for all ages past than I could ever hear of, or pretended to'. Hooke concluded that his hypothesis had withstood its first test. Even critics 'who would have left no stone unturned to overthrow it' had at last 'come to grant that tis possible it may be soe as I supposed'. This letter also revealed an underlying reason for the bitterness of Hooke's reaction to Wallis' criticism. Wallis had sided with Hevelius in the famous dispute over telescopic sights, first at the time of the original controversy in 1674, and probably also in 1685, when Hevelius's new book, *Annus Climacticus*, was praised by an anonymous reviewer in *Philosophical Transactions*.

Hooke thought Wallis was the author of the favourable review, and referred to 'that Scandalous Pamphlet' several times in the second letter.

Hooke was a tenacious opponent, and used his Royal Society platform to keep the dispute with Wallis on the boil. On 25 May he renewed his attack in a lecture, castigating his opponents for twisting and falsifying his arguments. He was insistent that such unfairness would not discourage him from putting forward new conjectures, and to prove his point he went on to describe some observations he had made eighteen years earlier in his work to prove the motion of the Earth. He had noticed then an apparent irregularity in the pace at which the Earth moved during its daily rotation, and wondered whether this might be caused (assuming it happened at all) by the fact that one part of the Earth was denser than another, and thus moved more quickly (like a heavy pendulum) when it was facing the Sun. It might also be the case, he said, that the Moon always kept one face towards the Earth because that side was more solid than the other. Hooke's speculations rested on the possibility that 'the gravitating Power in the Sun and Moon' was not exactly the same as that of the Earth, and that 'there may be something Specifick in each of them'. That is, only a few weeks before the publication of Newton's *Principia*, a book which he claimed had stolen his own ideas, Hooke was still not sure that gravity was a universal and unchanging force.[16]

The last of these attacks on Wallis was delivered to the Royal Society on 29 June 1687. A week later, on the day that Halley and Hooke were told they were to be paid their annual salaries in unsold copies of Willughby's *History of Fishes*, the three books of Newton's *Principia* were published. This marvellous book, perhaps the greatest scientific work ever published, brought together in one great mathematical and celestial system many of the things that Hooke had speculated about, measured and guessed at for nearly thirty years – the tides, comets, the Moon, orbital mechanics, the laws of motion, falling bodies, universal gravitation, the shape of the Earth. The book was recognized as a masterpiece in London and all over Europe, and two of the greatest figures of the age, Huygens and Leibniz, greeted it with admiration. For those not skilled in mathematics, and even for the philosopher John Locke, much of *Principia* was incomprehensible. Much later, Newton told a friend that he had deliberately made Book

III abstruse 'to avoid being baited by little Smatterers in Mathematicks'.[17] It is often said that the 'little Smatterer' Newton wanted to shut out of the debate was Hooke. This is probably untrue, and would have been very unjust to Hooke. Although he could not have written *Principia* Hooke could certainly understand it, and was probably one of a handful of people who appreciated its full importance. This is exactly why he found its publication, and its pointed failure to recognize what he saw as his own contribution to the subject, so hard to bear.

Hooke spent some of July 1687 dissecting eyes with his friend Richard Waller, the Society's new secretary, and in a dispute with Denis Papin over the best way to convey force or motion over a distance. At the end of June Papin had spoken to the Society about using a tube attached to a vacuum pump for lifting weights in mines, and Hooke, reasonably enough, had pointed out that it would not work as well as a rope. Over the next two weeks he gave the Society a demonstration of the advantages of using a long pole or a length of thread to communicate a force across a long distance. Waller saw the experiment, in which a length of packthread was run around a pulley at the other side of Gresham College quadrangle and back to the meeting room. Hooke showed how motion or weight applied to one end of the thread was transmitted immediately to the other, and used long taut wires to demonstrate the rapid transmission of sound: 'the sound was propagated instantaneously, even as quick as the motion of Light, the sound convey'd by the Air coming a considerable time after that by the Wire.' Papin responded by showing how he could stop his vacuum pipe leaking by laying it in a trench and covering it with coarse turpentine, and Hooke came back on 20 July with a paper arguing that it would be impractical to keep a long vacuum pipe free of leaks, especially if it were buried, and that in any case the springiness of the air made it the worst of all media for conveying power. Water or some other liquid would be a much better choice.[18]

Hooke's main scientific task in the summer of 1687 was probably to read through his collection of classical texts in search of evidence to support his hypotheses on the history of the Earth. Although he had dismissed Wallis' criticisms, he was apparently persuaded that his ideas would not win acceptance unless he could show that the changes he described had happened within the accepted span of human history, and that there was evidence for them in ancient texts. In short, he had to find a way of squeezing the vast bulk of geological time into a biblical or classical corset.

As a result of this work Hooke was able to deliver a long series of lectures between November 1687 and February 1688 which concentrated almost entirely on showing that there was ample evidence in the Bible and in Greek and Roman texts, if their meaning was properly understood, to support a cataclysmic account of the history of the Earth. By doing this Hooke hoped to 'take off the odium of Novelty' from his theory, and answer Wallis' main objection to it. From Plato's *Timaeus* he drew an account of 9,000 years of Egyptian history full of disastrous floods, and of the destruction of the Atlantic island of Atlantis in a great earthquake. In the travels of Hanno the Carthaginian he found a description of a voyage into the Atlantic with stories of volcanoes, rivers of fire and islands in flame. One of these, Hooke was sure, was the famous Pike of Teneriffe in the Canaries. Ovid's *Metamorphoses*, which Hooke took to be a history of the world since its creation 'wrapped up in Mythology and Mascarade', was one of his favourite sources. The Gorgon's head represented the process of petrification, the battles between Gods and Titans were metaphors for earthquakes and eruptions, Perseus' rescue of Andromeda referred to great inundations, and so on. These myths and fables, Hooke argued, were careful symbolic representations of the ancient history of Egypt and Greece, as it had been passed down and remembered through poetry and song. In each lecture he brought in new authorities to bolster his case. Herodotus wrote of the flooding of Egypt, Aristotle's Book of Meteors spoke of the Earth being once covered in water. If men of such wisdom believed these things, he argued on 29 February, perhaps we should take more notice of the evidence of rotting shells than we are accustomed to do.

> And tho' it must be granted, that it is very difficult to read them, and to raise a *Chronology* out of them, and to state the intervals of the Times wherein such, or such Catastrophies and Mutations have happened; yet, tis not impossible, but that, by the help of those joined to other means and assistances of Information, much may be done in that part of Information also. And tho' possibly

> some may say, I have turned the World upside down for the sake of a Shell, yet, as I think, there is no one has reason for any such assertion from any action I have hitherto done; yet if by means of light and trivial Signs and Tokens as these are, there can be Discoveries made and certain Conclusions drawn of infinitely more important Subjects; I hope the attempts of that kind do no ways deserve reproach.[19]

Although Hooke was obviously quite prepared to abandon the biblical story of Creation and the Flood in favour of a longer and more complex account based on the fossil and classical record, he seems to have realized by the time he gave his last lecture in the series, on 29 February 1688, that this was asking his audience to swallow too much. Perhaps he recognized that he had strayed too far from Christian teaching, but it seems more likely, in the light of his earlier lectures, that he simply realized, as Darwin did almost two centuries later, that the gulf between his ideas and those of his audience was too wide to be bridged. Conceding more ground to Wallis and his supporters, Hooke used most of his last lecture trying to show that the stories of the Creation and Noah's Flood might themselves be a part of this ancient attempt to tell the story of the Earth in simple terms. Perhaps the words of the second verse of Genesis, 'and darkness was upon the face of the deep', meant that the Earth as first created was covered in water, not for a day or two, but for an age, and perhaps Noah's flood lasted long enough to return the planet to its primitive condition. With this rather weak attempt to reconcile his fossil-based view of the Earth's history with traditional Christian belief, Hooke suspended his series of lectures, and moved on to other things.

24. A Revolution and Old Battles

(1688–1690)

According to Richard Waller, who knew Hooke well in the 1680s, Hooke's health was even worse than usual in 1688. For a great part of the year he was 'very weak and ill, being often troubl'd with Head-achs, Giddiness and Fainting, and with a general decay all over, which hinder'd his Philosophical Studies'. Illness seems to have kept him away from Royal Society meetings in March and April, and limited his contributions in May and June to book summaries and brief interjections. Waller believed that Hooke's unresolved dispute with Cutler, who was still refusing to pay his annual £50, made him uneasy and increased his illness. Hooke, on the other hand, was inclined to wonder whether his ill-health, like so many other things, was due to recent earthquakes and volcanic eruptions. A more obvious explanation for his illness was the fact that the land next to his lodgings was again occupied by a stableyard, turning what had been a pleasant garden into a 'noysom and unquiet place', with dungheaps piled against the wall under his windows.[1]

Hooke's surviving diary resumes on 1 November 1688, four days before William of Orange's landing at Torbay began the so-called 'Glorious Revolution' which drove the Catholic James II into exile. The tiny diary, which measures about four inches by two, is briefer and more cryptic than that of the 1670s, but it gives us another chance to spy into Hooke's life beyond the Royal Society and the lecture hall, which has been generally hidden from us since 1683. The diary shows us that Henry Hunt, the Royal Society's Operator, was Hooke's almost constant companion. He took morning tea with Hooke almost every day, and often dined with him too. Theodore Haak, now in the last years of his long life, was still a close friend, and so were John Aubrey the antiquary and Edward Tyson the anatomist. The tea and chocolate Hooke shared with his regular visitors was made and served by Martha, Hooke's housekeeper, who was impudent or free-spirited, depending on your point of view. Hooke was always complaining about her rudeness and cheating, but when she ran off at various times to Nell Young, to Captain Paggen or to Edward Tyson he made sure that he got her back. Hooke's favourite coffee house was still Jonathan's, and the friends he met there nearly every evening included some old companions, Francis Lodwick ('Lod'), Sir John Hoskins ('Hosk') and Edmond Halley ('Hall'), and some newer ones, Richard Waller ('Wall'), Alexander Pitfield ('Pif') and Mr Currer ('Cur'). They were all fellows of the Royal Society. His friend John Godfrey was still clerk of the Mercers' Company, and seems to have dabbled in luxury imports. In October 1689 he sold Hooke twelve pounds of tea for £3 7s, and the following February he supplied him with some East Indian sugar (Hooke called it 'candy'), which was 'clammy and brown'. Hooke still worked with Sir Christopher Wren and met him quite often, but Wren was frequently out of town, rebuilding Hampton Court for the new King and Queen, William and Mary. And he often visited his old friend Thomas Tompion, the greatest of watchmakers. On 12 October 1689, at Tompion's, he saw 'a most curious gun beyond any for neatness I had ever seen'. The main women in Hooke's life, apart from his troublesome housekeeper, were Nell Young, who had been his servant and mistress in the early 1670s, and Mrs Moore, with whom he often took tea, sometimes with cheesecake or 'much discourse'.[2] Hooke occasionally dined with Samuel Pepys, who had once been so impressed by *Micrographia*. In June Hooke lent him some petrified wood and told him about the drawing device he called a picture box, and in October he showed him some optical experiments and demonstrated the workings of a telescope.

Although Hooke occasionally noted that he was 'melancholy', the tone of his diary was not that of a broken or dispirited man, embittered by the death of his niece or by his conflict with Newton. He hardly ever spent long periods alone, and he met friends and fellow scientists, who apparently found him good company, every day. He condemns his opponents as 'rogues', 'rascals' and 'dogs' much less than he used to do. The neglect of his appearance which characterized the last years of his life was not yet evident. In

November and December 1688 he was busy restocking his wardrobe – a new black coat from Richardson the tailor for 17s 6d, new drawers (cut by himself), new stockings for 5d, new shoes after Christmas. In February he bought six ells (over seven yards) of 'Holland' linen and cut out two shirts for Martha to sew. He was fifty-three in 1688, but his general health seems to have been no worse than it had been in his late thirties. Although average life expectancy was low, it was not uncommon for people who reached their middle years to live to an impressive age. Newton, Haak and Halley lived to their mid-eighties, and Wren made it to ninety-one. No doubt Hooke was feeling his age, but his diary is not the catalogue of intestinal pain, emetics and regular vomiting that it had been in the early 1670s. It is possible that he was simply not recording these incidents as he used to, but there is no sign that he had lost interest in his ailments, and he was careful to note that he was suffering from 'colick and stopage' in November 1688, and that his urine was 'high colourd'. His sleeping pattern was a little unorthodox, since the combined effects of strong coffee, afternoon naps, old age and interesting work often kept him up past three o'clock in the morning. He was certainly fit enough to take quite long walks, usually with Godfrey or Lodwick, northwards across tenter grounds and open fields to Kingsland, a small village just over two miles away, where Dalston Junction now stands. Business could be combined with pleasure. In May 1689 he set off with Lodwick and Godfrey to 'view' Mr Bignoll's 'strangely built house', and earn himself 18s. Sometimes, for a change, Hooke and Godfrey set off to the north-west, across Finsbury Fields to Islington, a five-mile round trip.

Hooke's eyes had taken the place of his sinuses and intestines as the main focus of his ill-health. Robert Boyle, who was nearing the end of a life that had been ruined by chronic illness, took a helpful interest in his health problems, and had tried many remedies for his own eye troubles. He suggested an 'infusion of Crocus metallorum to clear the sight' in May 1689, and recommended clarified honey with rosemary flowers in June. Whether or not Hooke tried these remedies, his eyesight was getting worse in August and September. On 18 August he became 'quite blind' in his right eye, but recovered the same day. His sight failed again on 3 September, but four days later he experienced a 'strange clearnesse of sight', followed by another deterioration, and another recovery on 14 September. At least he knew London's best opticians, and he often turned to them to replace spectacles that he had lost or broken. Working too late in poor light put a strain on his middle-aged eyes, and he sometimes recorded (as on 22 December) that his eyes were very weak after a long evening's work. These intermittent problems were a warning of what was to come, but for the time being he could see well enough to carry on with his work, unlike his unlucky friend Thomas Henshaw, who lost his sight, Hooke recorded, early in 1690.

Hooke was still interested in every aspect of craft and manufacture, frequently reading books or having conversations about such things as mapmaking, glass-painting, lacquering, papermaking and fire-fighting, and he maintained his interest in astronomy and microscopy. He was as keen as ever to discover new manufacturing secrets, and to demonstrate his discoveries to his Royal Society colleagues. In April 1689, on a visit to one of his favourite booksellers, Caillou, Hooke saw 'strang Marbled paper', and noted in his diary 'I know how to make it'. On 20 May he 'tryd marbling paper', and nine days later he told the Royal Society that it was done 'by laying specks or stroacks of severall Colours on one side of a sheet of paper, and then folding it against the other and pressing in the colours . . ., which makes a very Pretty appearance'. Hooke was still in touch with John Dwight, the greatest English potter of his day. Working in his Fulham pottery, Dwight had discovered or invented many craft secrets that Hooke would have loved to know, but seems to have guarded them well. They spoke of his 'transparent china', the imitation porcelain on which he held the patent, in September 1689, but Hooke never learned enough about its manufacture to explain the process to the Royal Society.

Hooke's lonely and tenacious campaign to persuade scientists to consider the history of the Earth in a non-biblical way gave him a great interest in classical texts and travellers' tales, and in February 1689 he tried to add to his impressive list of foreign languages – French, Dutch, Latin, Greek and a scrap of Chinese – by learning Portuguese. To serve all these interests, and no doubt to satisfy his collector's urge, he went to bookshops and book auctions every week, sometimes almost every day. Books were his main extravagance, costing far more than his other everyday purchases, coffee, cacao beans (for making drinking chocolate), tea, beer, bread, and candles. There is a paper in Hooke's hand listing fifty-seven auctions and catalogues between August 1686 and August 1689, but in fact he went to auctions far more often than that. Some books were cheap enough: 3d for a Spanish grammar, 2½d for a *Voyage to Siam*. But he paid 15s for Milton's *Areopagitica* and £8 15s for the twelve-volume works of Aldrovandi, an Italian naturalist. Sometimes he combined business with pleasure by acting as an agent for collectors who lived outside London, and for the Royal Society library. Hooke repaired the Aldrovandi volumes, and on 5 July 1689 he paid a carpenter, Kettle, 3s to crate them, then took them to the Swan Inn on Holborn Bridge, where a carrier delivered the books to Sir John Long FRS of Wiltshire, one of Hooke's regular customers.

The famous auctioneer Edward Millington, who sold Hooke's collection after his death, had an auction house in Ave Maria Lane, but most book auctions were held in City coffee houses: Hargrave at Tom's, Hussey at Sam's, Nathaniel Rolls at Will's. All these were in the streets and lanes around St Paul's Churchyard, where the London book trade had re-established itself after the disaster of September 1666, when its whole stock had been lost when St Paul's collapsed into the chapel of St Faith. Around Moorfields, the open space just to the north of Hooke's Bedlam Hospital, there were open-air bookstalls along the railings. Hooke usually came back from his Moorfields book-hunts with nothing ('MF0'), but there was always the chance of a lucky find: in March 1693, about fifteen months after Boyle's death, he came across nearly a hundred of his old friend's Dutch chemistry books 'exposed in Moorfields on the railes'. Hooke was a valued customer to the booksellers of St Paul's and Cornhill, and they often lent him books to study and return. He exchanged books with friends, too, and borrowing, lending and returning books, and discussing them in Jonathan's or the Royal Society, was an important part of his social and intellectual life.

Hooke seems to have received his last payment from the Royal Society in 1688, but he still had his regular salary from Gresham College. His City work was not as demanding as it had been in the years after the Great Fire, but he met his fellow surveyor, John Oliver, from time to time, and they still had duties connected with the Fleet Ditch. Hooke still carried out occasional views on properties in the

City, providing professional adjudications in disputes between property owners or builders, usually for a fee of 10s. In May 1689 he dealt with the case of a 'Rascall bricklayer' who was trying to build a party wall that encroached on a neighbouring property, and in October he looked at the new building being constructed in the Strand by Sir Francis Child, and judged that 'it could not be hindered'. Child was at this very moment deserting the goldsmith's trade to become London's first full-time banker, and the new building was almost certainly connected with his bank at the Temple Bar, where Fleet Street meets the Strand. So Hooke, by chance, was involved in the very beginning of specialist commercial banking in London.

Hooke's visits to the sites of new City churches, which had been so frequent in the 1670s, were much rarer in 1689. Most of the new churches had been completed in the 1670s or 1680s, and probably only five were being constructed in 1689–90. After eighteen years of large and regular payments, Hooke received no salary for work on City churches in 1689 or 1690. He was often at St Paul's, either looking for Wren or meeting John Oliver, who had been Wren's assistant surveyor at the Cathedral since 1676. Hooke was an architect and builder of distinction, and commissions were as plentiful as ever. He was often called for by his old teacher and friend Dr Busby, the Prebendary of Westminster Abbey, to carry out repairs on the Abbey or School: the great North Window of the Abbey in November 1688, the cloister house in June 1689, the cloister house vault in December, the stairs and museum in February 1690. In January 1690 his position was formalized, and he became Surveyor to the Dean and Chapter of Westminster at a salary of £20 a year. At the end of 1689, he began work on almshouses in Willen, where he had built Busby's church. He took on other repair and redecorative work in London churches. Between May and July he was employed to refurbish a 'temple', possibly the Dutch Church of Austin Friars. His diary for 10 July mentions a 'Rayle and Ball, 3 ft 8 in. high; plinth, base ballester and rail at 8 sh per foot runing in Dutch Church'.

A much more important job came his way in March 1689, when the leading liverymen of the Haberdashers' Company took him to dinner at Romer's coffee house to ask him to design their new school and almshouses, to be paid for from the bequest of Robert Aske, a silk merchant. In the spring Hooke and the Haberdashers went in search of a large site north of the City, and settled on one in Hoxton, near Pitfield Street, less than a mile from the City Wall. Hooke finally presented his design to the wardens of the Company in January 1690, and supervising the building kept him busy for the next four or five years. When the school and houses were finished they provided homes for twenty poor freemen of the company and education for twenty sons of Haberdashers. The new chapel was consecrated in 1695 by Hooke's old friend John Tillotson, who became Archbishop of Canterbury in 1691.

Hooke always had several jobs on the go at once. In March 1689, when he had just taken on the Haberdashers' almshouses, he was approached by Mr Gould, who wanted him to design a house in Highgate. Hooke met Gould and his wife several times over the next month, and prepared a design and model of the house in May. On 1 June, the day he showed the model to the Goulds, he was approached by an Essex landowner, Richard Vaughan, who wanted a new house in the country. Two days later, starting out at six o'clock in the morning, he went with Vaughan by coach and four to Burntwood (now known as Brentwood) to look over the proposed site. For his trouble he received a gold sovereign and a plate of bread and butter. Within three days of his visit he had designed a forty-foot-square house with a courtyard, and over the summer and autumn of 1689 he received a string of visits from Vaughan and his craftsmen discussing amendments to the plans and seeking advice on practical details. The house Hooke built for Vaughan is Shenfield Place, a brick house with a hipped roof and well-preserved wood panelling, about a mile north of Brentwood.

Although Hooke was a famous and controversial scientist, most of his Gresham lectures were as poorly attended as ever. He often recorded 'noe auditors' for his Thursday morning and afternoon sessions, and once found 'one fat man' with a coat and stick waiting for him to start. But when Mr Paget brought a party of schoolboys along from Christ's Hospital Mathematical School, Hooke would lecture to an audience of forty. Perhaps they did not understand what he was talking about, but the boys' behaviour did not give him cause for complaint. And he still gave his Cutler lectures, usually to the Royal Society, in the hope that Sir John Cutler would one day pay him the promised £50 a year.

Hooke supplemented his earnings as a surveyor, architect and scientist with a small rental income. Several of the stables and coach-houses that had been built without permission in Gresham College had been demolished as a fire hazard in 1681, but Hooke still had a stable, and let it in 1688 to Robert Foot for £4 6s a year. Foot's replacement, a man called Fips, left in July 1689 because his carriage was stolen, but Hooke found another tenant, Pargiter, who was prepared to pay £8 a year and mend the stable doors into the bargain. Hooke also had property on the Isle of Wight, which was leased in January 1689 to a reliable new tenant, Wey. He also received rent for Isle of Wight property from his cousin Robert Giles, to whom he sent newspapers and a telescope. Giles, whose son Tom had died in Hooke's care in 1676, lost a second son in July 1689, and died in 1693.

The months covered by the early part of the diary were dramatic and anxious ones, and Hooke, picking up news in coffee houses and from printed gazettes and street proclamations, recorded them all. He got news of William's landing at Torbay on 7 November 1688, two days after it had happened, and he noted the worrying news that the 'Rabble' was rioting a few days later. He recorded King James' unsuccessful attempt to rally the army to him at Salisbury on 19 November, and the sense of confusion in London coffee houses between 24 and 27 November, when James' closest allies, including his own daughter Anne, deserted him. This uncertainty and confusion, mostly created by James' panic and indecision, caused Hooke and his friends some anxiety. On 5 December Hooke wrote that there were 'jealousies of much danger to the City. Great confusion of reports, noe certainty.' James fled Whitehall in the early hours of 11 December, but he was captured and brought back to London in a sort of triumph five days later. He was allowed to escape again on 18 December, the day that William of Orange finally arrived in London. Hooke watched James' three coaches passing by on their way to Greenwich, escorted by some Yeomen of the Guard. He seems to have been pleased by the change of government, and especially by the fact that it all happened without much civil disorder or bloodshed. His attitude generally was that of a detached and well-informed observer, not a partisan and certainly not a participant. The most explicit comments on the revolution to be found in Hooke's diary is a T.D.L. (*Te Deum Laudamus* – praise God) when the Convention Parliament recognized William as Regent and his point of view in a discussion with 'the usual company' in Jonathan's in March 1689 – 'I pleaded against Division and Revenge'.

William of Orange's friendly invasion and James' flight created a constitutional crisis but not a social one, and Hooke's work as a scientist and architect was not seriously interrupted. The Royal Society met every Wednesday as usual, though its annual elections were postponed from St Andrew's Day to St George's Day (30 November 1688 to 23 April 1689), giving Hooke an extra five months on the Council. In the early weeks of November, when William's army was making its way virtually unopposed through the West Country and down the Thames valley to London, and anti-Jacobites were seizing power in many provincial towns and cities, Hooke was discussing repairs to the North Window of Westminster Abbey with Dr Busby, and organizing suitable craftsmen to carry out the work. In late November, when James' political and military position was falling apart, Hooke was reading the second part of Thomas Burnet's *Sacred Theory of the Earth*, which had just been published. Burnet's theory was that the Earth had been created like a giant egg, and that the Flood had released its internal waters and turned its smooth shell into jagged mountains. These ideas, which seem laughable now, had a much better reception in the 1680s than Hooke's radical speculations. Hooke's lectures on the various books of Burnet's work were his main contribution to Royal Society meetings in December 1688.

In early December, when city after city was going over to William and there were rumours of French landings on the south coast, Hooke gave the Royal Society a lecture on the value of simple inventions (a regular theme) and told them of an easy way of making wooden cranks and attaching them to an axle. This, he said, would improve the performance of engines where 'the difficulty and dearness of Iron Cranks had made them impracticable'. Later in the month, when James was making his two escapes from Westminster, Hooke entertained the Society with a series of lectures on Burnet's theory.

Hooke was anxious to re-establish himself as a salaried Royal Society official, and saw Halley's desire to reduce his commitments as a way to achieve this. In January 1689 Hooke offered to take over the editorship of *Philosophical Transactions* from Halley, for an annual fee of £60. Hooke lost his position on the Council in the April 1689 elections, and in November 1689 the Council rejected his proposal. Even Abraham Hill, one of his coffee-house cronies, voted against his scheme. Hooke got a sort of revenge a few days later, when he was elected to the Council again, despite the campaign (as he saw it) of a coterie of his opponents: 'The gang tryd utmost effort, but faild.' Hooke talked with Halley about taking over as the Society's clerk, and with Richard Waller about joining him as secretary, but others remembered Hooke's performance after Oldenburg's death, and they were not tempted to make the same mistake twice.

Hooke was not the original scientist he had once been, but now that he had recovered from the ill-health of 1688 his lectures and demonstrations were a mainstay of Royal Society meetings again. He tried to provoke and invigorate the Society with ideas and speculations: the growing tendency of modern artists to paint in perspective, the mysterious ability of some spiders and insects to walk across the surface of ponds, the possibility of pickling meat in brandy for long voyages. He did his best to keep the spirit of invention alive, and to encourage his listeners to contemplate the enormous possibilities of improvement as scientific knowledge grew. In May 1689 he tried to revive interest in the origins of the Earth, and at end of the month he gave a far-sighted talk which was intended to answer objections to his theories. He concentrated on justifying his hypothesis that the petrified shells of unknown sea creatures suggested that some of the species God had created were now extinct. Hooke accepted that some apparently extinct species might still exist in little-known regions, and he called for new natural histories of the Americas and the Far East. But he insisted that to suppose that some species no longer existed was not a denigration of God's handiwork, as his critics had claimed.

> We find nothing in Holy Writ that seems to argue such a constancy of Nature; but on the contrary many Expressions that denote a continual decay . . . if the Body of the Earth be accounted one of the number of the Planets, then that also is subject to such Changes and final Dissolution, and then at least it must be granted, that all the Species will be lost; and therefore, why not some at one time and some at another?[3]

Once again Hooke was pointing his fellow scientists along the road that might lead to a new understanding of the development of species, but none of them seemed interested in following him.

Hooke often went over old ground, perhaps because he was running out of new ideas, but especially when he had to assert his claim to a forgotten idea or innovation that someone else was introducing as their own. In February 1689 the Society heard a letter from Denis Papin describing a valved cylinder in which a small explosion would create a vacuum, which would drive down a piston and lift a weight. Hooke declared that this was the 'new Invention in Mechanicks of prodigious use' that he had described in code in *Helioscopes* in 1675. This was a danger of using codes, but even when Hooke had presented an idea to the Royal Society years ago in plain English this seemed to count for nothing, unless he could dig through the Society's records and prove that he had said it. Hooke's friend Halley, who was almost as inventive as Hooke had been in his prime, was often the unwitting culprit. On 6 March 1689 (according to Hooke's diary) Halley 'read a paper of Walking under Water. The same with what I shewd the Society 25 years since. Hoskins belied me, as he does every time.' Sir John Hoskins often chaired Royal Society meetings, and Hooke had not forgotten his tactless reaction to the first book of Newton's *Principia* in April 1686, when he had ignored Hooke's earlier work and provoked the outburst that led to the terrible clash with Newton. To assert his priority over Halley, Hooke got Waller to read his 1660s work on diving bells and air supply to the next Royal Society meeting. He took the opportunity to remind the Society of his depth sounder, which was now much improved, and of his submarine. The same thing happened in February 1690, this time over telescopic sights in quadrants, the subject of Hooke's famous dispute with Hevelius. In a conversation in Jonathan's, 'Hally pretended to the glasse sights of Sea quadrant, though it was printed in *Hist. of the Royal Society* before he went to school.'

From time to time, Hooke was oppressed by the feeling that all his work for the Royal Society and the increase of natural knowledge had been forgotten or disregarded. On 21 June 1689 he went with Hunt and Lodwick to see Mr Molt, a chemist in Fish Street, to buy a supply of alcohol for some experiments on congruity and the penetration of liquors (an old favourite) that he was carrying out for the

Royal Society. Hooke used his lecture on the mixing of liquids, which was given to the Society on 26 June, to set out his grievances against the Royal Society and the scientific community in general:

> I have had the misfortune either not to be understood by some who have asserted I have done nothing, or to be misunderstood or misconstrued (for what ends I now enquire not) by others who have secretly suggested that their expectations – how unreasonable soever – were not answered . . . And though many of the things I have first Discovered could not find acceptance yet I finde there are not wanting some who pride themselves on arrogating of them for their own – But I let that passe for the present.[4]

It would have been easy, he told them a week later, for him to have devoted his life to making money, as most other men of learning did. While most had joined the ranks of 'Divines, Lawyers or Physicians, where their way to Canaan is already chalked out', and others had devoted their skills to a mercenary search for perpetual motion or the Philosopher's Stone, he had followed the lonelier and less well-rewarded path of true scientific enquiry, and had endured mockery and name-calling because he had not been 'contented to tread in the comon tracks and to goe round in the same way like the horse in the mill'. Yet it had been him, he reminded the Society, who had first discovered 'the true cause of the motions of the Cœlestiall Bodys' and the oval figure of the Earth, and who would, in time, show that the Earth's axis had shifted.[5] On 24 July, shaking off this melancholy and complaining mood, he read a lecture on 'the Limits of Possible Discovery', in which he discussed inventions that might one day be possible, but were 'not yet known by the Limitation of our understanding'. He urged the Society to press on with new enquiries into light, colour, sound, tastes, odours and 'all tangible qualitys, such as heat and cold, drynesse and moysture, Gravity & levity, Fluidity & solidity, expansion & contraction, Elasticity'. We had a superficial knowledge of plants, he said, 'but as to their vertues or qualities for medicine or for many other physicall & mechanical uses we are yet in the Dark'. The Society should not be deterred by past failures or the world's ingratitude, 'for the limits of Naturall Knowledge are as infinite and boundlesse as the quantitys of Natures productions'.[6]

Hooke's feelings about his career and reputation would have been very different if he had been given credit, along with (or instead of) Newton, for the explanation of planetary motion. It would be quite inaccurate to say that he was obsessed with this issue, but he was certainly not prepared to forget it. Newton was elected as one of Cambridge University's representatives on the Convention Parliament in January 1689, and spent almost the whole of 1689 in London. Halley, probably trying to act as a peacemaker, brought the two adversaries together at his house on 15 February 1689, but they had a difficult and inconclusive conversation: 'At Halley's met Newton; vainly pretended claim yet acknowledged my information. Interest has no conscience: A posse ad esse non valet consequentia. [Being cannot be deduced from possibility].' Hooke and Newton met again at the Royal Society on 12 June, when Hooke's other old rival, Huygens, was also there. This was probably the only time the three were all in one place together. Hooke showed an experiment of staining a stone in imitation of the branched treelike pattern found on moss agate (or 'mochus'), and the meeting seemed to pass without incident. His aim in this experiment was to show that such plantlike formations were not 'designs of nature' (as traditionalists said fossils were) but 'productions of chance'. Nature did not make copies of shells or trees to delight us or to test our faith. At a meeting three weeks later things went less smoothly ('Newton and Mr Hamden came in, I went out: returnd not till 7'), but the two both attended a meeting on 7 August without conflict. Three times in late August, on the 20th, 24th and 30th, Hooke's diary recorded the word 'Newton', suggesting that the two met privately that month, when Newton was certainly in London. Perhaps their relationship was just a little easier than is usually believed, or possibly Hooke wrote down Newton's name for some other reason.

Whatever the significance of the meetings in August, it is clear that Hooke still believed that Newton and the world at large had cheated him of credit for his ideas on gravity and planetary orbits. In 1689 the antiquary Anthony Wood was coming to the end of his research for a biographical dictionary of Oxford writers and clergymen, *Athenæ Oxoniensis*. Hooke's close friend John Aubrey, who lived in Oxford and helped Wood gather information for his book, alerted him to the fact that Wood had not acknowledged his work on gravity. On 15 September 1689 Hooke got together with Aubrey to draft a letter to Wood putting his claim to be the true author of Newton's theory. The letter reproduced his account of circular motion from *An Attempt to Prove the Motion of the Earth*, and explained that he had told Newton of this in 1679. It recounted the story of the 1679–80 correspondence on falling bodies, and claimed that Newton had learned of the inverse square law from it (which Hooke thought was the case). All these ideas had been printed by Newton without any mention of Hooke, Aubrey said. 'Likewise Mr Newton has in the same booke printed some other theories and experiments of Mr Hooke's, as that about the oval figure of the earth and sea: without acknowledging from whom he had them.' The letter, which had given a fairly straight summary of Hooke's claims, ended on an emotional note:

> Mr Wood! This is the greatest discovery in nature that ever was since the world's creation. It never was so much as hinted by any man before. I know you will doe him right. I hope you may read his hand. I wish he had writt plainer, and afforded a little more paper.
>
> Tuus,
>
> J. Aubrey.
>
> Before I leave this towne, I will gett of him a catalogue of what he hath wrote; and as much of his inventions as I can. But they are many hundreds; he belives not fewer than a thousand. 'Tis such a hard matter to get people to doe themselves right.[7]

This letter is in Aubrey's hand, with Hooke's amendments. Wood may not have received it, and he was apparently not influenced by it. It was Wood's misfortune that he listened to Aubrey on another subject, and printed his story about the late Earl of Clarendon's selling of offices. Clarendon's son had Wood prosecuted, and Hooke recorded the outcome in one of his last surviving diary entries, on 3 August 1693: 'Woods booke cond[emned] to be, and was, burnt at Oxon, July 31 93, and he banisht the University'.

Hooke is remembered today mainly for his conflict with Newton, and this letter was a colourful and eccentric contribution to their long-running dispute. It is reprinted in full in John Aubrey's *Brief Lives*, the most widely known account of Hooke's life, where it takes up almost half of the space devoted to him. As a result, its impact on the general impression of his life and character has been too great. Once we begin to see Hooke's life in the round, this letter and his other skirmishes with Newton fall into place as important incidents in a busy and varied life. If we take the four days either side of 15 September 1689, the day the letter was written, it at once becomes clear that his life was not devoted to fighting and refighting his war with Newton. On Wednesday 11 September he set out the foundations of a house with Godfrey, visited Knox's ship at Deptford, met his friends at Jonathan's, went to a book auction and did a view for 5s. On Thursday he lost his ring dial and asked Hunt, Halley and Waller if they had seen it, met friends at Man's and Jonathan's coffee houses, visited Boyle, went with Sir Edmund King to see Boyle's new portrait, and bought several books at an auction. On Friday he sent a collection of books to Sir John Long, received Tycho Brahe's globe from Boyle, visited his old friend Thomas Henshaw, walked with Lodwick, met the usual crowd at Jonathan's and looked for his ring dial. On Saturday he did another view, called on his cousin Hannah Giles (who was out), treated his troublesome eyes with apple juice, returned a book to the Master of the Mint, and spent the evening at Jonathan's discussing the news of Prince Louis of Baden's crushing defeat of the Turks two weeks earlier. On Sunday, the day of the letter, he showed Brahe's globe to Tompion. On Monday he found his ring dial, 'demonstrated equal motion in a parabola', worked with Hunt on fitting and soldering a brass microscope, and spent another evening at Jonathan's with Halley, Godfrey, Lodwick, Currer and others rejoicing over the utter rout of the Turks. On Tuesday he worked with Hunt on the brass microscope, dealt with the carpenters who were working on the new house he had designed for Mr Vaughan, and went to Jonathan's. On Wednesday he worked on Henshaw's microscope, showed him the new brass one, and had a 'good discourse' with him. The two went to Jonathan's to meet Halley, Sloane and the usual Royal Society set, and then he sat up until five in the morning watching an eclipse of the Moon. Whatever else it may be, this is not the life of a solitary and embittered monomaniac, wasting his energies on a single quarrel.

Hooke always replenished his stock of ideas and information through conversations in coffee houses and visits to workshops.

Without question, his best informant in late 1689 was Captain Knox, who arrived back from his latest voyage in September. Hooke visited him on the 11th, and they met at Jonathan's a fortnight later, and Knox, supplied with refreshments at Hooke's expense, started to talk: 'good Discourse of Mauricius, the Cape, Bombay, etc. Bespoke Shooes. Ebony very black when cut: true musk to be had at the Cape of Good Hope: I saw 3 cods of it. The Malabar arithmetick the same with ours, characters only differing.' Knox gave Hooke books in Arabic and Malabar, and (a few days later) three preserved fish for the Royal Society. Hooke met Knox at Jonathan's again on 24 October, and heard about the trees that grew on Mauritius, and about the island of 'Diego Rodriges', 'as big as the Isle of Wight: an excellent Island uninhabited'. Two days later Hooke bought him a drink of chocolate, and heard the most interesting story of all. There was in India 'a strange intoxicating herb like hemp' known as gange, which 'takes away understanding and memory'. Less interestingly, Hooke learned that rice was 'not injurious to the sight nor harmf. to the teeth'. Knox told him more on 5 November. Gange was also called Consa and Bangue, and it was 'accounted very wholsome'. Knox never ran out of stories, as long as Hooke kept his cocoa cup full. Over the winter he told Hooke about Nova Zembla and Sumatra, of sapan wood, jewels and nautical instruments.

Hooke was an excellent listener, and made good use of what he had learned in Jonathan's. Throughout the autumn of 1689, and into the new year, he brought Knox's gifts before the Royal Society, and used Knox's information to enliven his discourses. One week it was a leaf of the taliput palm tree and the three small fishes, another the leaves of the mana and cinnamon trees of Mauritius and Malabar, or advice on dyeing leather or ripening grapes. On 4 December 1689 he spoke more generally of 'the great advantages that might accrue to the Nation, if People were Incouraged to Discover the Secrets of the several Artifices used in India and Elsewhere, whereby the foreign manufactures are made, that are brought to us'. As examples he explained the basics of Indian arithmetic from a book in the Brachman (Brahmin) character, described the fine Indian gold thread used in silk-weaving, and demonstrated the method used by North American Indians to weave a previously unknown coarse cloth, using a technique he had learned in a coffee house the previous August.

Before the next meeting on 11 December he borrowed a Dutch map of Tartary and translated Muller's *Relation of Tartary* from the French. With Lodwick's help he translated Backhoff's *Voyage into Cathay*, and went with Halley to see a map of the South Seas in Dr Hans Sloane's impressive collection. Much later Sloane's collection, which Hooke had first seen in August 1689, was moved to Montagu House (Hooke's burned and repaired building) as the founding collection of the new British Museum. Using all these sources Hooke lectured the Royal Society on voyages through Siberia to China, and told them that it was 'performed in old times as well as lately'.[8]

In case the fellows were not convinced by all these advantages of opening their minds to the wider world, Hooke saved his best argument until the end. On 18 December 1689, and again three weeks later, he described the many advantages of the drug known to the Indians as Bangue and to us as cannabis, marijuana or Indian hemp. Hooke's talk, which was quite new to the fellows of the Royal Society, survived in the papers found in his rooms after his death. It is a good example of his clear and informative lecturing style, and an early proposal for the use of cannabis as a medicine in the West:

> It is a certain Plant which grows very common in *India*, and the Vertues, or Quality thereof, are there very well known; and the Use thereof (tho' the Effects are very strange, and, at first hearing, frightful enough) is very general and frequent: and the Person, from whom I received it, hath made very many Trials of it, on himself, with very good Effect. 'Tis called, by the *Moors*, *Gange* by the *Chingalese*, *Comsa*, and by the *Portugals*, *Bangue*. The Dose of it is about as much as may fill a common Tobacco-Pipe, the Leaves and Seeds being dried first, and pretty finely powdered. This Powder being chewed and swallowed, or washed down, by a small Cup of Water, doth, in a short Time, quite take away the Memory and Understanding; so that the Patient understands not, nor remembereth any Thing that he seeth, heareth, or doth, in that Extasie, but becomes, as it were, a mere Natural, being unable to speak a Word of Sense; yet he is very merry, and laughs, and sings, and speaks Words without any Coherence, not knowing what he saith or doth; yet he is not giddy, or drunk, but walks and dances, and sheweth many odd Tricks; after a little Time he falls asleep, and sleepeth very soundly and quietly; and when he

wakes, he finds himself mightily refresh'd, and exceeding hungry. And that which troubled his Stomach, or Head, before he took it, is perfectly carried off without leaving any ill Symptom, as Giddiness, Pain in the Head or Stomach, or Defect of Memory of any Thing (besides of what happened) during the Time of its Operation. And he assures me, that he hath often taken it, when he has found himself out of Order, either by drinking bad Water, or eating some Things which have not agreed with him. He saith, moreover, that 'tis commonly made Use of, by the Heathen Priests, or rambling Mendicant Heathen Friars, who will many of them meet together, and every of them dose themselves with this Medicine, and then ramble several Ways, talking they know not what, pretending after that, they were inspired. The Plant is so like to Hemp, in all its Parts, both Seed, Leaves, Stalk, and Flower, that it may be said to be only *Indian* Hemp. Here are divers of the Seeds, which I intend to try this Spring, to see if the Plant can be here produced, and to examine, if it can be raised, whether it will have the same Vertues. Several Trials have been lately made with some of this, which I here produce, but it hath lost its Vertue, producing none of the Effects before-mentioned . . .

. . . this I have here produced, is so well known and experimented by Thousands; and the Person that brought it has so often experimented it himself, that there is no Cause of Fear, tho' possibly there may be of Laughter. It may therefore, if it can be produced, possibly prove as considerable a Medicine in Drugs, as any that is brought from the *Indies*; and may possibly be of considerable use for Lunaticks, or for other Distempers of the Head and Stomach, for that it seemeth to put a Man into a Dream, whilst yet he seems to be awake, but at last ends in a profound Sleep, which rectifies all; whereas Lunaticks are much in the same Estate, but cannot obtain that, which should, and in all Probability would, cure them, and that is a profound and quiet Sleep.[9]

To an invalid who had been dosed with emetics, poisons and placebos by the best physicians in London, bangue must have seemed a most welcome alternative therapy. To complete his account of its virtues Hooke added that bangue could 'ease the sense of hard labour' and 'remove all sense of Grief or Pain', and that in Linschoton's *Voyages* it was 'said to be likewise used to excite Venery, of which quality the Captain is silent'. The medicinal uses of hemp were mentioned in some sixteenth-century herbals, and a book by the Portuguese physician Garcia da Orta, *Colloquies on the Simples and Drugs of India* (published in Portuguese in 1563 and English in 1577), spoke of a 'kind of ecstasy' induced by a herb called bang. Robert Burton's *The Anatomy of Melancholy* (1621), a massive compendium of information on melancholy and its remedies, briefly quoted da Orta's words, but such a full and graphic account of the many effects of taking cannabis had not been published in England before Hooke's paper.[10] Hooke's suggestion that cannabis might have medical uses was not taken up. Berlu's *Treasury of Drugs*, published in 1690, described it as having 'an infatuating quality and pernicious use', and its therapeutic properties were not systematically investigated by Western doctors until Dr O'Shaughnessy's trials in India in the late 1830s. Whether the effects of bangue might explain the length of time it took Knox and his friends to organize their escape from captivity is a question we cannot answer.

Hooke quickly moved on to other things. On 14 December 1689 a French watchmaker, M. Grillet, came to see him claiming that he had invented a new way to establish longitude, but Hooke found him 'unintelligible'. Four days later Grillet put his proposal to the Royal Society, and Hooke was chosen to lead a committee to examine the new method. After several delays, the committee – Hooke, Waller and Pitfield – met Grillet on 8 January 1690 in the Artillery Ground, a walled space between Moorfields and Bunhill Fields which had previously been used for artillery practice. There then took place a demonstration which must have appeared quite ludicrous to onlookers who were not aware of its scientific purpose. Grillet marked the walls of the Artillery Ground into degrees and minutes with chalk, then placed a paper box over his head so that he could not see where he was going, and wore a long canvas coat so that he could not feel his way with his legs. He tied his legs together with a two-foot string so that he could take measured strides, and carried what appeared to be a compass. Then, striding across the ground in two-foot steps, 'he seemed to perform what ships do in dead reckoning'. Grillet made two attempts but neither was a success, since his method took no account of the sea currents that would throw a ship off course. Hooke reported all this to the Royal Society that afternoon, and a second trial was arranged for the end of the month. The second trial took place in Gresham College quadrangle on 29 January 1680, with Grillet now using a hoop around his ankles instead of a string to regulate his paces. The result was another embarrassing failure, and Hooke reported to the Royal Society that afternoon that Grillet had 'erred so considerably, that he was made to understand that what he could not doe, would not suffice to answer his Proposition'.[11]

Hooke's scientific work often involved him in pleasant excursions to London's open spaces with his Royal Society friends. On 1 February 1690, two days after the Grillet fiasco, he went with Harry Hunt to try out a new model sailing ship on a pond in some nearby fields. After this outing, which recalled his childhood interest in model ships, he warmed himself up in Jonathan's, in the company of Godfrey, Lodwick, Waller and the talkative Captain Knox. On 10 February he went to a brickpond with Hunt and Waller to try the model again, but it was too windy. This did not prevent him delivering a paper to the Royal Society on 5 March arguing that taut or flat sails would convert the force of the wind into speed against the resistance of the water more efficiently than the bellying sails universally preferred by sailors and navigators. The lecture began with a self-justifying passage which shows him at his worst. He mentioned Huygens's work on gravity, which accepted the inverse square law, 'the Theory which I had the happiness first to invent', and went on to remind his listeners of his troubles and triumphs as a scientific pioneer, making the boldest claims to priority over Newton's work on gravity:

> I have experimentally verify'd the effects of propounding new Inventions to improve such as are present now in vogue; witness the improvement of Astronomical Instruments, the Spring-watches, the universality of Gravity, and the Motions of the Heavens, according to the Rules and Laws of Mechanical Motions. And yet after all the obloquy and reproaches, and unhandsome treatment I have met with for making these discoveries, I find the things themselves, in tract of Time, become to be approv'd, and come to be of general use. There are, I believe, but very few in the World now that will adhere to *Hevelius* his magnify'd Contrivances for

> Instruments with plain Sights, tho' at the same time they joyn with him in the Aspersions he hath cast upon me.[12]

In the rest of the lecture Hooke argued the mathematical case for straight over bellied sails in a way that did not convince his contemporaries any more than it convinces modern commentators. Even if he had properly understood the work of Newton and Huygens on force, impact and fluid resistance (which he had not), Hooke would not have been able to solve this apparently simple but actually very complex problem with the knowledge available to him at the time. Practical sailors would not have understood the weakness of Hooke's dynamics, but they knew which way up they wanted their ships to sail, and that practical experience was a better guide than Royal Society lectures. The best place for Hooke's taut sail would have been on Sir William Petty's embarrassingly unseaworthy catamaran.[13]

Hooke saw another opportunity to return to his old battles with Huygens and Newton in February 1690, when he was asked to give the Society an account of Huygens's newly published *Treatise on Light*. He bought the book for 3s on 8 February, and over the next ten days, when much of his time was taken up with appearances at the Court of Aldermen in a lengthy case involving one of his clients, he read the book and prepared two lectures on it for the Royal Society. In the first of these, given on 19 February, he defended his own theory of light, propounded in *Micrographia*, against Huygens's, and in the second, a week later, he focused on Huygens's theory of gravity, and used the occasion to repeat his claim that he was the source of Newton's most important ideas.

> Those proprietys of Gravity which I myself first discovered and shewed to this Society many years since . . . of late Mr. Newton has done me the favour to print and publish as his own Inventions . . . and particularly that of the Ovall figure of the Earth was read by me to this Society about 27 years since upon the occasion of the carrying the Pendulum Clocks to Sea and at two other times since, though I have had the ill fortune not to be heard. And I conceive there are some present that may very well Remember and Doe know that Mr Newton did not send up that addition to his book till some weeks after I had read & shewn the experiments and demonstrations thereof in this place.

Newton had promised a correspondent that he would do him justice, Hooke said, but until now he had done 'Just nothing', and had instead made 'a pretence of falling out and being much offended with me who should dare to chalenge what was my owne'. Unfortunately, once again Hooke made the mistake of basing his claim on the inverse square law, which he had not originated or taught to Newton, rather than the mechanics of circular orbits, which Newton had used, albeit in a much more sophisticated form, in *Principia*.[14]

Newton was irritated but not threatened by Hooke's persistent claims. When William Derham, who later edited Hooke's *Experiments and Observations*, told Newton of another of Hooke's accusations, he replied 'with greater warmth & peevishness than was usual in him . . . That he believed Dr Hook could not perform that wch he pretended to: let him give Demonstrations of it: I know he hath not the Geometry enough to do it'. Newton made a point of showing those interested in the dispute a manuscript of his dating from before 1669 which contained the essentials of his ideas on the gravity of the Moon, Earth, planets and Sun. If Newton's dating was correct, his paper predated Hooke's letters of 1679–80, and even his Cutler Lecture (*An Attempt to Prove the Motion of the Earth*) of 1670, but not his seminal paper on 'the inflection of a direct motion into a curve by a supervening attractive principle', which was read before the Royal Society on 23 May 1666. So Hooke's claim to priority, in this matter at least, was not overturned.[15]

25. The Fear of Being Forgotten

(1690–1693)

Hooke's diary is missing between 10 March 1690 and 5 December 1692, a period in which he received a doctorate from John Tillotson, the new Archbishop of Canterbury. He was still drawing a regular income as a Gresham professor, but his attempt to persuade Cutler to pay the dues that had built up since the 1683 court case was a failure, although he enlisted the support of two ecclesiastical heavyweights, Tillotson and Thomas Sprat, Bishop of Rochester.[1] The replacement of his defiant housekeeper Martha with the churchgoing Mary also took place in these years.

Hooke was by now well into that phase of life in which deaths and funerals were more common among his friends than marriages and baptisms. On 30 December 1691 he lost his old patron and friend Robert Boyle, who was at last carried off by the disorder that had kept him virtually housebound for years. He died in Lady Ranelagh's house in Pall Mall at the age of sixty-five, only a week after his sister. Hooke no doubt went to Boyle's funeral in St Martin's in the Fields, and heard the famous Bishop Burnet, his favourite preacher, praise Boyle's many virtues. Death brought indignities even to the greatest names. Just over a year after his death Boyle's collection of chemistry books was for sale on an open-air bookstall in Moorfields. Perhaps a sadder personal loss for Hooke was the death of his great friend and chess partner, Theodore Haak, who was buried in St Andrew Holborn in May 1690. He left unpublished a collection of six thousand proverbs that he had translated from German and Spanish into English.

Hooke was still supervising the Haberdashers' school and almshouses in Hoxton, which were not finished until 1695. This was not a project for which he simply provided a plan and a model, leaving the construction itself to others. He was responsible for employing and paying craftsmen, checking their work and making regular visits to the site, which was less than a mile from Gresham College. When the carpenter Matthew Banks refused to sign the contract that the Haberdashers had prepared for him in June 1691, it was Hooke's job to sort the problem out and find terms acceptable to both sides. Banks would interpret his obligations in his own way, and he might cost more than other carpenters, Hooke warned Sir Richard Levett, Master of the Haberdashers' Company, 'but Sir we shall suddenly have occasion for a Carpenter, or be forced to delay the work'. His advice was to prepare to sign with another craftsman, to bring Banks into line. A further delay was threatened by the lack of stone for some small columns, but Hooke intended to redesign them in 'rubbed and gaged brickwork'. Meanwhile, bricklayers were at work on the walls of the cellar storey, and hoped to be almost finished by the end of the next week.[2] Building was well advanced by the end of 1692, but the Company was not yet ready to pay Hooke anything for his work.

As Surveyor to the Dean and Chapter, Hooke had a steady flow of repair work at Westminster Abbey and the various buildings connected with it. Other jobs that he took on between 1690 and 1692, and which were still current when the diary reappears in December 1692, included some flood-prevention work for Sir Robert Southwell, the Royal Society's President from 1690 to 1695, on his low-lying estate in the Severn valley, and some buildings across the river in Lambeth. It is also possible that at around this time Hooke drew some plans for the Plymouth Royal Dockyards, including the officers' houses, the great storehouse and the ropehouse. The plans are among his drawings in the British Library, and though the building was supervised by Edmund Dummer, in the view of Sir Nikolaus Pevsner and Bridget Cherry 'the design is closest in style to that of Robert Hooke'. The officers' houses and offices, which survived the destruction of most of the dockyard in the Second World War, are 'the first example in any of the English dockyards of a unified approach for officers' housing . . . a remarkably early date for a terrace of individual houses treated as a long palace front.'[3]

In the spring of 1690 Hooke came across the writings of the notorious Elizabethan 'magician' and international secret agent, John

Dee. His first instinct when he saw Dee's *Converse with Spirits* years earlier was to ignore it as a superstitious and irrelevant work, but in 1690 he discovered a pamphlet of Dee's, dedicated to the Archbishop of Canterbury, which declared Dee's honest search for truth, and listed the titles of all his works. Hooke's interest was aroused, he told the Royal Society on 11 June, 'by the Titles whereof he seeming to be an extraordinary Man, both for Learning, Ingenuity and Industry, I had a desire to peruse the Book with a little more Attention than I had formerly Thoughts of'. Its elaborate and superstitious title and its 'long and frighting preface' could be ignored, he said, 'since I conceived both these to have been the Ingenuity of the Publisher, to make the Book sell better'. On the face of it Dee's *Converse with Spirits* was about the Devil and his imps, and had nothing to do with natural philosophy. Much of it seemed to be 'a Rhapsody of incoherent and unintelligible Whimsies of Prayers and Praises, Invocations and Apparitions of Spirits, strange Characters, uncouth and unintelligible Names, Words and Sentences, and Relations of incredible Occurrences', and he often threw it aside and forgot it. But finally he decided to ignore the fact that it had been condemned by learned men and forgotten for thirty years, and try to unlock its secrets. His surprising conclusion was that it was a work of cryptography, a coded account of Dee's travels and discoveries on Queen Elizabeth's behalf in Germany and Poland, devised to make it possible to send intelligence to the English court without arousing suspicion. Hooke knew that Dee was an experienced cryptographer, and thought he could see hidden meanings in some apparently meaningless passages. The key to the whole text, he guessed, was the Book of Enoch, of which Dee had always taken special care, and since Dee was skilled in optics and mechanics, its hidden subject matter might be of interest to natural philosophers.[4]

Some modern writers have suggested that Hooke's interest in Dee shows that he, like Newton, was fascinated by medieval magic and believed that inexplicable or 'occult' forces played an essential part in the working of the natural order.[5] Gravity itself, in Newton's philosophy, was an ultimately inexplicable occult force. Hooke, on the other hand, was sure that it could eventually be understood, and that it would prove to be a form of matter in motion, like light, sound, heat, magnetism and everything else in the universe. Several times in his 1689 diary he suggested that he had discovered its secret ('Satisfyd of Gravity', 'Perfected Theory of gravitation'), and both Hooke and Wren thought that Newton's belief in gravity as an occult and mysterious force was laughable.

Although Hooke was re-elected to the Royal Society's Council in December 1690 he was still feeling angry and undervalued. Many of his innovations, he thought, had been ignored or inadequately understood when he first proposed them, generally because they conflicted with prevailing beliefs at the time. Thus he had not been urged to develop them, and sometimes, as in the case of his work on the shape and history of the Earth, he had been actively discouraged from doing so. Now, when others had caught up with his thinking, his long-forgotten inventions were being reinvented and credited to younger scientists. He made his resentment known in a series of Cutler lectures delivered to the Royal Society in December 1690.

First, he spoke of his proposition, 'made known to this Society near thirty Years since', that gravity was greater near the Poles and weaker at the equator, a view which was supported by the behaviour of pendulum clocks at sea. Though Hooke had often repeated the proposition the fact that it contradicted the opinions then in vogue had prevented its further development, he said, and it was only when Newton was converted to the idea in 1687 that any notice was taken of it. Now, thanks to the cooperation of Captain Knox, Hooke intended to put the question to a thorough test at last. 'I have got a Pendulum Watch fitted for that Purpose, which I now produce'.

There was another instrument that he hoped to send on Captain Knox's next voyage. His innovations in the design of the barometer, an instrument that had undergone so many changes and refinements since 1660, were especially liable to be forgotten, or to be confused with improvements associated with others, including Sir Samuel Morland and the London clockmaker Daniel Quare. Hooke's two-liquid barometer, invented in 1668, had been 're-invented' by Huygens four years later, and his three-liquid version was almost unknown in France, where Guillaume Amontons proposed the idea as his own in 1685. The problem with traditional barometers, he said, was that their readings indicated the combined effects of air pressure and temperature, but in 1668 he had made a combined thermometer and barometer calibrated for accuracy at all temperatures.

This simple device, which he claimed could have saved thousands of pounds and hundreds of lives, had been ignored because it was based on unfamiliar principles, but now, he said, 'it ought not to be any longer neglected, but rather to be made and try'd as soon as possible'. It was his nature to remember all these wasted opportunities as the Society's fault rather than his own, but looking back over Oldenburg's minutes for January 1668 we can see that it was the Society that had urged Hooke to construct his marine barometer and organize its sea trials, and Hooke who had let the moment pass.

In his next lecture, on 10 December 1690, Hooke pursued his tale of neglect and injustice. It was only too common, he said, for his inventions to be dismissed as impractical, and then, some years later, for them to be rediscovered and produced in the name of another, and praised for their utility. He began with the example of the backstaff, which he had fitted with a lens to improve the clarity of the shadow which the navigator had to read, only to find the innovation attributed to his younger rival John Flamsteed. The lens is known to this day as the 'Flamsteed glass'.

> The Instrument which I shew'd the Society, some Years before the Sickness, by making use of a Telescope-glass, instead of the small hole or slit of the Shadow-vane of a Back-staff, was not made use of 'till about ten Years after, and yet now it meets with general approbation, and is in constant use, and pretended to be the invention of another, tho' my shewing thereof was Printed in the History of the Royal Society. It cannot well be expected that I should spend my Time and Studies in inventing Instruments, and be at the expense of making and putting into practice those which I have contriv'd, without receiving any Benefit or Assistance from those to whom they may be of use, or any other. It may, I conceive, be judg'd sufficient, by reasonable Men, for the Inventor to contrive and describe the Means and Ways how such as have occasion, or desire of experimenting the thing, may, with ease enough, put the same in practice; . . . his Reward . . . is commonly ill Treatment, and not only rough Usage from those that envy his acquists, but even from the Artificers themselves, for whose sake he has labour'd; whilst another that adds some small matter to it, is enrich'd thereby, but the first discoverer is dismist with Contempt and Impoverishment.

Undeterred by what he saw as years of ingratitude and obstruction, Hooke announced a new invention to enable seamen to find their latitude accurately even at night or in mist, when existing methods, which depended on a good view of the horizon, were useless. This was a quadrant with a reflecting plate and a telescope lens (both of which were Hooke's innovations in the 1660s) and with a plumb line hanging between the mirror and the lens to enable the navigator to hold the instrument level without using the horizon. To steady the plumb line and make readings easier, its weight (the plumbet) should hang into a little pot of water. What mariners also needed, Hooke said, was a small telescope accurate enough to see the eclipses of Jupiter's moons, which they could use in conjunction with the astronomical tables produced by Cassini and Halley (Flamsteed was pointedly ignored) to find their position at sea. He made bold promises: 'I shall procure an ingenious Workman, who shall provide such Telescopes fit for that purpose; so that nothing will then be wanting to compleat the use of those Eclipses for the discovery of the Longitudes of such parts as they shall be observable in, save only an exact Time-keeper to observe the precise times of such appearances, which I can also accommodate them withal.' Hooke protested that he 'should have done it long since, if I had not been discourag'd by undeserved Troubles. But I see it to be the general Fate of all such as make any new discoveries, and therefore bear it with more patience'.

There was a particular problem, Hooke argued in his next lecture, for those who invented apparently simple devices. 'There are many things, that before they are discover'd, are look'd upon as impossible, which yet, when they are found, are said to be known by every one, the inventor only excepted, who must pass for an Ignoramus.' Many years ago, he said, he had demonstrated a marine waywiser, which would measure the speed and the distance travelled by a ship and all the twists and turns of its course, though not the effects of currents. 'Whether it has been since try'd I know not, yet I have many years since that, heard of one or two who were getting a Patent for a like Instrument; whether they succeeded or not, I have not inquir'd, for I freely imparted it for the general Good'. He now had a new device, based on the wind gauge of his weather clock, which could be fixed to a mast to measure wind speed and send the information to the navigator below. This would replace guesswork with measurement, and allow the performance of two ships, or two types of sail – flat or bellied – to be compared fairly, taking wind conditions into account. Hooke had presented both these devices, the waywiser and the wind gauge, to the Royal Society in November 1683.[6]

In 1691 Hooke came close to entering the London craft and manufacturing world that he knew so well. His involvement with church-building, and especially his work on Westminster Abbey, had given him a particular interest in the production of coloured glass. In February 1689 he had met Giles, a celebrated glass-painter from York, who made 'the true red glasse', and a year later he was still trying to find out how the glass was made. Mr Jed, whom he met in Jonathan's in January 1690, knew the secret, and Hooke tried to extract it from him: 'Red glass made with antimony, Query how?' When he showed a piece of the glass to the Society on 15 January 1690 he had still not discovered the secret of its manufacture, but could only tell them that it was 'clear glass that upon the blast of a Lamp turned a fair Ruby redd, which was made by one Mr Jedd by a preparation of Antimony'. But over the next year, in the period not covered by his diary, Hooke must have discovered the secret, or extracted it from Jed. On 19 May 1691 he was granted a ten-year patent, in partnership with Christopher Dodsworth, 'for the sole use of mixing metal, so as to make glass for windows of more lustre and beauty than that heretofore made in England, red crystal glass of all sorts'. Along with this, the pair were granted a patent over 'the art of casting glass, particularly looking-glass plates, much larger than ever blown in England or foreign parts'. How much Hooke contributed to this joint enterprise is not known, but since glass manufacture does not feature in his diary for 1693 it is likely that he was simply staking a claim with the intention of exploiting it if the opportunity arose. In October 1691 preparations were made for the incorporation of a new City guild or livery company, the Company of Glass Makers, with Hooke as one of its wardens, but nothing came of the idea.[7]

Hooke continued his campaign to rescue his early inventions from oblivion in December 1691, when he lectured on the depth-sounding device he had invented in the early 1660s. The original depth-sounder was a buoyant wooden ball attached to a heavy stone by a spring-hook in such a way that when the device hit the seabed the stone would be uncoupled, allowing the ball to return rapidly to the surface.

The time taken for the ball to reappear would indicate the depth of the water. The practical problem that sailors had found with the depth sounder was that it rose to the surface in unpredictable places, and that it was therefore impossible to measure the time it had spent under water. Hooke's refinement, which he had announced many years ago, was to add a container to the wooden ball, 'a Cylinder, Cone, or Hyperbolick Trumpet', which was pierced to take in a small amount of water when it was submerged. Since more water would enter this vessel the longer it was submerged and the deeper it went, anyone weighing it when it was recovered, even after a long delay, would be able to calculate the depth to which it had sunk. With such an instrument, Hooke went on, the contours of the deepest seas could be mapped. He was sure of the scientific basis of his device, because he had established by experiment that water pressure, and thus the speed at which the punctured vessel would fill up, was proportionate to depth, but he admitted that variations in temperature or water density at depth, and the unknown ways in which gravity changed nearer the centre of the Earth, might distort his calculations.

To solve these problems, he had invented a set of related instruments, which he described to the Society in detail. First, he had a depth sounder which would not be affected by temperature, water density, pressure or gravity. Instead of a simple wooden ball with its detachable stone weight, he had designed a hollowed ball which contained two vanes which were turned by the flow of water as the ball sank. Each revolution of the vanes turned a counting wheel, which would record the total number of revolutions. When the weight was released at the seabed this would also release a spring that would shut a small door in the hollowed ball, stopping the vanes turning any more. For greater accuracy, he devised a more complex device with two boxes containing revolving vanes and counting wheels, one of which opened at the seabed when the other closed. However complicated this marine waywiser sounded, Hooke was sure it could be made for under a crown (five shillings), and designed so that almost anyone could operate it. In addition, he had designed a minimum-temperature alcohol thermometer in which a small piston would be drawn down a strong tube as the alcohol inside it contracted in the cold at the bottom of the sea, but in which a valve would allow the alcohol to pass through the piston as it expanded again. In case pressure affected the results, the perforated cone could be sent down with the thermometer and the waywiser, so that temperature, pressure and depth could be measured against each other. Finally, he explained that the mechanical waywiser could be used in association with the self-closing bucket to take samples of seawater from any depth, not just at the seabed, to test its salinity or other qualities. The rotations of the counting wheels could be set like an alarm clock to open a spring catch and release the weight at a certain number of fathoms, and the whole device would rise, sealing the bucket, from that point. Pressure or temperature could be taken at a desired depth in exactly the same way, using his perforated cone or thermometer in conjunction with his mechanical waywiser. So here was Hooke, motivated by a combination of personal ambition and a commitment to the expansion of natural knowledge, doing his best to establish the science of oceanography more than a century before its time. Scientists and mariners took little interest in the subject for the next hundred years, and when oceanographic study intensified in the mid-nineteenth century, mainly because of the laying of submarine telegraphic cables, hardly anyone remembered the debt that was owed to Robert Hooke.[8]

Although Hooke found it difficult, as many people do, to welcome the achievements of younger men that threw his own work into the shade, he was also worried that the great work that he and others had started earlier in the century was not being continued and improved upon by a younger generation. In a retrospective lecture on the history of optical instruments in February 1692, Hooke described the discoveries made with telescopes and microscopes, 'many of which he shewed to have been made by himself'.[9] He reminded his listeners that he had discovered the permanent spot on Jupiter, which proved that it rotated on its axis, that he had pioneered the study of comets, and used a vertical telescope to see stars in daylight and to establish a stellar parallax that proved the motion of the Earth. But lately, he thought, interest in the use of telescopes had cooled, and beliefs which should have been discredited by the new philosophy, including astrology, the Aristotelian elements and the Cartesian Vortices, seemed to be finding new converts and defenders. The present use of microscopes was especially disappointing. They were 'now reduced almost to a single Votary, which is Mr Leeuwenhoek; besides whom, I hear of none that make any other use of that Instrument, but for

Diversion and Pastime'. Why had this come about? The early pioneers were 'gone off the stage', and among the present generation it was believed that there was nothing left to discover, or that what was left would bring no profit. The fact that Leeuwenhoek was working virtually alone was 'not for Want of considerable Materials to be discover'd, but for Want of the inquisitive Genius of the present Age'. Hooke, who had been unusual among his colleagues in seeing natural philosophy as a way of making a living, felt able to condemn a new generation of mercenaries who preferred cash to knowledge: 'all other notions are insipid with them, besides such as bring ready Money'. He was, as always, confident that knowledge and natural philosophy would triumph in the end: 'notwithstanding all this, there is a real Beauty and Allurement in Truth, that will produce some Votaries in the worst of Times, and shine out, and dispel the Clouds of Error that encompass it'.[10]

To hear an old scientist recounting his achievements might have been tiresome, but Hooke was naturally anxious that his forgotten work should not be credited to others. When Wallis suggested using a concave mirror to reflect light onto an object under a microscope, Hooke protested that he had done this 'twenty years since', and when the prolific and inventive Halley produced a draft in March 1692 of a quadrant with a telescopic sight and a mirror 'so that both the Sun and Horizon . . . coalesce in the same point', and an adjusting screw 'whose parts would divide the Angle most minutely' Hooke felt compelled to remind the Society (as he had told Halley walking home from Jonathan's exactly ten years earlier) that his own identical invention had been described in the 1660s.[11] Despite an occasional dispute, Hooke and Halley seem to have remained good friends, sharing the same interests, the same coffee houses and a dislike of John Flamsteed. Hooke's new work in the spring of 1692, a study of the construction of the Tower of Babel from classical sources, was less likely to attract interlopers.

In December 1692, when the diary reappears, Hooke had just been confirmed as a member of the Royal Society's Council, and was engaged in several building projects. A job he had recently taken on for the President of the Royal Society, Sir Robert Southwell, involved an interesting combination of building and natural philosophy. Southwell's estate at King's Weston, near Bristol, had been inundated by a very high Severn tide in 1687, and was vulnerable to further flooding. Southwell, along with other landlords in the district, was responsible for his flood defences, and he turned to Hooke, a great expert on water pressure, rates of flow, the strength of building materials and related topics, to make sure that the sluices, quays, wharves, breakwaters and waterwheels he was intending to construct would be designed to the best modern standards. So we find Hooke on 9 December 1692 'explicating the key, the wharfing, the mole, and the sluces' to Southwell and his son, and discussing the qualities of mortar. On 2 January Southwell came to Hooke's rooms to collect models of the sluice and the mill-gates Hooke and Hunt had made for him, and together they tested the holding power of different types of wall. Hooke provided Southwell with papers explaining how to calculate water pressure and the necessary strength of retaining walls, and on 10 January he spoke to him and his son at length about the best system of drainage. 'I diswaded from rake and captston, advised ditching and wall with buried trunks to carry off the water; explained to Mr Southwell all the papers of pressure, resistance of wall, centre of gravity.' In February 1693 Hooke provided Southwell with a design for sluicegates, and an estimate of the quantity of timber he would need, and finally, in May, he checked and corrected the designs for Southwell's dyke and flood gates, and drew up the timber contract. So Southwell was given a course of lessons in hydraulics by one of the greatest experts on the subject, and had his flood defences designed on scientific principles. In return, Hooke got at least twenty guineas, and a friend in the upper ranks of the Royal Society.

Though Hooke's great days as a scientist were over, his career as a builder and architect was as busy as ever, with important projects in Westminster, Lambeth and Hoxton. Work on the Haberdashers' school and hospital was well advanced, and Hooke often conducted negotiations with builders, discussed their bills with the company's master and wardens, and inspected the site. Even the carving of the inscription on Aske's statue was closely supervised by him. He visited Hoxton often, generally walking across the fields with John Godfrey, to check that his orders were being followed. On 20 December 1692 he discovered that almost every craftsman had ignored his instructions: 'both gable ends brickd up without order; also a flat about cupola, Lucern lights backward; Iron cramps etc without reason, salt mouldring stone: Plumbers laying high gutters, bricklayers tiling front etc: all without my order'. From time to time the Haberdashers committee added refinements to the building – a slated front, lead-covered pediments – which added to the cost of the palatial structure Hooke had designed. Later, when his friend Waller asked him why the Hospital cost so much more than he had estimated it would, he blamed the Haberdashers for not giving him control over the design and the builders' contracts. It was not his fault, he told Waller, 'but partly by new additions and alterations of the first Design, and chiefly by his not procuring and agreeing with the Workmen himself, which if he had done, as he said, he would have ingag'd it should have come to little or no more than his first propos'd Sum.'[12]

There was a difficult meeting at Haberdashers' Hall on 10 March 1693, when Hooke confronted the bricklayer and carpenters in front of the Committee, 'spak plain' and apparently resisted their efforts to 'lay all on me'. Good craftsmen knew their worth and stood their ground. When the wardens met Nicoll, the bricklayer, two weeks later he 'asserted that the usuall prise for Rubbed and Gaged work [where the bricks were rubbed to a uniform size] was 8d per foot and that he would doe it soe'. Despite the efforts of Hooke and the Haberdashers to keep the cost of the work down the original design had been too ostentatious. By December 1693 the Hospital and its fittings had cost £11,786, and the company, believing they had been overcharged by £600, refused to pay. The Haberdashers tried to economize in 1694 by abandoning Hooke's planned cupola, but the total cost, when the Hospital took in its first residents in 1695, was so great that there was not enough left in Aske's generous bequest to maintain the building or its occupants properly. As a home for twenty retired haberdashers and twenty schoolboys, Hooke's building, which would fit nicely along one side of Trafalgar Square, was completely out of proportion. In 1708 *A New View of London* described the finest building in Hoxton:

> The new Hospital here is a sumptuous Edifice built of Brick and Stone, with a Piazza at the Front, where is an Ambulatory 340 feet in length, constituted by the Elevation of that part in Stone Columns of the Tuscan Order . . . under the Pediment is

> a Nich wherein stands the Figure of the Founder carved in full proportion.

Seventy years later, when the consequences of Hooke's and the Haberdashers' extravagance were clearer, Entick's *History of London* took a harsher view: 'But a moiety of the sum being shamefully squandered in erecting an Edifice fitter for a Palace than an Alms-house to the great reproach of those concerned, the Company were obliged to turn off the boys for several years.' Without the money to maintain the vast building, it fell into disrepair, and in 1824 it was demolished and replaced for £13,500 by the building that still occupies the west side of Aske Square, next to the Haberdashers' housing estate, today.[13]

Hooke's work in the little riverside settlement of Lambeth, which seems to have included a 'house, greenhouse and brewhouse', as well as a moat, garden and birdcage, probably near Lambeth Palace, also ran into budgeting problems. Between January and April 1693 Hooke did his best to persuade the builder, Hayward, to adjust his bills before they were presented to the client. Hooke was used to these wrangles, but this project seems to have troubled him: at the end of April he had a 'Strange frightfull dream of Lambeth house'.[14]

In Westminster Abbey and School, Hooke earned his annual £20 by acting as a maintenance manager and building supervisor, hiring glaziers, plumbers, plasterers, bricklayers, masons and smiths, checking their work and arguing over their inflated bills. Among other tasks commissioned by Dr Busby, Hooke undertook to repair the Cloister House and build a wall around all or part of Tothill Fields, the large area of pasture and playground south-west of Horseferry Road. More importantly, the roof, walls and windows of the Abbey itself were in need of repair. On 4 May 1693 Hooke dined with Busby and 'viewd Henry 7 chapel, Abby butteress, mend key, viewd chimny in little Cloyster', and three weeks later he 'viewd Abby Window and considerd of its repair'. Hooke's plans to take down the battlements over some of the windows and to brick up and plaster the gable end at the north of the Abbey were approved by the Chapter of the Abbey at the end of May. On 10 June work began on holding the North Window in place with cramps (metal bars), and Dr Birch, the Prebend of the Abbey, approved Hooke's choice of plasterer for the gable end. After discussions with Wren, it was decided to model the plastering of the Abbey wall and some carving on its ribs or arches on what had recently been done in St Paul's. There was some sort of difficulty on 20 June, when Hooke met the Westminster Chapter with the Bishop of Rochester: 'To Sr Chr. Wren, then to the chapter at Westminster. I stayd expecting mason and carp[enter] to confront: the face of them changed. Rochester shy, Orles malapert [impudent], Birch sly; I spoke to Sr Chr. Wren. Busby cool.' Hooke's close working relationship with Wren, who had to approve and countersign all these Abbey works, helped him through most of these difficult moments. On 11 July Wren signed the note the Chapter needed, and Hooke got the work under way: 'Directed Tufnell and Collins about window gable for cramps, 2 stone, 4 punchions [supporting posts], place brick etc . . . with Dr B viewd and measured wall Tuttle feild'. Later that month Dr Birch ordered work on the faulty arches to begin, and the scaffolding started going up. One of Hooke's last surviving diary entries, on 29 July 1693, shows the work in progress, and new faults in the Abbey's lead roof coming to light: 'Dobleday saw gable end brickd to 2nd cramp, scaffolding for the rest: South gable end worse: S.E. leads very bad: H7 Chappel leading.'

Hooke was often called upon to inspect dilapidated churches or houses and to recommend repairs that others would carry out. On 24 May 1693 a messenger told him he was needed at the medieval and pre-Fire church of All Hallows, Barking, next to the Tower of London. Hooke went at once, and met the minister and the vestry. 'I found beams, plates, corbells all very much rotted. Informd it necessary to be repaired forthwith. 20s.' On 8 July he inspected the towers of two City churches which had been patched up after the Fire, St Vedast and St Dunstan. His opinion of their condition is not recorded, but both were demolished and replaced with Wren or Hawksmoor steeples within a few years. Hooke's other work in the City included his usual views or inspections of houses and building sites, and extensive rebuilding of the Fleet wall, part of which collapsed in May 1693.

Hooke was not a man of great influence, but through his work for the Abbey, the City and the Livery companies he made some powerful friends, whose help he was able to call on in times of difficulty. He had not been paid by Cutler since the court case of 1683, but when his benefactor died on 15 April 1693 Hooke saw his chance to recover his arrears and secure a steady payment in future. Cutler left most of his very large fortune to his daughter, Lady Radnor, and her husband, a substantial sum to his nephew, Edmund Boulter, and nothing to Hooke. Hooke knew Boulter, and hoped that he would persuade Lord Radnor to resume the payments. Rather than relying on the Royal Society, Hooke asked Sir Richard Levett, Master of the Haberdashers' Company, to use his influence with Boulter. After one or two reminders Levett spoke to Boulter in mid-May, but Hooke decided on a second line of attack. He persuaded Mr Moor, a City official and the husband of his good friend Mrs Moor, to go with him to Boulter's house on 20 May. 'With Mr Moor to Mr Boulter, met him at his Door; he sayd he would consider of it soe soone as some other Acc[ounts] were past with Ld Radnor, to whom he was going, that he knew of it, and had payd me the last £200.' Levett and Hooke maintained a steady pressure, but Boulter found it easy enough to put them off. On 7 June he promised Levett 'he would shortly consider of it', a month later he told Hooke 'he had not yet had time to settle it with Ld Radnor, but would speedily', and on 31 July he said he was waiting for a report on the affair. The diary ends on 8 August with Hooke not a penny richer as a result of Cutler's death.[15]

While this was going on a more embarrassing problem arose. On 20 May Hooke heard that the bankrupt bookseller and map-maker Moses Pitt, who was free again after two terrible years in the Fleet debtors' prison, had accused Hooke of owing him £400. This was almost the opposite of the true position, since Pitt had never paid Hooke the £400 he had promised him for his work on his atlas in 1678. Hooke was worried about the threat to his good name, and perhaps to his pocket, and he turned to his influential friend for help. 'I acquainted Mr Moor with it, he sayd he would secure me against his claim for a dish of Coffe.' Ten days later Moor and Hooke visited Carbonell, who was spreading Pitt's story, and told him what had really happened over Pitt's Atlas. Apparently Carbonell was unconvinced, and Moor saw him again on 7 June to 'shew Pitts praetence'. By the time Hooke 'met rasc. Pitt' on 20 June the matter seems to have been dropped.

Hooke's old employer, the Royal Society, had paid him nothing since 1688, though he looked after the publication of *Philosophical Transactions* between March and July 1693 and often presented papers to their meetings. Hooke thought that he was still entitled to a salary, and spoke to Southwell, the President, about his 'arrears' at the end of January. Two weeks later he asked the Vice-President, Sir John Hoskins, to pay his arrears, but 'Hoskins shuffled it off'. This was no great surprise. As Sir Robert Southwell told him the next day, the 'Pontac Counsell' – the influential group of richer Royal Society members who met and dined at Pontac's in the 1690s – had an 'aversion' to Hooke, and they knew that he was no longer the Society's indispensable source of new ideas and inventions.

Hooke's presentations to Society meetings in the spring and summer of 1693 had a familiar and retrospective ring – discussions of Burnet's work on the history of the Earth, speculations about ancient Chinese printing, reminiscences of his battle with Hevelius over telescope sights, a microscopic display of a gnat and mussels in coral, and a series of lectures interpreting Ovid's fables as codified accounts of ancient earthquakes. When he asserted his 'invention of Cœlestiall Motions' once more on 26 July it was not even thought worthy of recording in the Journal Book. If we look beyond the Royal Society's journals, it is clear that the range of Hooke's scientific and technical interests was still wide, even though his activity had diminished with age. In December 1692 and January 1693 he organized the printing and distribution of Simon de la Loubère's *New Historical Relation of the Kingdom of Siam, wherein a full account is given of the Chinese way of Arithmetick and Mathematick*, which his friend Alexander Pitfield had edited and translated from the French. Hooke, who was strongly in favour of the publication of travellers' accounts of far-off places, saw the book through the press, and delivered twelve copies to Pitfield on 11 January 1693. Another of Hooke's closest friends, Francis Lodwick, was also a linguist, and for years they had both toyed with the idea of studying Chinese as a model for their projected universal language. In summer 1693 they got an opportunity to try their linguistic skills on some real Chinese visitors. Hooke heard about them at Jeremy's coffee house on 2 June, and met them on 10 July, when they came to Jonathan's with Sir James Houblon, a City alderman. Two days later Hooke's diary shows him taking action to track the men down: 'Sought out Chinese: visited

Lod ... With Lod to Chinese: I shewed him Repository, Atlas Sin[ensis], Chin Almanack etc'. Finally, on 31 July, they met them over tea, with disappointing results: 'With Lod at the Chinese, tea: I could learn little, 8 or 10 characters pronounced all alike but of differing signification.' And this is the last we hear of Hooke's ambition to learn Chinese, and almost the last we hear of Francis Lodwick, who died in 1694.

Hooke still worked occasionally with his old friend and ally Thomas Tompion, whose thriving business now supplied clocks and barometers to William and Mary. Tompion's best ideas now came from Edward Barlow, the inventor of the repeating watch and the cylinder escapement, but Hooke still plied him with advice and employed him to make components for scientific instruments, including a wheel for his new 'logarithm Compasse'. Hooke also worked with John Marshall, one of London's leading lens makers. Marshall made two lamps for Sir Robert Southwell, and showed Hooke how he could make great concave lenses, twenty inches across, and an object lens that could be fitted into a window. With his worsening eyesight, Hooke also had a personal interest in Marshall's skills as an optician. In July 1693 he chose a new pair of spectacles from the eight pairs Marshall had made, 'all ground by his new way', and paid him 18d for them. At the same time he left his pocket lens to be 'new polisht'. The next month Marshall showed Hooke his new grinding tools, for lenses of different focal lengths, perhaps in the hope that Hooke would help him win the Royal Society's endorsement of his new technique.[16]

Hooke's consecutive diary entries end on 8 August 1693, and with them ends our intimate knowledge of the gossiping, coffee-drinking, building, bargaining, cajoling, enquiring, walking, book-buying, reading and measuring that filled his long and busy days. But he still had ten years to live, and the diary does not leave the impression that he was broken in health or spirit when he entered his fifty-ninth year in July 1693. His eyes were still sharp enough for him to proofread Pitfield's *History of Siam*, and the constant ill-health that afflicted him in the 1670s was either gone or ignored. True, he complained of giddiness in August, but a walk to Kingsland with Lodwick and Godfrey refreshed him. There were troublesome incidents, too: he cracked his forehead on an overhanging shop sign in the Old Bailey in July (an everyday hazard in London), his hat blew into a gutter in June, he was harassed by 'a cheating quack' as he walked to the Temple, and he was sometimes upset by 'spies' or intruders who wandered into his rooms. On 21 March, the day he saw the melancholy sight of Robert Boyle's chemistry books for sale in Moorfields, 'an impudent cheesemonger' rushed into his closet, along with 'a Rabble of boys'. This is a reminder that Gresham College was far from being a tranquil refuge in a noisy city. Over ten years earlier the Gresham committee had complained about 'a rude multitude of Boyes and fellows with footballs & stones wilfully thrown', and the problem had not been solved.[17]

The diary does not suggest that Hooke was unduly morose or bad-tempered, though there were times when he used his little book to express his irritation with the stupidity of others: 'Royal Society met: little said: nonsense about infinities', and 'At Jon. Lod, Spen, Vincent, Ashby. Not a farth[ing] for a cartload of Hypotheses.' But the prevailing impression is of pleasure in the variety of his work, a love of company, a circle of devoted and interesting friends, and an addiction to coffee and coffee-house conversation, whether it was 'venturous', 'smut', or just 'chat'.[18]

26. Hooke's Last Years

(1693–1703)

LET US NOT HURRY Robert Hooke into his grave. In the middle of the 1690s he remained an important architect and builder, and was probably involved in some private commissions which are unknown to us. The Haberdashers' hospital was not completed until 1695, and Hooke's work as surveyor of Westminster continued until January 1697, when James Broughton took over. Sir Christopher Wren became surveyor of the Abbey in 1698. But as a City Surveyor and churchbuilder it is likely that 1693 was his last really active year. He was paid £50 a year for his work on City churches in 1691, 1692 and 1693, but little after that, and his last known report on a view of a disputed property in the City is dated 14 December 1693.[1] In the world of natural philosophy he was still an active figure, defending his reputation as an inventor, struggling to convince a sceptical or indifferent Royal Society that the Earth had been shaped by floods and earthquakes, reporting on the work of others, and contributing some new ideas of his own. He no longer had the energy or originality to drive the Society's experimental programme as he had done for so many years, but at least he insisted on raising serious issues in a club which seemed to get more pleasure from gawping at seven-legged puppies and pigs with no snouts. In spite of his growing infirmity and cantankerousness he was elected to the Royal Society's Council every November from 1689 to 1695, and again in the elections of 1697, 1698 and 1700.

On 1 November 1693 Hooke demonstrated John Marshall's ten-inch wide telescope lens, which could be fitted into a window and viewed from twenty-five feet away, to the Royal Society. Marshall had shown it to him a year earlier, and the two had often worked together since then. Marshall appeared before the Society later in

November, and explained his new way of making multiple spectacle lenses, which involved grinding several lenses at once while they were cemented to the same surface. Hooke was asked, along with Halley, to report on Marshall's method to enable the Society to decide whether to grant his request for a testimonial. Since Hooke had been wearing a pair of spectacles made by this method for five months, it is not surprising that his report was positive, and Marshall got his certificate. Hooke's support probably helped him to retain this privilege in November and December 1694, when rival London spectacle makers complained to the Society that Marshall's method was neither new nor effective.[2]

In the winter of 1693–4 Hooke lectured on microscopes, telescopes and combustion. But among the reminiscences and self-justification there were still some new ideas, for those who were listening. In 1694 Captain Robert Knox was dismissed from the service of the East India Company and turned up in London again. Hooke met him in the spring, and spent several weeks in May and June entertaining the Society with stories and curious objects provided by the seafarer. He displayed the utensils Knox had carried and the clothes he had worn when he escaped from Ceylon, showed a specimen of pickled pepper, and gave a full account of the shape and uses of the mighty taliput tree, whose gigantic leaves, Hooke assured them, were meant to nourish the tree, not to provide shade for humans. This may seem obvious to us, but it was not to most thinkers of Hooke's day, who agreed with Richard Bentley FRS, the first Boyle lecturer, that God had created all things 'principally for the benefit and pleasure of man'. The general view, even in the early eighteenth century, was that the ape's antics, the nightingale's song, the cow's flesh, the horse's strength, the herb's culinary and medicinal value, the palm tree's shade, were provided for human enjoyment or benefit, not for their own.[3]

Hooke was worried by the fact that the French Academy was better at publishing its achievements than the Royal Society, and that academicians were taking credit for English or Italian ideas. Thinking of his own work on petrified wood and sea creatures, he complained that public and Royal Society disapproval had discouraged him from publishing his ideas. But now the French Academy had published some work by de la Hire in which he described petrified wood, just as Hooke had observed and described it in *Micrographia* thirty years earlier. Hooke felt enormous frustration that while the French could accept the reality of petrified organic matter the Royal Society, after so many years, could not. 'Must these be questioned or rejected, only because such Substances are found in places where we cannot give particular Histories of their pristine Estate, and how they come to be there placed and transformed; or because possibly we are not able to produce patterns of Creatures now at hand, and in being, which are of exactly the same Shape and Magnitude'? Would the absence of documentary evidence for their construction lead us to call the Pyramids 'tricks of nature'? Hooke gave the Royal Society a lesson in the mutability of species that had more in common with the thinking of the later eighteenth century than that of his own day, and issued a challenge to his timid and conservative colleagues to live up to the ideals of observation and open-mindedness that had motivated the founders of their Society:

> Certainly there are many *Species* of Nature that we have never seen, and there may have been also many such *Species* in former Ages of the World that may not be in being at present, and many variations of those *Species* now, which may not have had a Being in former Times: We see what variety of *Species*, variety of Soils and Climates, and other Circumstantial Accidents do produce . . . when we consider how great a part of the preceding Time has been . . . unrecorded, one may easily believe that many Changes may have happened to the Earth, of which we can have no written History or Accounts. And to me it seems very absurd to conclude, that from the beginning things have continued in the same state that we now find them, since we find everything to change and vary in our own remembrance; certainly 'tis a vain thing to make Experiments and collect Observations, if when we have them, we may not make use of them; if we must not believe our Senses, if we may not judge of things by Trials and sensible Proofs, . . . but must remain tied up to Opinions we have received from others, and disbelieve every thing, tho' never so rational, if our received Histories doth not confirm them; . . . But this is contrary to the *Nullius in Verba* [Bound to no man's words] of this Society, and I hope that sensible Evidence and Reason may at last produce a true and real Philosophy.[4]

Hooke spent the winter presenting papers on one of his favourite subjects, navigation. In December 1694 he explained once again the construction of his marine barometer which gave an accurate reading regardless of temperature changes, and the Society offered him £2 towards the cost of making it. This was familiar material, but later that month he described a drawing machine, which would enable seafarers to produce much more accurate and useful pictures of coastlines, mountains, towns, houses, animals, fashions, carriages, weapons, and so on, than the crude and exaggerated drawings that were usually found in books of travellers' tales. 'If we enquire after the true Authors of those Representations, for the Generality of them, we shall find them to be nothing else but from some Picture-Drawer, or Engraver, here at Home, who knows no more of the Truth of the Things to be represented, than any other Person, that can read the Story . . . So that instead of giving us a true Idea, they misguide our Imagination, and lead us into Error, by obtruding upon us the Imaginations of a Person, possibly more ignorant than our selves.' Therefore Hooke had devised 'a portable Picture-Box or Camera Obscura to be put upon the head of the Person that designs' that would enable anyone to make a quick accurate sketch of any object or view, for the benefit of natural philosophers and mariners. It was a device he had described to the Royal Society long ago, and had once urged the governors of Christ's Hospital to buy for the use of the children. When he tried to demonstrate the picture box on 2 January it was too dark to see anything but the candles that lit the room, but the next week it worked better, so long as the object being drawn was strongly lit.[5]

Pursuing the theme of the accurate recording of heights and distances for the use of mariners, in February 1695 Hooke told the Royal Society of a new way for sailors to measure the distance or height of coastal features, lighthouses, and so on. Two accurate sextants should be placed at opposite ends of a ship and precise readings on a particular coastal feature should be taken simultaneously, by two men communicating by tugging on a thread. These two readings, together with the known distance between the two sextants, would enable the distance or height of the chosen feature to be calculated accurately. Those making coastal maps would find this method easy to use, he said, and much better than the guesses they had previously relied upon.[6]

Hooke was still, in his own view at least, a Cutler Lecturer, and had given a Cutler lecture on proving the movement of the Earth in July 1693. In May 1694 he had abandoned friendly persuasion and obtained a writ to compel Edmund Boulter, Cutler's executor, to pay ten years' arrears, and in October 1694 he brought his case before the Court of King's Bench. Boulter's argument was that the £475 arrears Hooke had received in 1684 was a part of the £1,000 penalty Cutler had to pay if he defaulted on Hooke's regular £50, and that Hooke had already been offered the rest of the £1,000 and refused it. This was probably untrue, and the court rejected Boulter's argument and allowed Hooke's Chancery case to continue.[7]

The case came before the Court of Chancery on 23 February 1695. Boulter's lawyer asked for an injunction to stop Hooke's claim, because no money was due, and Hooke's responded that since Hooke had read the lectures as promised he was owed £550 (eleven years' arrears) with costs. The Court ordered an interim payment of £100 to Hooke from Cutler's estate (which was duly paid), and a resumption of the case after Easter. To strengthen Hooke's case, the Council of the Royal Society put the Society's seal on a certificate, partly written in Hooke's hand, confirming that Hooke had carried out his duties under the Cutler agreement 'to the full of his undertaking'. Boulter and his lawyers did their best to delay the case for as long as possible, perhaps hoping that Hooke would die before it was settled. When the case resumed on 10 May 1695 they got a month's delay by claiming that there had been a process of arbitration, and in June they argued that they needed to study the Royal Society's records to see whether the lectures had been delivered. Hooke now had the Attorney-General, Thomas Powys, on his team, but Boulter's lawyers managed to win further postponements in November and again in January 1696.[8]

In May 1695 Hooke returned to an invention he had proposed ten years earlier, a device for drawing huge circles. The instrument was simple enough – a 100-foot wire fixed to a central point, and with a 'small Truckle-wheel fix'd in a thin Ruler, so that the Axis of the Truckle kept parallel to the extended Wire Radius', at its other end. This simple but pleasing device would keep the tension of the wire steady, Hooke assured his audience, and draw a truer circle than the sixty-foot beam compass that Wren had used for marking out the dome of St Paul's. He reminded them, too, that this invention, like movable type or the pendulum, only seemed obvious once someone had thought of it. In June he returned to the subject to explain how a huge circle could be drawn even when the centre of the circle was inaccessible, using two truckle-wheels fitted with telescopes. His lecture is more interesting, though, for the further evidence it gives us of his resentment of the Royal Society, and his anger at the way he (and inventors in general) had been treated during his career. Why were instruments such as this, which were easy to understand and easy to make, and seemingly obvious once they had been thought of, so slow to appear? The true cause was

> the unwillingness that Men generally have to be at the Trouble of thinking and meditating, especially when they observe that those that are so, do generally reap nothing for all their Labour, but either Contempt, and the Nicknames of Madmen and Projectors, or the Emulations of others, which creates them continual troubles. Nor is there less difficulty in procuring the Instruments and Apparatus necessary to put a new Invention into use and practice, than to invent and contrive the same; for Workmen are generally very unwilling to be put out of their common Road of working, and make a hundred Objections before they will undertake, and very often make as many mistakes in the performing, before they will rightly execute what is desir'd; and the inventor must be content not only to afford them his patience, but his Purse also, otherwise no further progress is to be expected . . . but supposing him at last to have executed his design and made his purchase, what has he got but some *Difficiles Nugae* [difficult trifles], some new Swing Swangs, which were the names that the Barometer for the weather, and the Pendulums for Clocks did for a long time bear; but when the Truth at length doth prevail, and the usefulness of the Invention appears, then everyone claims it for his own, though possibly he never had the thought of it, 'till all the World knew it. These may be some of the reasons why inventions have come so thin and seldom into the World: And why many parts of useful Knowledge do yet remain undiscover'd.[9]

Perhaps this was the story of Hooke's scientific life, as it seemed to him when he was sixty. His inventions had been ignored or undervalued, and the best of them had been claimed by others. If he had won a reputation for himself, it was as an eccentric inventor of useless or impractical gadgets, who gave lectures that nobody came to and had to chase spies and schoolboys from his rooms. Still, love of science or force of habit drove him on, and he continued to present experiments and ideas to the Royal Society, some useful, others, like his suggestions that the day should be divided into twenty-nine hours and that advanced cancer might be cured by smoking tobacco, laughable.[10]

Science still gave him pleasure. In May 1695 Halley, Wren, Hooke and two others met in the Roman tavern in Queen Street to investigate the rate at which water flowed from cisterns of different heights through pipes of different lengths, to help Sir John Hoskins develop a fire engine. After dining at the Roman they measured the time it took for water from the waste pipe to fill a barrel, and took an average from four observations. Then they walked to the Fleet Bridge, to a cistern near the Exchange, and to Charing Cross and measured the time taken to fill a barrel in all these places. Hooke's brief report the next week focused on the flow from different lengths of lead pipe, without giving a hint that his investigation had involved a pleasant scientific jaunt around London with his friends.[11]

As a successful and experienced experimental scientist, Hooke gave lessons to the Royal Society in the art of invention and discovery, using his own career as a model. Some discoveries are made by chance, he told them in July 1695, others by design and careful thought, and some by a combination of the two. Observations could lead to theories, but theories in turn guided our observations and experiments, and could be confirmed by this process. His own proposition that the location of the Earth's poles changed over time, which had been greeted with opposition and contempt, only needed 'some lucky chance to prove it positively', and he still hoped that one day he would find the decisive demonstration. His discovery of the spot on Jupiter only happened, he said, because of his conjecture that Jupiter turned on its axis: 'therefore, tho' the Observation were the Demonstration, yet the Theory was the occasion of seeking after it'.

At the end of July 1695 Hooke spoke to the Society about his work on the structure of the eye – probably the dissection he did with Waller in July 1687 – and told them he had found 'certain muscles in the eye serving to alter the figure thereof', and that he was sure there must be a mechanism in the eye to expand and contract the pupil, to 'give a greater or lesser curiosity thereto'. This might be connected with the 'small white nervous line surrounding the Ball of the Eye, with severall smaller Filaments therefrom, which he conceived had a further use to Nourish the Exterior Coates of the Eye, and convey a Nervous Juice to them.'[12] And in the winter of 1695–6 Hooke was still active, giving a series of talks on the history of barometers and reminding the Society that he had been the first to notice a relationship between barometric pressure and the weather, and following up with a lecture on a possible north-eastern route to China.

Hooke's case against Boulter and the Cutler estate was finally resolved in the Court of Chancery on 18 July 1696, Hooke's sixty-first birthday. The court ordered that Hooke's arrears since 1684 be calculated and paid, and that the £50 annuity should be continued so long as the Royal Society certified that the Cutler lectures were still being given. This victory lifted a burden that had oppressed Hooke for several years, and he was overjoyed at the result. Waller, his biographer, quoted an entry from a page of his diary which is now lost, expanding the letters DOMSHLGISS:A into the Latin words of praise he understood them to represent: 'Deo Opt. Max. Summus Honor, Laus, Gloria in secula seculorum, Amen. I was born on this Day of July 1635, and God has given me a new Birth, may I never forget his Mercies to me; whilst he gives me Breath, may I praise him'.[13]

Hooke knew well enough that this new life he had been given might not last very long. If there was one thing he wanted to do before his death it was to establish an irrefutable record of the Royal Society's (and his own) achievements in natural philosophy before it was too late. In June 1696, according to Waller, the Royal Society offered him money to make a full account of all his experiments, observations and inventions, but 'by reason of his increasing Weakness and a general Decay, he was absolutely unable to perform it, had he desir'd it never so much'. Yet it was quite clear to him that if he did nothing to establish his authorship of his many ideas and inventions they would soon be credited to others, and his work would be forgotten. On 29 July 1696 he spoke to the Society about Cassini's recent discourse on the history of astronomy and navigation, which attributed many of the Royal Society's achievements, including several of Hooke's inventions, to the French Academy. This showed what might happen when he was no longer alive to defend himself and the Royal Society. The discovery of the variation in the time of a pendulum swing in different parts of the world, which Cassini attributed to France, had been announced by Hooke to the Royal Society '32 or 33 years since', in December 1664. England had been well ahead of France and Italy in making good seventy-foot telescopes, thanks to the work of Sir Christopher Wren and Richard Reeve, and Hooke had been the first to use such object lenses in tubeless telescopes. Cassini had even claimed the application of telescopic sights and micrometers to astronomical instruments, and Hooke's well-known invention, the clockwork-driven telescope, for France, when in every case they had been pioneered in England. Hooke was especially hurt by Cassini's claim that Huygens had invented the watch with a spring-regulated balance. Something had to be done, he concluded, to defend the achievements of the Royal Society against Cassini's ill-informed claims, to stop the mouths of those who still said that 'this Society invented or improved nothing of real use'.[14]

In the autumn of 1696 Hooke spoke of red window glass, the coconut shells that were often washed up on Isle of Wight beaches and the numbing effects of opium, and in December he gave three entertaining lectures on the subject of the mysterious and unique shellfish called the nautilus, the living relative of the much larger creatures (ammonites) whose fossilized shells he had often discussed before the Royal Society. Hooke only knew the nautilus by its flat spiral shell, which was divided into many separate cells. He described the way the nautilus moved between deep and shallow water, and guessed that it had the ability to fill its shell's compartments with water or air in order to produce the desired movement. The problem of how the submerged nautilus made the 'air' it needed for buoyancy led Hooke to recollect his own success in making 'artificial air' with Boyle in April 1673. He noted, in a typical aside, that the experiment had been ignored at the time and published by someone else as their own work some years later. He then moved on to assert that many fish lived on the deep seabed, and that it was very likely that there was rich vegetation there for them to feed on. He reminded his listeners that he had designed deep-sea 'messengers' to reveal the secrets of the ocean floor, as well as diving bells and underwater spectacles: 'I shall be able to produce divers others, for other Purposes, if God spare my Life so long as to see the Seas free from Rovers, and that the Study of Arts does succeed the Study of Arms'. He recalled that many unknown places had been thought barren and uninhabitable, but turned out on investigation to be fertile, and the deep ocean floor might not be the wasteland most believed it to be. 'Probably, most people will treat me as Columbus was, when he pretended the Discovery of a New World to the Westward: But I have been accustomed to such Kind of Treatments, and so the better fitted to bear them'.[15] He seemed to find it almost impossible in his later lectures to avoid some reference to the mistreatment and neglect he now fancied he had suffered throughout his career with the Royal Society.

Hooke's health was failing, but he had to keep delivering Cutler lectures if he wanted his £50 annuity. In the spring of 1697 he gave three talks on amber, which were prompted by the publication of a book on the subject by Philip Hartman. Hooke contended, in opposition to Hartman, that amber was the resin of an ancient tree, which had been petrified by submersion in the sea and deposited in layers of sand, especially in Prussia, when much of the world was under water. He promised to give a better account of how and when this had happened 'if God restore my health'. He dismissed Hartman's suggestion that amber had been carried from the sea to Prussian hilltops through 'subterraneous passages' as yet another evasion of the obvious explanation for this and many other phenomena. He did not try to conceal his frustration with all those, including most of his listeners, who could not accept his explanation for the existence of fossil deposits on land. It was 'as if a Mariner discovering an Island in some great Ocean, and finding some House on it, but no Inhabitants, should conclude that this House had there grown of itself, or else had been brought thither thro' the Air by some violent Hurricane'. He urged scientists to investigate the process of petrification, and the origins and nature of chalk and flint, since he no longer had the health, the money or the assistance to do the work himself.[16]

Hooke gave the last of these lectures on 19 May 1697, and the following two weeks he pursued his theme with papers on the recent discoveries of the remains of unknown or strangely located mammals

– Siberian mammoths, Kentish hippopotami, Norfolk elephants – and a story of a fully manned Roman ocean-going vessel buried in a Swiss mountain. Once again, he implored his audience not to content themselves with the absurd idea that Noah's Flood explained all these wonderful finds, and instead to embrace the rational hypothesis he had been offering them (he said) since 1664. Ask yourselves, he told them, 'whether the latitudes of places might have been changed, . . . whether Ireland and America might not have been formerly joyned – whether the bottom of the Sea might have been dry land and what is now dry land might not have been sea &c.'[17]

In April 1697 Hooke began to write the story of his own life, 'with all my Inventions, Experiments, Discoveries, Discourses. &c. which I have made, the time when, the manner how, and means by which, with the success and effect of them, together with the state of my Health, my Employments and Studies, my good or bad Fortune, my Friends and Enemies', but got no further than his adolescence. Hooke's health had been poor for years, but in the summer of 1697, according to Richard Waller, it deteriorated sharply:

> About July 1697, he began to complain of the swelling and soreness of his Legs, and was much over-run with the Scurvy, and about the same time being taken with a giddiness he fell down Stairs and cut his Head, bruis'd his Shoulder, and hurt his Ribbs, of which he complain'd often to the last. About September he thought himself (as indeed all others did that saw him) that he could not last out a Month.[18]

Hooke seems to have recovered a little from this low point, but his legs 'swell'd more and more, and not long after broke, and for want of due care Mortify'd a little before his death'. So there would be no more companionable walks across the fields to Kingsland or Islington. Let us hope that he could still struggle down Broad Street to meet his friends in Jonathan's coffee house and look for bargains in the Cornhill bookshops for a few more years. To judge from the Royal Society journals, he continued to buy books and report on them to the Royal Society, or offer them for the Society's library, until the end of 1701.

Death was also removing old and trusted friends from his coffee-house circle. Francis Lodwick died in 1694, Dr Busby in 1695, John

Godfrey in July 1697, and John Aubrey a month earlier, after a visit to London.[19] One of Aubrey's last letters to Hooke, written on 6 November 1695, captures the melancholy of infirmity and old age, and reminds us that grumpy and ill-favoured as he was, Hooke was loved by some good-hearted friends:

> Honoured Sir;
>
> I give you my hearty thanks for all your favours & kindness to me. I am I thank God in Good health: and want nothing but your good company & our Wednesday meeting at Gresham College . . . I wish you all happinesse, and shall pray for yr Illuminations; my eies grow dimmer: The next summer about June I intend (God willing) to see you once more. I am
>
> Sr your truly affectionat friend to serve you
>
> Jo. Aubrey.[20]

Like a dying emperor, Hooke wanted to revisit remote corners of his vast scientific empire, reminding his listeners of ancient and half-forgotten triumphs. Since the Royal Society still met in Gresham College, Hooke's immobility did not force him to give up going to the regular Wednesday afternoon meetings, and he attended and spoke at almost every one of them in 1697. He regained his place on the Council that November, and in December he read a series of lectures criticizing those who imagined limits to the potential power of telescopes. One day, he said, 'if Money were not wanting', telescopes or other artificial aids would enable astronomers to see animals or similar objects on the Moon, just as he had predicted in *Micrographia*.[21] In March 1698 he gave a lecture arguing that the shape of the Sun was a flattened sphere, and explaining how to make a helioscope with four reflex planes in a twenty-four-foot tube. At the end of June he lectured on the fire burning within the Earth, and in July he discussed Huygens's book on the possible inhabitants of other planets, *Cosmotheoros*, which was published in English in 1698, three years after its author's death. Unlike Huygens, Hooke said he thought there was a lunar atmosphere and areas of water, making it possible that the Moon was inhabited. And in 1699 and 1700, despite his failing eyesight, he managed to give several lectures on earthquakes and the age of the Earth, in which he also talked of the special qualities needed in a natural philosopher. The required combination of experimental accuracy, observational acuity and theoretical understanding, he now believed, was only given to one in a hundred, or even one in a thousand.

As a man with access to the best medical attention, but whose legs were being eaten away by decay, and whose precious eyesight was failing, Hooke had especially harsh words to say about physicians and their medicines. The 'observations made or pretended to be made by young Physicians' were nearly all useless, and most were only published 'as Advertisements to make themselves the more known, and so get Practice; and tho this or that Symptom may be true and matter of Fact, yet the true Cause of the Distemper, and the reason of the Cure or Miscarriage of the Patient possibly was really quite different from those assign'd by them'.[22] In his next lecture, in July 1699, Hooke argued again that the Earth had gone through various stages of development, moving from its youthful fluid and smooth state to its present senility, when (like him) it had grown rough, scarred and stiff, and its inner fires were burning with diminishing power. Hooke believed that, given time, he could answer all the objections to this view, but he knew that there would always be plenty who would not be convinced, or not admit that they were. 'All I can say, is *Valeat quantum valere potest*, let every one enjoy his own freedom'.

The wit Ned Ward, the 'London Spy', visited Gresham College, 'the Wiseacres' Hall', at about this time, and saw an old philosopher walking in measured steps around the quadrangle, with his eyes fixed on the pavement. 'He seemed to scorn gloves . . ., crossing his arms over his breast, and warming his hands under his armpits; his lips quaked as if he'd ague in his mouth, which tremulous motion, I conceived, was occasioned by his soliloquies, to which we left him.'[23] If this was Hooke, perhaps he was muttering about the Mercers' Company's problems, which threatened to expel him from his familiar College rooms at the end of his life. The company's difficulties stemmed from the fact that they had rebuilt the Royal Exchange in such an ostentatious and costly way after the Great Fire that the building, which had previously been the source of much of their income, now burdened them with debt. Hooke knew this only too well, because in 1666–7 the Company had ignored his advice to repair and reconstruct the Exchange using old materials from the site. The Gresham Joint Committee called the lecturers in on 11 August

1699 and told them that the failure of the Royal Exchange to attract enough tenants meant that Sir Thomas Gresham's bequest could no longer be paid. The Committee wanted to demolish and rebuild the College, combining rooms for lecturers with rooms for rent, and asked the professors to support their petition to win Parliament's approval for the plan. Discussions dragged on, and Hooke stopped coming to the meetings in November 1699, probably because he was too ill to do so. The other professors finally agreed to petition for the plan in February 1702, but the Royal Society, which would have lost its weekly meeting place, mobilized its friends in the House of Lords, and the Gresham Bill was lost when Parliament was dissolved.[24]

Knowing that his time was almost up, Hooke was determined that his reputation as an inventor and innovator would survive him. No doubt his repeated reminders to the Society that ideas they took to be new had in fact been proposed by him thirty years earlier became tedious or irritating. On 24 May 1699 he told the Society that a 'new' way of measuring the distance of the Moon which they had recently heard had been demonstrated by him in 1664, when it had been disregarded. Hooke said that he intended to repeat several of his early experiments, 'in order to make them fit for the Publique view, which cannot but be imperfectly remembrd by me after soe long a space of time', if the Society would cover his costs. On Sir John Hoskins' orders, this whole paper was excluded from the Journal Book, and ignored.[25] But six weeks later Hooke borrowed the second volume of the Society's register, and used it to prepare 'a discourse in Vindication of his Astronomical works' which he delivered in August. When Newton showed the Society an improved sextant or quadrant later that month Hooke came to the next meeting (in October) to declare that this was 'his own invention before the year 1665'. He probably meant the seventeen-inch quadrant he had shown the Society on 22 February 1665, which was 1664 in the old Julian calendar, then formally in use, in which the new year began on 25 March. Hooke's career in the Royal Society stretched back so far that his earlier achievements were unknown to most members in 1699. When he told them in December that 'he had invented an instrument to tell exactly the distance saild upon every Tack, and Rhumb, and gave a true calculation of every League', they asked the obvious question 'of this instrument of Dr Hook's, where it is'. Hooke's reply the next week was that he had made a model, but that the device had never been used at sea. A search of the Society's books was ordered, looking also for 'a kind of waywiser to describe the space drove, and Turnings of a Road'.[26] Hooke's claims were not always greeted with scepticism. In 1701 Edmond Halley, who had taken Hooke's marine barometer (with a thermometer to enable temperature changes to be discounted) on his great Atlantic voyage in 1698–1700, wrote an article in *Philosophical Transactions* giving the instrument his wholehearted approval.[27]

Despite his age and infirmity, the Society still relied on Hooke's skill and experience. In July 1699 the Society was given a Japanese peacock, and Hooke 'was asked to provide for it'. Luckily the opossum they acquired at the same time was given to Harry Hunt, who apparently kept it alive until 1704, when the Society had the pleasure of examining its severed penis. In the early months of 1700 Hooke was asked to devise some experiments to establish the nature of ambergris and to report on a new portable weighing scale, and was appointed to a committee to examine and catalogue some ancient Etruscan inscriptions sent by the Italian antiquary Raphael Fabretti. He was inactive in most meetings between April and October 1700, but he was elected to the Council on 30 November, and over the winter of 1700–01 his voice was heard more frequently, often describing some new Eastern titbit supplied by his old friend Robert Knox. In May 1701 he suggested making an incombustible cloth out of asbestos and fine glass fibres, and in July he declared that the Moon was probably inhabited. His last recorded contributions to Royal Society meetings were on 10 June 1702, when he reported on an unsatisfactory machine for measuring a ship's course, and two weeks later, when he spoke briefly on earthquakes and glass-grinding.

Richard Waller tells us that from the late 1690s Hooke's eyesight was fading, so 'that at last he could neither see to Read or Write'. But he had vision enough in 1701 and 1702 to read out his lectures, which were written in his tiny but clear hand, and to draft a memorandum on measuring the diameter of the Sun in December 1702. His last letters to Sir Robert Southwell, in December 1701 and January 1702, suggest an active mind, serviceable eyesight, and an undiminished ability to offend. In the first, Hooke criticized Mr Gillmore's design for Southwell's new fountain, and in the second he apologized for the offence he had caused: 'I am sorry he took my expression soe heinous, as if I had designed it as an affront to him . . . I doe therefore most earnestly beg your pardon, and desire you will ascribe it to my zeale to expresse my self as far as I am able'. Southwell died in September, but the two old friends probably met in January 1702.[28]

By that time Hooke was in a sorry state. He had an £8,000 fortune in his great iron chest, but like many very old people he lacked the energy or will to spend any of it. In short, he had become a miser. Years after his death those who had known him dined out on stories of his miserliness. A Royal Society friend, Sir Godfrey Copley, told people that in his old age Hooke was 'much concerned for fear he should outlive his estate', and consequently would 'endanger himself to save sixpence for anything he wants'. To improve his story, which later found its way into the *Gentleman's Magazine*, Copley added that Hooke had starved at least one housekeeper to death. In 1706 Lord Somers, who became President of the Royal Society in 1699, treated the Bishop of Carlisle to some 'easy and well-bred reflections' on Hooke's 'unaccountable covetousness . . . not daring to sleep in the same Room with his Hoord of Guinneas till the Neighbourhood were risen'. This was not just unkind gossip. His good friend Captain Knox recorded that old Hooke 'lived Miserably as if he had not sufficient to afford him foode & Rayment'.[29] Richard Waller, who knew Hooke well in his later life, said that his stoop grew 'very remarkable' in his last years, and his face, 'always very pale and lean', was 'nothing but Skin and Bone' at the end. At least Hooke had decided, around 1700, to cut off his hair, which he had worn 'very long and hanging neglected over his Face uncut and lank', and replace it with a periwig. Waller watched Hooke's condition deteriorate in his final months:

> Thus he liv'd a dying Life for a considerable time, being more than a Year very infirm, and such as might be call'd Bed-rid but kept in his Cloaths, and when over tir'd, lay down upon his Bed in them, which doubtless brought several Inconveniences upon him, so that at last his Distempers of shortness of Breath, Swelling, partly of his Body, but mostly of his Legs, increasing, and at last Mortifying, as was observed after his Death by their looking very black, being emaciated to the utmost, his Strength wholly worn

> out, he dy'd on the third of March 1702/3, being 67 Years, 7 Months, and 13 Days Old.[30]

As it happens, we have another account of Hooke's death. When he knew that his last moments had arrived he called his maidservant, and told her to run and fetch his dear old friend Captain Knox. Knox arrived too late to see Hooke alive, but he was able, with Harry Hunt, to lay out his body. Knox made a record of the events in his diary:

> The 2th March 1702/3 This night aboute 11 or 12 of the Clock my Esteemed Friend Dr Robert Hooke Professor of Geometry & Naturall Philosophy in Gresham College Died there, onely present a Girle that wayted one him who by his order (just before he died) came to my Lodging & called me. I went with her to the Colledge where, with Mr Hunt the Repository Keeper, we layed out his body in his Cloaths, Goune & Shooes as he Died, & sealed up all the Doores of his appartment with my Seale & so left them.[31]

Hooke's death was announced in the Royal Society meeting on 3 March, and his funeral took place three days later in the great medieval church of St Helen, where Tom Giles and Grace Hooke had been buried. He was 'decently and handsomely interred' 'about the Midle of the South Eyle of the Church', accompanied to his grave by every available member of the Society. Captain Knox was there too, and noted in his diary that Hooke's executors 'made a Noble finerall giving Rings & gloves & Wine to all his friends thare, which were a great Number for he left a very Considerable Estate all in Mony besides some Lands'. Most of those who hoped for a share of Hooke's fortune were disappointed, since in the absence of a will the whole lot went to Hooke's closest living relative, Elizabeth Stephens. Fortunately many of Hooke's scientific papers, which had little commercial value, soon came into the hands of Richard Waller, who used them to compile his life of Hooke and published many of them as Hooke's *Posthumous Works* in 1705. Waller used scraps of Hooke's diary in writing his life, but three years after this he was given Hooke's main diary by Mr and Mrs Dillon, Hooke's niece and her husband. When Waller died in 1714 the diary and papers were passed on to William Derham, who eventually published a second collection of Hooke's papers in 1726. Hooke's 1672–83 diary then disappeared until 1891, when it was bought by the Corporation of the City of London, its present owner.

Within three weeks of Hooke's death the Mercers' Company had demanded the keys to Hooke's lodgings, and ordered the Royal Society to clear out its library and repository and find a new place to meet. Wren led a deputation which secured a stay of execution, and the Society continued to use Gresham College until 1711. But reminders of the curator who had impressed, entertained and irritated the Society in these rooms for forty years were soon removed. His only portrait, which seems to have hung on the wall of the Society's room, was removed and lost, and a German visitor in 1706 found Hooke's instruments and inventions in a shocking state, dirty, neglected, or dismantled.[32]

Nobody had served the early Royal Society longer or better than Robert Hooke, but it is likely that his death came as something of a relief. His most creative days were long gone, and in his last years his habit of complaining about his treatment at the Society's hands, and claiming so many new inventions as his own, must have been trying. The 1660s, when Hooke (alongside Henry Oldenburg) had kept the Royal Society alive with his almost limitless energy and intellectual fertility, and produced in *Micrographia* one of the greatest and most exciting scientific books of the century, seemed a very long time ago. For the twenty or so fellows who saw Isaac Newton as the best future leader of the Royal Society, Hooke's death opened the way to a new era. While Hooke was a grumbling presence at nearly every meeting Newton hardly ever attended the Society, but with his death it became possible to elect Newton to the Council and the presidency in November 1703. As for Newton, he felt able at last to publish his work on light and colours, *Opticks*, which had been written in 1672 but held back for fear of controversy. All references to his debt to Hooke's work on light and colours were removed.

Newton achieved an extraordinary dominance over the Royal Society, of which he was President for almost twenty-four years, and over the whole scientific community in England. His personal hegemony, along with the triumph of his highly mathematical approach to science, almost completely eclipsed Hooke's posthumous reputation. Not only did Hooke's experimental and mechanical approach to science appear outdated after the Newtonian revolution, but Newton's virtual deification cast Hooke in the role of a false prophet who had resisted Newton's triumph and dishonestly claimed credit for his work. This does not suggest that there was a deliberate attempt on Newton's part to belittle Hooke's achievements – next to him, almost any scientist looked a diminished figure. The fact that during his long career Hooke had conducted acrimonious disputes with several scientists (just as Newton did) was counted against him, and even his wonderful versatility was treated as a weakness, a reason for his failure to achieve greatness in any one field. The significance of his speculations on combustion, light waves, elasticity, fossils, the formation of continents and the development of species was not grasped in the early eighteenth century, and by the time scientists had moved in his direction they had all but forgotten his work in these fields.

The enduring image of Hooke as a difficult and unhappy man was inadvertently reinforced by the work of his two friends, John Aubrey and Richard Waller, one because he focused on the dispute with Newton, the other because of his account of Hooke's final years. Hooke's odd appearance, his disputatiousness, his fear of poverty, his very inventiveness made him a figure of fun. Jonathan Swift's satire on scientists and the Royal Society in *Gulliver's Travels*, published in 1726, is a joke at the expense of Boyle, Newton, Grew and all the scientists whose essays Swift read in *Philosophical Transactions*. But anyone who still remembered Robert Hooke would have recognized him as one of the professors in the Academy of Projectors in Lagado, which Gulliver visited in his Voyage to Laputa. In one of the academy's many rooms Gulliver met 'the universal Artist', who 'had been Thirty Years employing his Thoughts for the Improvement of Human life. He had two large Rooms full of wonderful Curiosities, and Fifty men at work. Some were condensing Air into a dry tangible Substance, by extracting the Nitre, and letting the aqueous or fluid Particles percolate: Others softening Marble for Pillows and Pincushions: others petrifying the Hoofs of a living Horse to preserve them from foundring.' Hooke had never tried to extract sunbeams from cucumbers or plough a field with hogs, but the astronomer combining a sun dial and weathercock, the physician pumping up a dog with bellows, and the creators of a wordless universal language which involved carrying huge sacks of *things*, which they showed to each other to convey their meaning, could all have been drawn from Hooke's life.

If we followed Swift and Shadwell in thinking of Hooke as the archetype of the unworldly scholar, fiddling about in his laboratory on projects which never had the slightest chance of doing any good to anyone, we would be utterly misunderstanding Hooke's life and work. More than any other scientist of his day, except perhaps Wren, Hooke turned his skills to practical ends, directing the rebuilding of the centre of one of Europe's greatest cities, designing and constructing colleges, hospitals, churches, suburban mansions and West End town houses, and discovering and publicizing a range of important craft skills in pottery, glassware, metalwork and textiles. The mighty Monument, still standing straight on its clay and gravel foundations after over three hundred years, was not the work of a laboratory-bound eccentric. Hooke was one of the most prolific and innovative builders of country houses and civic institutions of his day. He has been unlucky in the loss of all of his biggest London buildings, but those who want to judge for themselves Hooke's talents as an architect and builder can still do so. Three of the City churches with which he was closely associated, St Benet Paul's Wharf, St Martin Ludgate and St Edmund the King, are in excellent condition, as are Willen church in Buckinghamshire, Ragley Hall in Warwickshire, Ramsbury Manor in Wiltshire, Shenfield Place in Essex, and Buntingford Almshouses.

Hooke was driven by the desire to make his name, to make money, to promote the advance of natural philosophy on almost every front, and to prove to a doubting world that science could bring practical benefits to mankind. Though some of his endeavours failed, and others were ignored or misunderstood, he could claim that through his work he had made a real difference to the comfort and safety of his fellow citizens. Sailors could travel more safely with his quadrants, sextants, backstaffs and marine barometers, his spring-regulated watch ticked in hundreds of pockets, and the London watch and clock industry was entering a golden age, partly on the strength of his innovations. If the medical profession was still stuck in the Middle Ages, where it remained for another century or more, this was no fault of Hooke, who had done as much as any other scientist to demonstrate the purpose of respiration and the lungs, and more than any other Englishman to show what might be learnt about disease by careful microscopic observation. His suggestion that doctors could learn more from listening to their patients' chests than by tasting their urine, if anyone had picked it up, might have advanced diagnosis by a century. Hooke sowed the seeds of several new sciences, though in many cases the world had forgotten the sower by the time the plant had appeared. In the study of insects, friction, the strength and elasticity of materials, meteorology, oceanography, comets, evolution, geology, crystals, light, heat, sound, combustion and respiration, his work was seminal. His insights into light and planetary motion were important in preparing the way for a greater scientist, and though Newton's famous 'shoulders of giants' compliment may not have been sincerely meant it was nonetheless true. Because Hooke combined outstanding mechanical skills with a commitment to accuracy in measurement and observation, he was able to make an almost unparalleled contribution to the development of scientific instrumentation. He transformed or improved all the important scientific instruments of his age: thermometers, barometers, reflecting and refracting telescopes, single-lens and compound microscopes, quadrants, sextants, pendulums, spring watches, precision balances, vacuum pumps, micrometer eyepieces, and dividing engines. His gadgets and mechanical devices, some adopted in his lifetime, others forgotten for a century or more, played an important part in later technology: the universal joint, the punched-paper record keeper, the iris diaphragm, the worm gear, the optical telegraph, the deep-water sampler, the wind gauge, perhaps the velocipede, the wheel-cutting engine, and the sash window.

Hooke had a vision of the endless possibilities of scientific and technical advance, and spent his life trying to teach the fellows of the Royal Society, some of whom had a limited grasp of what he was talking about, to 'throw off that lazy and pernicious principle, of being contented to know as much as their Fathers, Grandfathers, or great Grandfathers ever did'. This optimism about what science might achieve was typical of the age, but it reached its highest point in Hooke, whose mind was remarkably free of the magical and religious superstitions that muddled and constricted some of the best brains of his day. This freedom enabled him to make suggestions about the changing shapes of continents, climatic change, the formation of rocks and the development of species which anticipated and perhaps influenced the ideas of James Hutton in the eighteenth century and Charles Lyell and Charles Darwin in the nineteenth. His unlimited confidence in the problem-solving potential of the new science usually translated into a similar confidence in his own powers, and for some people this made him a difficult man to like. But those who were not offended by his vanity or his sharp tongue enjoyed his intelligence, his curiosity, his energetic pursuit of all kinds of knowledge, and his ability to stay up and talk nearly all night. This is why Robert Hooke enjoyed the sustained friendship of such a wide variety of Londoners: the Mercers' clerk John Godfrey, the needlewoman Nell Young, the linguists Theodore Haak and Francis Lodwick, the sea captain Robert Knox, the 'operator' Henry Hunt, the clergymen John Wilkins and Seth Ward, the watchmaker Thomas Tompion, the antiquarian John Aubrey, the schoolmaster Dr Richard Busby, the landowner Sir Robert Southwell, and the scientists Robert Boyle, Edmond Halley, Thomas Henshaw and Christopher Wren. Hooke's charm was not in his face or his unfortunate body, perhaps not in his manners or his social sophistication, but in his mind. On 15 February 1665 Samuel Pepys went to his first meeting as a member of the Royal Society, and afterwards joined Lord Brouncker and the rest for a club supper at the Crown tavern, behind the Royal Exchange. In this sophisticated and well-dressed company one man, a poorly formed and socially insignificant young scholar, impressed him more than anyone else: 'Mr Hooke, who is the most, and promises the least, of any man in the world that ever I saw'.

Notes

1. 'The History of My Own Life'

1. M. Hunter, et al., 'Hooke's Possessions at his Death: a hitherto unknown inventory', in M. Hunter and S. Schaffer, *Robert Hooke: New Studies* (Woodbridge, Suffolk, 1989), pp. 287–94.
2. P. Earle, *The Making of the English Middle Class* (London, 1989), p. 109.
3. R. Bud, D. J. Warner (eds), *Instruments of Science* (New York and London, 1998). The book was produced by the Science Museum of London and the National Museum of American History.
4. R. Waller, *The Life of Dr Robert Hooke*, p. 2, in R. T. Gunther, *Early Science in Oxford*, vol. 6 (Oxford, 1930).
5. John Hooke's inventory and will are available on the Isle of Wight History Centre Web site, http://freespace.virgin.net/roger.hewitt/iwias/home.htm.
6. Waller, *The Life of Dr Robert Hooke*, pp. 3–4.
7. J. Aubrey, *Brief Lives* (ed. O. Lawson Dick, Harmondsworth, 1962), p. 242.
8. H. Nakajima, 'Robert Hooke's family and his youth', *Notes and Records of the Royal Society*, 48 (1994), pp. 11–16.
9. S. Inwood, *A History of London* (London, 1998), pp. 157–9.
10. Aubrey, *Brief Lives*, pp. 242–3.
11. Waller, *The Life of Dr Robert Hooke*, p. 64.

2. The Revolution in Science

1. B. J. Shapiro, *John Wilkins, 1614–1672* (Berkeley and Los Angeles, 1969), p. 25.
2. C. Webster, *The Great Instauration* (London, 1975), pp. 153–72. Shapiro, *John Wilkins*, pp. 118–47.

3. This story is repeated, in a fictionalized form, in Iain Pears' *An Instance of the Fingerpost* (London, 1997), a book which features many of the men who were important in Hooke's life.

3. Hooke at Oxford

1. *Oxford English Dictionary*, 1971.
2. S. Shapin, 'Who was Robert Hooke?', in M. Hunter, S. Schaffer (eds), *Robert Hooke: New Studies* (Woodbridge, Suffolk, 1989), pp. 263–4.
3. J. Agassi, 'Who discovered Boyle's Law?', *Studies in the History and Philosophy of Science* 8 (1977), p. 202.
4. A. Chapman, 'England's Leonardo: Robert Hooke (1635–1703) and the art of experiment in Restoration England', *Proceedings of the Royal Institution of Great Britain*, 67 (1996), pp. 239–75.
5. R. Waller, *The Life of Dr Robert Hooke*, p. 9, in R. T. Gunther, *Early Science in Oxford*, vol. 6 (Oxford, 1930).
6. M. Hunter, 'The life and thought of Robert Boyle', http://www.bbk.ac.uk/bb/biog.html.

4. The Royal Society

1. M. Hunter, *Establishing the New Science* (Woodbridge, Suffolk, 1989), pp. 245–60. M. Hunter, *Science and Society* (Cambridge, 1981), pp. 49–54. Oldenburg's correspondence has been edited and published in thirteen volumes by A. R. and M. B. Hall (Madison, 1965–86).
2. M. Hunter, *The Royal Society and its Fellows, 1660–1700: the Morphology of an Early Scientific Institution* (Chalfont St Giles, 1982), pp. 7–11, 22–5, 115–17.
3. Hunter, *The Royal Society*, p. 18.
4. Hunter, *Establishing the New Science*, pp. 279–89. Thomas Birch, *The History of the Royal Society of London* (London, 1756–7), vol. ii, p. 142.
5. Hunter, *Establishing the New Science*, ch. 9.
6. M. A. R. Cooper, 'Robert Hooke, City Surveyor. An assessment of the importance of his work as a Surveyor for the City of London in the aftermath (1667–74) of the Great Fire', City University (London) Ph.D. thesis (1999), pp. 32–6, 83. R. T. Gunther, *Early Science in Oxford*, vol. 6 (Oxford, 1930), p. 179.
7. R. Waller, *The Life of Dr Robert Hooke*, pp. 10–12, in Gunther, *Early Science in Oxford*, vol. 6 (Oxford, 1930). A. R. Hall, 'Horology and

Criticism: Robert Hooke', *Studia Copernicana*, vol. 16 (1978), pp. 261–81. D. S. Landes, *Revolution in Time: Clocks and the Making of the Modern World* (Cambridge, Mass., 1983), pp. 114–28. M. B. Hesse, 'Hooke's Philosophical Algebra', *Isis*, 57 (1966), pp. 67–83.

8. M. Wright, 'Robert Hooke's Longitude Timekeeper', in M. Hunter, S. Schaffer (eds), *Robert Hooke. New Studies* (Woodbridge, Suffolk, 1989), pp. 63–118, especially appendix, pp. 103–18. I have tried to quote Hooke's first draft, omitting later additions that may not represent what he had achieved by 1664.
9. Wright, 'Robert Hooke's Longitude Timekeeper', pp. 108–9.
10. Wright, 'Robert Hooke's Longitude Timekeeper', p. 110.

5. 'Full of Employment'

1. S. Pumfrey, 'Ideas above his station: a social study of Hooke's curatorship of experiments', p. 6, in *History of Science*, xxix (1991), pp. 1–44. Wren's letter is quoted in J. A. Bennett, 'Hooke and Wren and the system of the world', *The British Journal for the History of Science* 8 (1975), no. 28, p. 55.
2. Royal Society Classified Papers, vol. xx (Hooke Papers), f. 67.
3. S. Shapin, S. Schaffer, *Leviathan and the Air-Pump* (Princeton, 1985), pp. 235–56.
4. Thomas Birch, *The History of the Royal Society of London* (London, 1756–7), vol. i, pp. 174–8.
5. M. Deacon, 'Founders of Marine Science in Britain: the work of the early fellows of the Royal Society', *NRRS*, 1965, vol. 20, pp. 28–50.
6. R. T. Gunther, *Early Science in Oxford*, vol. 6 (Oxford, 1930), p. 204, letter from Hooke to Boyle, 6 October 1664.
7. Robert Hooke, *Micrographia* (London, 1665), pp. 139–40.
8. *Encyclopædia Britannica*, 11th ed., 1910–11, vol. 19, pp. 970–1.
9. Birch, *History of the Royal Society*, vol. ii, p. 18.
10. Gunther, *Early Science in Oxford*, vol. 6, pp. 270–5. Royal Society Classified Papers, vol. xx (Hooke papers), ff. 2, 24.
11. Birch, *History of the Royal Society*, vol. i, pp. 411–12, 13 April 1663.
12. W. E. K. Middleton, *Invention of the Meteorological Instruments* (Baltimore, 1969), pp. 26–8.
13. Gunther, *Early Science in Oxford*, vol. 6, p. 203, Hooke to Boyle, 6 October 1664.
14. Gunther, *Early Science in Oxford*, vol. 6, pp. 180–1.
15. Birch, *History of the Royal Society*, vol. i, pp. 202–4, 367.

16. Royal Society Register Book, vol. 2, p. 193. R. G. Frank, *Harvey and the Oxford Physiologists* (Berkeley, 1980), pp. 154–60. Gunther, *Early Science in Oxford*, vol. 6, pp. 216–18, Hooke to Boyle, 10 November 1664.
17. H. D. Turner, 'Robert Hooke and theories of combustion', *Centaurus*, 4 (1956), pp. 297–310.
18. Birch, *History of the Royal Society*, vol. i, pp. 449, 454–6, 460–2. Gunther, *Early Science in Oxford*, vol. 6, pp. 186–191, 200–1. P. J. Pugliese, 'The scientific achievement of Robert Hooke: method and mechanics' (Harvard University Ph.D. thesis, 1982), pp. 393–8.
19. Royal Society Classified Papers, vol. xx (Hooke papers), f. 7.
20. Gunther, *Early Science in Oxford*, vol. 6, pp. 193–5, Hooke to Boyle, 8 September 1664.
21. Birch, *History of the Royal Society*, vol. i, pp. 505–7.
22. Birch, *History of the Royal Society*, vol. i, pp. 141–4.
23. Hooke, *Micrographia*, pp. 218–19, 228–30.
24. John Aubrey, *Brief Lives* (ed. O. L. Dick, Harmondsworth, 1962), p. 243.
25. R. Waller, *The Life of Dr Robert Hooke*, in Gunther, *Early Science in Oxford*, vol. 6, pp. 64–5.
26. Waller, in Gunther, *Early Science in Oxford*, vol. 6, p. 139, Hooke to Boyle, 3 July 1663.
27. H. C. King, *The History of the Telescope* (London, 1955), p. 71.
28. D. J. Oldroyd, *Thinking About the Earth* (London, 1996), p. 327.
29. Hooke, *Micrographia*, pp. 242–6.
30. Gunther, *Early Science in Oxford*, vol. 6, p. 205, Hooke to Boyle, 6 October 1664.
31. Pepys's Diary, 1 May 1665; R. Straus, *Carriages and Coaches* (London, 1912), pp. 115–17.
32. Birch, *History of the Royal Society*, vol. i, p. 271.
33. Pepys's Diary, 1 February 1664.
34. Gunther, *Early Science in Oxford*, vol. 6, p. 223.

6. A Secret World Discovered

1. Pepys's Diary, 2 January and 20 January 1665.
2. J. A. Bennett, *The Mathematical Science of Christopher Wren* (Cambridge, 1983), p. 73. Robert Hooke, *Micrographia* (London, 1665), Preface.
3. On the problem of witnesses and the idea of 'virtual witnessing' see M. A. Dennis, 'Graphic Understanding: Instruments and Interpretation in Robert Hooke's *Micrographia*', *Science in Context*, 3 (1989), p. 345.
4. *Micrographia*, Preface, and illustration facing p. 1. Royal Society Classi-

fied Papers vol. xx (Hooke Papers), f. 30, Hooke to Beale, 24 June 1664. M. Fournier, *The Fabric of Life* (Baltimore and London, 1996), pp. 13–18. B. J. Ford, *Single Lens. The Story of the Simple Microscope* (London, 1985), pp. 40, 62.

5. Ford, *Single Lens*, pp. 60–2.
6. *Micrographia*, Preface and pp. 203–4.
7. *Micrographia*, pp. 47–67.
8. *Micrographia*, pp. 100–6, 109–12.
9. *Encyclopædia Britannica* (2001).
10. Hooke, *Micrographia*, pp. 82–8.
11. *Micrographia*, pp. 112–16.
12. *Micrographia*, pp. 142–5.
13. Pepys's Diary, 8 August 1666. *Micrographia*, pp. 169–74.
14. *Micrographia*, pp. 186, 211–13.
15. *Micrographia*, pp. 205–7.
16. *Micrographia*, pp. 185–91.
17. Huygens's letter is quoted in M. B. Hesse, 'Hooke's philosophical algebra', *Isis*, 57 (1966), p. 73.
18. Newton's notes on *Micrographia* are printed in G. Keynes, *A Bibliography of Robert Hooke* (Oxford, 1966), pp. 97–108. R. S. Westfall, *Never at Rest* (Cambridge, 1980), pp. 156–9. E. Shapiro, *Fits, Passions and Paroxysms* (Cambridge, 1993), pp. 49–59.

7. Falling Bodies

1. R. T. Gunther, *Early Science in Oxford*, vol. 6 (Oxford, 1930), pp. 247–9, Hooke to Boyle, 8 July 1665.
2. Royal Society Journal Book, vol. 7, p. 291, 25 June 1690.
3. Hooke to Boyle, 26 September 1665. Gunther, vol. 6, p. 252.
4. M. A. R. Cooper, 'Robert Hooke, City Surveyor. An assessment of the importance of his work as a Surveyor for the City of London in the aftermath (1667–74) of the Great Fire', City University (London) Ph.D. thesis (1999), p. 121.
5. J. A. Bennett, *The Mathematical Science of Christopher Wren* (Cambridge, 1993), pp. 67–8.
6. Thomas Birch, *The History of the Royal Society of London* (London, 1756–7), vol. ii, pp. 90–2.
7. P. Pugliese, 'Robert Hooke and the dynamics of motion in a curved path', in M. Hunter, S. Schaffer (eds), *Robert Hooke. New Studies* (Woodbridge, Suffolk, 1989), pp. 187–93.

8. A. Koyré, *Newtonian Studies* (London, 1965), Appendix H, pp. 180–4.
9. R. T. Gunther, *Early Science in Oxford*, vol. 6 (Oxford, 1930), pp. 253–5. Hooke to Boyle, 3 February 1666. M. Hunter, *Establishing the New Science* (Woodbridge, Suffolk, 1989), pp. 126–35.
10. J. A. Bennett, *The Divided Circle* (Oxford, 1987), is a good guide to all these instruments.
11. Gunther, vol. 6, pp. 249–50, Hooke to Boyle, 15 August 1665.
12. Cooper, 'Robert Hooke, City Surveyor', p. 228, n. 514. http://homepage.swissonline.net/dedual/wild_heerbrugg/heinrichwild.htm.
13. A. R. Hall, M. B. Hall (eds), *The Correspondence of Henry Oldenburg* (Madison, 1965–86), vol. 2, pp. 605, 610, Moray to Oldenburg, 12 and 16 November 1665.

8. 'London Was, But It Is No More'

1. J. Schofield, *The Building of London, from the Conquest to the Great Fire* (London, 1984), pp. 162–6.
2. There is a convenient selection of these early maps in F. Barker and P. Jackson, *The History of London in Maps* (London, 1990).
3. Mercers' Company, Gresham Repository 1626–69, p. 232, minutes of Joint Committee of 21 September 1666. A. Saunders, *The Royal Exchange* (London, 1997), pp. 124–7. Royal Society meetings were changed from Wednesdays to Thursdays in January 1667.
4. Thomas Birch, *The History of the Royal Society of London* (London, 1756–7), vol. ii, p. 115.
5. Hooke, *Animadversions on the First Part of the Machina Coelestis of Johannes Hevelius*, in R. T. Gunther, *Early Science in Oxford*, vol. 8 (Cutler Lectures) (Oxford, 1931), pp. 97–101. M. A. R. Cooper, 'Robert Hooke, City Surveyor. An assessment of the importance of his work as a Surveyor for the City of London in the aftermath (1667–74) of the Great Fire', City University (London) Ph.D. thesis (1999), pp. 128–30.
6. R. Waller, *Life of Dr Robert Hooke*, in Gunther, *Early Science in Oxford*, vol. 6 (Oxford, 1930), p. 51.
7. T. F. Reddaway, *The Rebuilding of London After the Great Fire* (London, 1940), chs. 3, 4.
8. Cooper, 'Robert Hooke, City Surveyor', pp. 43–6.
9. M. A. R. Cooper, 'Robert Hooke's work as a surveyor for the City of London in the aftermath of the Great Fire', part 3, *Notes and Records of the Royal Society* (1998), vol. 52 (2), pp. 205–20.
10. M. A. R. Cooper, 'Robert Hooke's work as a surveyor for the City of

London in the aftermath of the Great Fire', part 1, *Notes and Records of the Royal Society* (1997), vol. 51 (2), pp. 161–74. Cooper, 'The Legacy of Robert Hooke: Science, Surveying and the City', a talk given at Gresham College on 4 April 2001, and available at www.royalsoc.ac.uk./events/hooke.

11. A. Saunders, ed., *The Royal Exchange* (London Topographical Society, 1997), pp. 124–9. Cooper, 'Robert Hooke, City Surveyor', pp. 41–2, 187–9.
12. Cooper, 'Robert Hooke, City Surveyor', pp. 94–7.
13. Quoted in M. Warner (ed.), *The Image of London* (London, 1987), p. 117.

9. 'Noble Experiments'

1. R. T. Gunther, *Early Science in Oxford*, vol. 6 (Oxford, 1930), pp. 313–14. Hooke to Boyle, 5 September 1667.
2. R. G. Frank, *Harvey and the Oxford Physiologists* (Berkeley, 1980), pp. 197–201. Royal Society Classified Papers, xx, f. 45. Hooke's report, 24 October 1667.
3. Thomas Birch, *The History of the Royal Society of London* (London, 1756–7), vol. ii, pp. 282–3, 287–8, 14 and 28 May 1668.
4. Frank, *Harvey and the Oxford Physiologists*, pp. 208–20.
5. Birch, *History of the Royal Society*, vol. ii, pp. 189, 469–73.
6. Birch, *History of the Royal Society*, vol. ii, pp. 126–7, 133–4, 137, 157–60.
7. Gunther, *Early Science in Oxford*, vol. 6, pp. 313–14.
8. Hooke, *Animadversions on the First Part of the Machina Coelestis of Johannes Hevelius*, pp. 106–7.
9. Birch, *History of the Royal Society*, vol. ii, pp. 203–4. R. W. Symonds, *Thomas Tompion. His Life and Work* (London, 1951), pp. 131–2.
10. M. 'Espinasse, *Robert Hooke* (London, 1956), p. 67.
11. J. H. Leopold, 'The Longitude Timekeepers of Christiaan Huygens', in W. J. H. Andrewes (ed.), *The Quest for Longitude* (Harvard, 1993), pp. 102–14.
12. Birch, *History of the Royal Society*, vol. ii, pp. 246, 251, 283–6, 295–8.
13. Birch, *History of the Royal Society*, vol. ii, p. 139.
14. Birch, *History of the Royal Society*, vol. ii, pp. 188–9, 197, 204. Gunther, *Early Science in Oxford*, vol. 8 (Cutler Lectures) (Oxford, 1931), 'An Attempt to Prove the Motion of the Earth from Observations', p. 20.
15. A. R. and M. B. Hall, *Correspondence of Henry Oldenburg*, vol. iv, pp. 396–7, Oldenburg to Hevelius, 11 May 1668, pp. 447–8, Hevelius to

Oldenburg, 3 June 1668; vol. v, pp. 244–5, Hevelius to Oldenburg, 11 December 1668.

16. There is a good account of Bradley's work, and of the aberration of light, in the *Encyclopædia Britannica* (11th edition, 1910–11), vol. 1, in the article on Aberration by Otto Eppenstein. Hooke mentioned the possibility of 'variation in the perpendicularity' to the Royal Society on 27 April 1671.
17. E. G. Forbes, L. Murdin, F. Willmoth (eds), *Correspondence of John Flamsteed, the First Astronomer Royal* (Bristol, 1995), vol. 1, pp. 16–17, Flamsteed to Brouncker, 24 November 1669; pp. 81–2, Flamsteed to Oldenburg, 18 February 1671.
18. W. E. K. Middleton, *The Invention of the Meteorological Instruments* (Baltimore, 1969), pp. 35, 37–8. R. Waller (ed.), *The Posthumous Works of Robert Hooke* (London, 1705, reprinted with an introduction by R. S. Westfall (London, 1968)), pp. 555–6.
19. Middleton, *The Invention of the Meteorological Instruments*, pp. 25–6.
20. Birch, *History of the Royal Society*, vol. ii, pp. 235–313.
21. Hooke, *Posthumous Works*, p. 15. P. J. Pugliese, 'The scientific achievement of Robert Hooke: method and mechanics' (Harvard University Ph.D. thesis, 1982), pp. 9–10, argues for a date in 1668, but seems to ignore Hooke's earlier discovery of Jupiter's spot in 1664. M. B. Hesse, 'Hooke's philosophical algebra', *Isis*, 57 (1966), p. 68, prefers 1666.
22. Hooke, *Posthumous Works*, pp. 1–65, *A General Scheme or Idea of the Present State of Natural Philosophy*.
23. Hooke, *Posthumous Works*, pp. 320–2.
24. Hooke, *Posthumous Works*, pp. 327–8.
25. P. Rossi, *The Dark Abyss of Time. The History of the Earth and the History of Nations from Hooke to Vico* (Chicago, 1984), pp. 12–20. Hooke, *Posthumous Works*, Westfall's introduction, p. xxiv.
26. Birch, *History of the Royal Society*, vol. ii, pp. 449, 469.

10. A New Career

1. M. Hunter, *Establishing the New Science* (Woodbridge, Suffolk, 1989), pp. 156–82.
2. Royal Society Journal Book, vol. 7, pp. 217–18, 17 July 1689. John Evelyn's Diary, 25 September 1672.
3. W. Thornbury, *Old and New London* (6 vols, London, 1883–5), vol. 2, p. 431.

4. H. J. Louw, 'The origin of the sash-window', *Architectural History*, 26 (1983), pp. 49–72. H. J. Louw, R. Crayford, 'A constructional history of the sash-window c. 1670–1725', parts 1 and 2, *Architectural History*, 41 (1998), pp. 82–130, and 42 (1999), pp. 173–239.
5. M. A. R. Cooper, 'Robert Hooke, City Surveyor. An assessment of the importance of his work as a Surveyor for the City of London in the aftermath (1667–74) of the Great Fire', City University (London) Ph.D. thesis (1999), pp. 90–4.
6. C. Wren, *Parentalia* (London, 1750), p. 263.
7. The Wren Society, vol. x, *The Parochial Churches of Sir Christopher Wren, 1666–1728* (London, 1933), and vol. xix, *The City Churches* (London, 1942), pp. 1–56. S. Bradley, N. Pevsner, *London: the City Churches* (London, 1998), pp. 24–6, 69, 83, 100, 103, 122, 143. P. Jeffery, *The City Churches of Sir Christopher Wren* (London, 1996). A. Geraghty, 'Nicholas Hawksmoor and the Wren City Steeples', *The Georgian Group Journal*, 10 (2000), pp. 1–14, n. 23.
8. Cooper, 'Robert Hooke, City Surveyor', pp. 139–47, 238–9.
9. Jeffery, *The City Churches of Sir Christopher Wren*, p. 93.
10. Bradley and Pevsner, *London: the City Churches*, p. 83. In attributing churches to Hooke I have followed Bradley's revised edition of Pevsner, which in turn based some of its attributions on the research of Anthony Geraghty.
11. For a good account of Wren's scientific work, see J. A. Bennett, *The Mathematical Science of Christopher Wren* (Cambridge, 1983).
12. T. F. Reddaway, *The Rebuilding of London After the Great Fire* (London, 1940), pp. 200–21. Cooper, 'Robert Hooke, City Surveyor', pp. 78–86.
13. Cooper, 'Robert Hooke, City Surveyor', pp. 86–8.
14. Royal Society Journal Book, vol. 7, pp. 219–20, 24 July 1689.
15. Cooper, 'Robert Hooke, City Surveyor', pp. 88–90.
16. M. I. Batten, 'The Architecture of Dr Robert Hooke, F.R.S.', *Walpole Society*, 25 (1936–7), pp. 83–113.
17. London Topographical Society, *The A to Z of Restoration London* (London, 1992), introduction by R. Hyde, pp. v–xii. Cooper, 'Robert Hooke, City Surveyor', pp. 130–4. E. G. R. Taylor, 'Robert Hooke and the cartographical projects of the late seventeenth century (1666–1696)', *Geographical Journal* 90 (1937), pp. 529–40.

11. Physicians, Lovers and Friends

1. L. Mulligan, 'Self-scrutiny and the study of nature: Robert Hooke's diary as natural history', *Journal of British Studies*, vol. 35 (July 1996), pp. 311–42.
2. Guildhall manuscript 1758.
3. M. 'Espinasse, *Robert Hooke* (London, 1956), p. 152.
4. There is a good account of Hooke's health in L. M. Beier, 'Experience and experiment: Robert Hooke, illness and medicine', in M. Hunter, S. Schaffer (eds), *Robert Hooke: New Studies* (Woodbridge, Suffolk, 1989), pp. 235–52.
5. Hooke's Diary, 16–20 November 1672, 4–5 July 1673.
6. There is information on Grace Hooke on the Web site of the Isle of Wight History Centre, http://freespace.virgin.net/ric.martin/vectis/hookeweb/roberthooke.htm.
7. S. Shapin, 'Who was Robert Hooke?', in Hunter and Schaffer, *Robert Hooke: New Studies*, p. 268.
8. The best accounts of coffee houses are B. Lillywhite, *London Coffee Houses* (London, 1963) and A. Ellis, *The Penny Universities* (London, 1956).
9. Revealing diary references to Godfrey include 17 March 1679, 4 September 1680, 17 December 1680, 6 December 1688, 23 April 1689, 31 January 1690, 5 February 1690. Many thanks to the Mercers' Company's chief archivist, Ursula Carlyle, for her help in identifying Mr Godfrey.
10. Shapin, 'Who was Robert Hooke?', pp. 272–6.
11. R. Iliffe, 'Material doubts: Hooke, artisan culture and the exchange of information in 1670s London', *British Journal for the History of Science*, 28 (1995), p. 317.
12. Iliffe, 'Material doubts', is the best study of Hooke's relations with London craftsmen.

12. Heat and Light

1. A. D. C. Simpson, 'Robert Hooke and practical optics', in M. Hunter, S. Schaffer (eds), *Robert Hooke: New Studies* (Woodbridge, Suffolk, 1989), pp. 48–9. H. C. King, *The History of the Telescope* (London, 1955), pp. 68–74.
2. H. W. Turnbull (ed.), *The Correspondence of Isaac Newton*, vol. 1,

1661–75 (Cambridge, 1959), pp. 110–14, Hooke to Oldenburg, 15 February 1672.

3. A. I. Sabra, *Theories of Light from Descartes to Newton* (London, 1967), pp. 233, 249–50.
4. Turnbull, *Correspondence of Isaac Newton*, vol. 1, pp. 171–88, Newton to Oldenburg, 11 June 1672.
5. R. H. Westfall, *Never at Rest. A Biography of Isaac Newton* (Cambridge, 1980), p. 246.
6. Turnbull, *Correspondence of Isaac Newton*, vol. 1, pp. 198–203, Hooke to Brouncker, June 1672.
7. A. R. Hall, M. B. Hall, 'Why blame Oldenburg?', *Isis*, 53 (1962), pp. 482–91. Westfall, *Never at Rest*, pp. 249–52. King, *The History of the Telescope*, pp. 68–74.
8. Hooke, *Micrographia*, Preface. Turnbull, *Correspondence of Isaac Newton*, vol. 1, pp. 111–12, Hooke to Oldenburg, 15 February 1672. Simpson, 'Robert Hooke and practical optics', pp. 44–9. King, *The History of the Telescope*, pp. 144–8.
9. Simpson, 'Robert Hooke and practical optics', p. 53. Turnbull, *Correspondence of Isaac Newton*, vol. 1, p. 255, n. 17, Collins to Gregory, 26 December 1672.
10. W. Derham, *Philosophical Experiments and Observations of the Late Eminent Dr Robert Hooke* (London, 1726), pp. 269–70.
11. Nicholas Blake and Richard Lawrence, *The Illustrated Companion to Nelson's Navy* (London, 2000), pp. 9, 192.
12. Thomas Birch, *The History of the Royal Society of London* (London, 1765–7), vol. iii, p. 29.
13. R. G. Frank, *Harvey and the Oxford Physiologists* (Berkeley, 1980), pp. 255–6. H. D. Turner, 'Robert Hooke and theories of combustion', *Centaurus* 4 (1956), pp. 297–310. Hooke's Diary, November 1672 to March 1673, passim.
14. R. Iliffe, '"In the Warehouse": Privacy, Property and Priority in the Early Royal Society', pp. 37–8, *History of Science*, xxx (1992), pp. 29–68.
15. A. R. Hall, M. B. Hall (eds), *The Correspondence of Henry Oldenburg* (Madison, 1965–86), vol. 9, pp. 492–6, Leibniz to Oldenburg, 26 February 1673.

13. Measuring the Heavens

1. Mercers' Company, Gresham repository 1669–1776, pp. 136–8, 7 December 1673.

2. Thomas Birch, *The History of the Royal Society of London* (London, 1765–7), vol. iii, pp. 120–1.
3. Robert Hooke, *Animadversions on the First Part of the Machina Coelestis of Johannes Hevelius* (London, 1674), pp. 80–1.
4. Hooke, *Animadversions*, pp. 106–11. For a brief account of the development of dividing engines, see T. K. Derry, T. I. Williams, *A Short History of Technology* (Oxford, 1960), pp. 348–9.
5. Hooke, *Animadversions*, pp. 113–14.
6. R. Iliffe, 'Material doubts: Hooke, artisan culture and the exchange of information in 1670s London', p. 316, *British Journal for the History of Science*, 28 (1995), pp. 285–318.
7. E. G. Forbes, L. Murdin, F. Willmoth (eds), *Correspondence of John Flamsteed, the First Astronomer Royal* (Bristol, 1995), vol. 1, pp. 307–11, Moore to Flamsteed, 10 October 1674; Flamsteed to Moore, 13 October 1674; Flamsteed to Hooke, 13 October 1674.
8. M. Feingold, 'John Flamsteed and the Royal Society', in F. Willmoth (ed.), *Flamsteed's Stars. New Perspectives on the Life and Work of the First Astronomer-Royal* (Woodbridge, Suffolk, 1997), p. 38.
9. Forbes, et al., *Correspondence of John Flamsteed*, vol. 1, p. 356, Flamsteed to Towneley, 3 July 1675.
10. D. Howse, *Greenwich Observatory*, vol. 3, *The Buildings and Instruments* (London, 1975), pp. 17–18.
11. M. Feingold, 'John Flamsteed and the Royal Society', in F. Willmoth, *Flamsteed's Stars. New Perspectives on the Life and Work of the First Astronomer-Royal* (Woodbridge, Suffolk, 1997), p. 38.
12. Forbes, et al., *Correspondence of John Flamsteed*, vol. 1, pp. 465–71, Hevelius to Flamsteed, 14 June 1676.
13. Forbes, et al., *Correspondence of John Flamsteed*, vol. 1, pp. 465–72, Hevelius to Flamsteed, 14 June 1676; pp. 491–5, Flamsteed to Hevelius, 20 July 1676; pp. 516–25, Hevelius to Flamsteed, 23 December 1676; pp. 539–40, Flamsteed to Oldenburg, January 1677; pp. 546–8, Flamsteed to Oldenburg, 5 April 1677.
14. Birch, *History of the Royal Society*, vol. iii, p. 488.

14. The Coiled Spring

1. M. Hunter, *Establishing the New Science* (Woodbridge, Suffolk, 1989), pp. 232–9, from Royal Society Classified Papers, xx, fols 92–4.
2. Birch, *History of the Royal Society*, vol. iii, pp. 173–4.
3. Birch, *History of the Royal Society*, vol. iii, p. 191.

4. R. Iliffe, '"In the Warehouse": privacy, property and priority in the early Royal Society', *History of Science*, 30 (1992), pp. 29–68.
5. T. Sprat, *A History of the Royal Society of London for the improving of natural knowledge* (London, 1667), p. 247.
6. D. S. Landes, *Revolution in Time: Clocks and the Making of the Modern World* (Cambridge, Mass., 1983), pp. 127–8.
7. Waller, *The Life of Dr Robert Hooke*, in Gunther, *Early Science in Oxford*, vol. 6, pp. 16–17.
8. E. G. Forbes, L. Murdin, F. Willmoth (eds), *Correspondence of John Flamsteed, the First Astronomer Royal* (Bristol, 1995), pp. 329–30, Flamsteed to Towneley, 16 March 1675.
9. Forbes, et al., *Correspondence of John Flamsteed*, pp. 351–7, Flamsteed to Towneley, 8 June, 22 June, 3 July 1675.
10. S. Shapin, 'Who was Robert Hooke?', in M. Hunter, S. Schaffer (eds), *Robert Hooke: New Studies* (Woodbridge, Suffolk, 1989), pp. 253–86, is excellent on Hooke's behaviour and social status.
11. Iliffe, '"In the Warehouse"', pp. 46–7. A. R. Hall, M. B. Hall (eds), *The Correspondence of Henry Oldenburg* (Madison, 1965–86), vol. 12, pp. 1–3, Huygens to Oldenburg, 2 October 1675.
12. Iliffe, '"In the Warehouse"', p. 47.
13. Forbes, et al., *Correspondence of John Flamsteed*, p. 374, Flamsteed to Towneley, 22 September 1675.
14. Hall and Hall, *Correspondence of Henry Oldenburg*, vol. 12, pp. 28–9, Oldenburg to Huygens, 1 November 1675.
15. Robert Hooke, *A Description of Helioscopes and some Other Instruments* (London, 1676), in R. T. Gunther, *Early Science in Oxford*, vol. 8 (Cutler Lectures) (Oxford, 1931), pp. 151–2.
16. R. Waller, *The Life of Dr Robert Hooke*, in Gunther, *Early Science in Oxford*, vol. 6, pp. 51–2.
17. Forbes, et al., *Correspondence of John Flamsteed*, p. 383, Towneley to Flamsteed, 24 November 1675.
18. Hall and Hall, *Correspondence of Henry Oldenburg*, vol. 12, pp. 19–24, Huygens to Brouncker, 22 October 1675; Huygens to Oldenburg, 22 October 1675.
19. Gunther, *Early Science in Oxford*, vol. 8, pp. 434–5, Hooke to Aubrey, 24 August 1675.
20. M. B. Hall, 'Oldenburg, the Philosophical Transactions, and Technology', p. 39, in J. G. Burke (ed.), *The Uses of Science in the Age of Newton* (Berkeley and London, 1983), pp. 21–47. Hall and Hall, *Correspondence of Henry Oldenburg*, vol. 12, pp. 42–3, 25 October 1675.
21. Hooke's Diary, 11 November, 2 December 1675.

22. J. Andrews, A. Briggs, et al., *History of Bethlem* (London, 1997), pp. 230–5.
23. Andrews and Briggs, et al., *History of Bethlem*, p. 248.
24. Andrews and Briggs, et al., *History of Bethlem*, pp. 230–52.
25. R. Iliffe, 'Material Doubts: Hooke, artisan culture and the exchange of information in 1670s London', *British Journal of the History of Science* 28 (1995), pp. 299–303.

15. 'Oldenburg Kindle Cole'

1. Hooke's Diary, 18 January 1676.
2. P. Gouk, 'The role of acoustics and music theory in the scientific work of Robert Hooke', *Annals of Science* 37 (1980), pp. 575–605. For a full discussion of Hooke's musical ideas in their seventeenth-century context, see P. Gouk, *Music, Science and Natural Magic in Seventeenth-Century England* (New Haven and London, 1999).
3. R. Waller (ed.), *The Posthumous Works of Robert Hooke* (London, 1705, reprinted with an introduction by R. S. Westfall (London, 1968)), pp. 186–90.
4. Thomas Birch, *The History of the Royal Society of London* (London, 1756–7), vol. iii, pp. 268–9.
5. Birch, *History of the Royal Society*, vol. iii, pp. 298–9.
6. H. W. Turnbull, *The Correspondence of Isaac Newton*, vol. 1, 1661–75 (Cambridge, 1959), pp. 404–11, Newton to Oldenburg, 21 December 1675, 10 January 1676.
7. Turnbull, *Correspondence of Isaac Newton*, vol. 1, pp. 412–13, Hooke to Newton, 20 January 1676.
8. Turnbull, *Correspondence of Isaac Newton*, vol. 1, pp. 416–17, Newton to Hooke, 5 February 1676.
9. Hooke, *Lampas*, in R. T. Gunther, *Early Science in Oxford*, vol. 8 (Cutler Lectures) (Oxford, 1931), p. 180.
10. Hooke's Diary, 2, 3, 25 June, 1 July.
11. R. Iliffe, '"In the Warehouse": privacy, property and priority in the early Royal Society', pp. 51–2, in *History of Science*, xxx (1992), pp. 29–68.
12. British Museum Add. mss, 4441 ff. 58–9, 100, for Oldenburg's two drafts.

16. In Two Worlds

1. There is a recipe for making phosphorus in W. Derham, *Philosophical Experiments and Observations of the Late Eminent Dr Robert Hooke* (facs. edn., London, 1967), pp. 174–6.
2. J. A. Bennet, 'Hooke and Wren and the system of the world', *The British Journal for the History of Science*, 8 (1975), pp. 32–9.
3. M. Wright, 'Robert Hooke's longitude timekeeper', in M. Hunter, S. Schaffer (eds), *Robert Hooke: New Studies* (Woodbridge, Suffolk, 1989), p. 116.
4. *Survey of London*, xxix, *The Parish of St James, Westminster* (1960), part 1, pp. 61–2.
5. M. I. Batten, 'The Architecture of Dr Robert Hooke, F.R.S.', *Walpole Society*, 25 (1936–7), pp. 105–6, 109–10. N. Pevsner, B. Cherry, *The Buildings of England: Devon* (2nd edition, Harmondsworth, 1989), pp. 85, 356.
6. W. Kent, *An Encyclopaedia of London* (London, 1937), p. 24.
7. The Isle of Wight History Centre Web site (http://freespace.virgin.net/roger.hewitt/iwias/home.htm).
8. W. B. Bannerman (ed.), *The Registers of St Helens* (Harleian Society, Bishopsgate, 1904), p. 324.

17. Tiny Creatures and Springy Bodies

1. Thomas Birch, *The History of the Royal Society of London* (London, 1756–7), vol. iii, pp. 346–52. Hooke, *Microscopium*, pp. 298–9, in R. T. Gunther, *Early Science in Oxford*, vol. 8 (Cutler Lectures) (Oxford, 1931), pp. 297–320.
2. Hooke, *Microscopium*. B. J. Ford, *Single Lens. The Story of the Simple Microscope* (London, 1985), pp. 128–9. M. Fournier, *The Fabric of Life* (Baltimore and London, 1996), pp. 34, 167–77.
3. Birch, *History of the Royal Society*, vol. iii, p. 365.
4. Birch, *History of the Royal Society*, vol. iii, pp. 364–5.
5. Birch, *History of the Royal Society*, vol. iii, p. 371. Hooke's Diary, 20 December 1677, 3 January 1678.
6. R. T. Gunther, *Early Science in Oxford*, vol. 7 (Oxford, 1930), pp. 463–4.
7. M. B. Hall, *Promoting Experimental Learning. Experiment and the Royal Society, 1660–1727* (Cambridge, 1991), pp. 107–9.
8. E. G. Forbes, L. Murdin, F. Willmoth (eds), *Correspondence of John*

Flamsteed, the First Astronomer Royal (Bristol, 1995), vol. 1, p. 600, Flamsteed to Bernard, 8 February 1678. The last two words refer to Horace's joke, 'The mountains are in labour and bring forth a ridiculous mouse'.

9. R. G. Frank, *Harvey and the Oxford Physiologists* (Berkeley, 1980), pp. 275–8.
10. There is useful information on Holmes' affair with Grace and John Hooke's suicide on the Isle of Wight History Centre Web site (http://freespace.virgin.net/roger.hewitt/iwias/home.htm).
11. E. G. R. Taylor, 'Robert Hooke and the Cartographical Projects of the Late Seventeenth Century (1666–1696)', *Geographical Journal*, 90 (1937), pp. 529–40. E. G. R. Taylor, 'The English Atlas of Moses Pitt, 1680–1683', *Geographical Journal* 95 (1940), pp. 292–9.
12. J. E. Gordon, *The New Science of Strong Materials* (2nd edn, Harmondsworth, 1976), pp. 36–42.
13. Robert Hooke, *Micrographia* (London, 1665), p. 12.
14. P. J. Pugliese, 'The scientific achievement of Robert Hooke: method and mechanics' (Harvard University Ph.D. thesis, 1982), pp. 234–48. M. Hesse, 'Hooke's vibration theory and the isochrony of springs', *Isis*, 57 (1966), pp. 433–41. A. E. Moyer, 'Robert Hooke's ambiguous presentation of "Hooke's Law"', *Isis*, 68 (1977), pp. 266–75. M. E. Ehrlich, 'Mechanism and activity in the Scientific Revolution: the case of Robert Hooke', *Annals of Science*, 52 (1995), pp. 127–51.

18. 'A Man of a Strange Unsociable Temper'

1. John Evelyn's Diary, 10 October 1683.
2. J. M. Crook, *The British Museum* (London, 1972), pp. 51–2, 136–8. M. I. Batten, 'The Architecture of Dr Robert Hooke, F.R.S.', *Walpole Society*, 25 (1936–7), pp. 94–6 and plate xxxviii.
3. Thomas Birch, *The History of the Royal Society of London* (London, 1756–7), vol. iii, pp. 463–4, 476–81, 486–8. W. E. K. Middleton, *The Invention of the Meteorological Instruments* (Baltimore, 1969), pp. 250–5.
4. Middleton, *The Invention of the Meteorological Instruments*, pp. 90–5, 250–5. W. Derham, *Philosophical Experiments and Observations of the Late Eminent Dr Robert Hooke* (facs. edn., London, 1967), pp. 41–7, Hooke's accounts of his weather clock and self-emptying rain bucket.
5. Royal Society Journal Book, vol. vii, p. 248, 18 December 1689.
6. Birch, *History of the Royal Society*, vol. iv, p. 277. Middleton, *The Invention of the Meteorological Instruments*, pp. 255–7.

7. P. J. Pugliese, 'The scientific achievement of Robert Hooke: method and mechanics' (Harvard University Ph.D. thesis, 1982), p. 99.
8. British Museum, Sloane mss 1039, f. 12.
9. Birch, *History of the Royal Society*, vol. iii, pp. 452–71.
10. Birch, *History of the Royal Society*, vol. iii, pp. 481–2, 486–7, 489.
11. John Evelyn's Diary, 12 April 1682.
12. Birch, *History of the Royal Society*, vol. iii, pp. 500–2.
13. Birch, *History of the Royal Society*, vol. iii, pp. 501, 504–6. M. Hunter, *Establishing the New Science* (Woodbridge, Suffolk, 1989), pp. 255–6, 275–6.
14. E. G. Forbes, L. Murdin, F. Willmoth (eds), *Correspondence of John Flamsteed, the First Astronomer Royal* (Bristol, 1995), vol. 1, p. 704, Flamsteed to Towneley, 25 October 1679.
15. R. Waller (ed.), *The Posthumous Works of Robert Hooke* (London, 1705, reprinted with an introduction by R. S. Westfall (London, 1968)), pp. 114, 178.
16. A. Koyré, 'A note on Robert Hooke', *Isis* 41 (1950), pp. 195–6.
17. M. F. Ashley Montagu, *Edward Tyson, M.D., F.R.S., 1650–1708* (Philadelphia, 1943), pp. 61–80. G. Keynes, *Bibliography of Robert Hooke* (Oxford, 1966), pp. 48–50, 95.
18. H. W. Turnbull (ed.), *The Correspondence of Isaac Newton*, vol. 2 (Cambridge, 1960), p. 297, Hooke to Newton, 24 November 1679.
19. R. S. Westfall, *Never at Rest. A Biography of Isaac Newton* (Cambridge, 1980), pp. 382–3.
20. R. S. Westfall, 'Hooke and the law of universal gravitation', *The British Journal for the History of Science*, 3 (1967), pp. 245–61, p. 260.
21. H. W. Turnbull (ed.), *The Correspondence of Isaac Newton*, vol. 2 (Cambridge, 1960), pp. 300–3, Newton to Hooke, 28 November 1679.
22. M. White, *Isaac Newton. The Last Sorcerer* (London, 1997), pp. 195–201.
23. Turnbull, *Correspondence of Isaac Newton*, vol. 2, Newton to Halley, 20 June 1686.
24. Birch, *History of the Royal Society*, vol. iii, pp. 515–18.
25. Turnbull, *Correspondence of Isaac Newton*, vol. 2, pp. 309–10, Hooke to Newton, 6 January 1680.
26. J. Lohne, 'Hooke versus Newton. An analysis of the documents in the case on free fall and planetary motion', *Centaurus*, 7 (1960), p. 33.
27. Turnbull, *Correspondence of Isaac Newton*, vol. 2, p. 447, Newton to Halley, 27 July 1686.
28. Turnbull, *Correspondence of Isaac Newton*, vol. 2, p. 447, Newton to Halley, 27 July 1686.

19. All Trades

1. Thomas Birch, *The History of the Royal Society of London* (London, 1756–7), vol. iv, pp. 3–17, 23, 29–32. P. J. Pugliese, 'The scientific achievement of Robert Hooke: method and mechanics' (Harvard University Ph.D. thesis, 1982), pp. 255–75. Hooke's handwritten paper is in the British Museum, Sloane mss 1039, f. 114.
2. British Museum, Sloane mss 1039, f. 120. 'An Account of some tryalls made about the mixture of metalls'.
3. Birch, *History of the Royal Society*, vol. iv, pp. 8–44. E. G. Forbes, L. Murdin, F. Willmoth (eds), *Correspondence of John Flamsteed*, vol. 1, p. 741, Flamsteed to Towneley, 20 May 1680.
4. M. I. Batten, 'The Architecture of Dr Robert Hooke, F.R.S.', *Walpole Society*, xxv, 1936–7, pp. 83–113, prints the whole correspondence.
5. M. A. R. Cooper, 'Robert Hooke, City Surveyor. An assessment of the importance of his work as a Surveyor for the City of London in the aftermath (1667–74) of the Great Fire', City University (London) Ph.D. thesis (1999), pp. 142, 237 (n. 581).
6. H. Colvin, 'Robert Hooke and Ramsbury Manor', in H. M. Colvin, *Essays in English Architectural History* (New Haven and London, 1999), pp. 191–4.
7. E. G. R. Taylor, 'Robert Hooke and the Cartographical Projects of the Late Seventeenth Century (1666–1696)', *Geographical Journal*, 90 (1937), pp. 529–40.
8. Mercers' Company, Gresham Repository, 1678–1722, pp. 53–8, 24 August 1680, 26 October 1680.
9. R. Waller (ed.), *The Posthumous Works of Robert Hooke* (London, 1705, reprinted with an introduction by R. S. Westfall (London, 1968)), pp. 71–82.
10. Hooke, *Posthumous Works*, pp. 83–95.

20. The Empire of the Senses

1. R. Waller (ed.), *The Posthumous Works of Robert Hooke* (London, 1705, reprinted with an introduction by R. S. Westfall (London, 1968)), pp. 156–9.
2. P. J. Pugliese, 'The scientific achievement of Robert Hooke: method and mechanics' (Harvard University Ph.D. thesis, 1982), p. 114.
3. Pugliese, 'The scientific achievement of Robert Hooke', pp. 111–16.

4. *Dictionary of National Biography*, entry on Robert Knox.
5. R. Knox, *An Historical Relation of the Island Ceylon* (1681; Glasgow, 1911), p. xlv.
6. Thomas Birch, *The History of the Royal Society of London* (London, 1756–7), vol. iv, pp. 64–72.
7. E. G. R. Taylor, 'Robert Hooke and the cartographic projects of the late seventeenth century (1666–1696)', *Geographical Journal*, 90 (1937), pp. 529–40.
8. Birch, *History of the Royal Society*, vol. iv, pp. 74–87. Hooke's Diary, Guildhall Library, 14, 16 May 1681.
9. Hooke, *Posthumous Works*, p. 118.
10. Birch, *History of the Royal Society*, vol. iv, p. 96.
11. P. Gouk, 'The role of acoustics and music theory in the scientific work of Robert Hooke', *Annals of Science* (1980), vol. 37, pp. 575–605, pp. 583–4. P. Gouk, *Music, Science, and Natural Magic in Seventeenth-Century England* (New Haven and London, 1999), p. 208.
12. M. A. R. Cooper, 'Robert Hooke, City Surveyor. An assessment of the importance of his work as a Surveyor for the City of London in the aftermath (1667–74) of the Great Fire', City University (London) Ph.D. thesis (1999), p. 239.
13. S. Bradley, N. Pevsner, *London, the City Churches* (London, 1998), pp. 69–70.
14. M. Feingold, 'Astronomy and Strife: John Flamsteed and the Royal Society', in F. Willmoth (ed.), *Flamsteed's Stars* (Woodbridge, Suffolk, 1989), pp. 31–48, p. 43.
15. R. Iliffe, 'Material doubts: Hooke, artisan culture and the exchange of information in 1670s London', *British Journal for the History of Science* (1995), pp. 316–17. A. Johns, 'Flamsteed's optics and the identity of the astronomical observer', pp. 81–5, in Willmoth, *Flamsteed's Stars*, pp. 77–106.
16. Birch, *History of the Royal Society*, vol. iv, pp. 98–103.
17. Birch, *History of the Royal Society*, vol. iv, pp. 102–5, 113–14.
18. Birch, *History of the Royal Society*, vol. iv, pp. 117–18, 120–2, 129.
19. E. G. Forbes, L. Murdin, F. Willmoth (eds), *The Correspondence of John Flamsteed, the First Astronomer Royal*, vol. 1, pp. 886–7, Flamsteed to Molyneux, 11 April 1682.
20. Hooke, *Posthumous Works*, pp. 130–7.
21. Hooke, *Posthumous Works*, pp. 139–47.
22. Birch, *History of the Royal Society*, vol. iv, pp. 153–4. John Evelyn's Diary, 20 June 1682. B. R. Singer, 'Robert Hooke on memory, association and time perception', *Notes and Records of the Royal Society*, 31 (1979),

pp. 115–131. J. J. Macintosh, 'Perception and imagination in Descartes, Boyle and Hooke', *Canadian Journal of Philosophy*, 13 (1983), pp. 327–52. D. R. Oldroyd, 'Some "Philosophicall Scribbles" attributed to Robert Hooke', in *Notes and Records of the Royal Society*, 35 (1980), pp. 17–32.

23. Birch, *History of the Royal Society*, vol. iv, p. 156.

21. A Curator Again

1. H. Colvin, 'Robert Hooke and Ramsbury Manor' in H. M. Colvin, *Essays in English Architectural History* (New Haven and London, 1999). H. J. Louw, 'New light on Ramsbury Manor', *Architectural History*, 30 (1987), pp. 45–9.
2. Hooke's Diary, Guildhall Library. R. T. Gunther, *Early Science in Oxford*, vol. 7 (Oxford, 1930), pp. 600–1.
3. R. Waller (ed.), *The Posthumous Works of Robert Hooke* (London, 1705, reprinted with an introduction by R. S. Westfall (London, 1968)), pp. 171–85.
4. Birch, *History of the Royal Society*, vol. iv, pp. 167–8, 174.
5. Birch, *History of the Royal Society*, vol. iv, pp. 196, 260, 455, 457.
6. Birch, *History of the Royal Society*, vol. iv, pp. 175–6, misdated 10 January 1684.
7. Birch, *History of the Royal Society*, vol. iv, 187–8.
8. S. Pumfrey, 'Ideas above his station: a social study of Hooke's curatorship of experiments', *History of Science*, 29 (1991), p. 34. M. Hunter, *Establishing the New Science* (Woodbridge, Suffolk, 1989) p. 330.
9. Birch, *History of the Royal Society*, vol. iv, pp. 207–10.
10. Hoskins and Colwall to Cutler, 19 December 1682, quoted in M. Hunter, 'Science, Technology and Patronage: Robert Hooke and the Cutlerian lectureship', in Hunter, *Establishing the New Science*, pp. 309–10. Hunter's essay unravels the Hooke–Cutler relationship brilliantly, and is my main source for these paragraphs.
11. Hooke's Diary, Guildhall Library, 22, 25 January 1683.
12. Hunter, 'Science, Technology and Patronage', pp. 316–25.
13. W. Derham (ed.), *Philosophical Experiments and Observations of the Late Eminent Dr Robert Hooke* (1726; reprinted 1967), pp. 107–8.
14. Hooke, *Posthumous Works*, p. 184.
15. Hooke, *Posthumous Works*, pp. 451–73. Gunther, *Early Science in Oxford*, vol. 7, p. 622, excerpt from Hooke's Diary, 22 September 1683.
16. Birch, *History of the Royal Society*, vol. iv, pp. 223–5. Derham, *Philosoph-*

ical Experiments and Observations, pp. 111–12. Hooke, *Posthumous Works*, p. 562.

17. Birch, *History of the Royal Society*, vol. iv, pp. 228–30, 237.
18. The dispute between Hooke and the chemists is unravelled and explained in S. Pumfrey, 'Ideas above his station', pp. 33–6.
19. Birch, *History of the Royal Society*, vol. iv, pp. 251–3, 261. Derham, *Philosophical Experiments and Observations*, pp. 128–9.
20. Birch, *History of the Royal Society*, vol. iv, pp. 261–71.
21. Derham, *Philosophical Experiments and Observations*, pp. 113–15.
22. R. S. Westfall, 'Robert Hooke, Mechanical Technology and Scientific Investigation', in J. G. Burke (ed.), *The Uses of Science in the Age of Newton* (Los Angeles, 1983), pp. 85–110.
23. John Evelyn's Diary, 24 January 1684.
24. Birch, *History of the Royal Society*, vol. iv, pp. 254–5, 263–5. Derham, *Philosophical Experiments and Observations*, pp. 130–42.

22. Newton's Triumph

1. M. Hunter, *Establishing the New Science* (Woodbridge, Suffolk, 1989), p. 331.
2. H. W. Turnbull (ed.), *The Correspondence of Isaac Newton*, vol. 2, 1676–87 (Cambridge, 1960), p. 442, Halley to Newton, 29 June 1686.
3. H. Colvin, *Biographical Dictionary of British Architects, 1600–1840* (New Haven and London, 3rd edn, 1995), pp. 508–10.
4. R. S. Westfall, *Never at Rest. A Biography of Isaac Newton* (Cambridge, 1980), pp. 402–10.
5. Turnbull, *Correspondence of Isaac Newton*, vol. 2, p. 442, Halley to Newton, 29 June 1686.
6. W. Derham (ed.), *Philosophical Experiments and Observations of the Late Eminent Dr Robert Hooke* (1726; reprinted 1967), pp. 150–67. D. Dowson, *History of Tribology* (London, 1979), pp. 145–7.
7. Thomas Birch, *The History of the Royal Society of London* (London, 1756–7), vol. iv, pp. 351–2.
8. Hooke, *Posthumous Works*, pp. 193–200.
9. Hooke, *Posthumous Works*, pp. 510–30.
10. P. J. Pugliese, 'The scientific achievement of Robert Hooke: method and mechanics' (Harvard University Ph.D. thesis, 1982), pp. 499–514.
11. Birch, *History of the Royal Society*, vol. iv, pp. 476–7.
12. Turnbull, *Correspondence of Isaac Newton*, vol. 2, 1676–87, pp. 442–3, Halley to Newton, 29 June 1686.

13. Turnbull, *Correspondence of Isaac Newton*, vol. 2, 1676–87, p. 431, Halley to Newton, 22 May 1686.
14. Turnbull, *Correspondence of Isaac Newton*, vol. 2, 1676–87, pp. 433–4, Newton to Halley, 27 May 1686.
15. Turnbull, *Correspondence of Isaac Newton*, vol. 2, 1676–87, pp. 435–7, Newton to Halley, 20 June 1686.
16. Turnbull, *Correspondence of Isaac Newton*, vol. 2, 1676–87, pp. 437–40, Newton to Halley, 20 June 1686.
17. Turnbull, *Correspondence of Isaac Newton*, vol. 2, 1676–87, pp. 441–3, Halley to Newton, 29 June 1686.
18. Turnbull, *Correspondence of Isaac Newton*, vol. 2, 1676–87, pp. 446–7, Newton to Halley, 27 July 1686.
19. Westfall, *Never at Rest*, pp. 444, 449.

23. The World Turned Upside Down

1. Thomas Birch, *The History of the Royal Society of London* (London, 1756–7), vol. iv, pp. 516–18.
2. M. Hunter, *Establishing the New Science* (Woodbridge, Suffolk, 1989), p. 332.
3. Birch, *History of the Royal Society*, vol. iv, pp. 524, 545.
4. Birch, *History of the Royal Society*, vol. iv, pp. 498–9.
5. Hooke, *Posthumous Works*, pp. 329–30.
6. Hooke, *Posthumous Works*, p. 338.
7. Hooke, *Posthumous Works*, pp. 340–8.
8. Hooke and the Royal Society used the word 'prolate' for a flattened sphere, but today this means an elongated sphere, and the word applied to a flattened sphere is 'oblate'.
9. Hooke, *Posthumous Works*, pp. 350–3.
10. Hooke, *Posthumous Works*, pp. 355–8.
11. D. Oldroyd, 'Geological controversy in the seventeenth century: "Hooke vs Wallis" and its aftermath', in *Robert Hooke: New Studies*, pp. 210–12.
12. The whole Wallis–Hooke correspondence is printed and analysed in Oldroyd, 'Geological controversy in the seventeenth century'.
13. W. Derham (ed.), *Philosophical Experiments and Observations of the Late Eminent Dr Robert Hooke* (1726; reprinted 1967), p. 206. Oldroyd, 'Geological controversy in the seventeenth century', pp. 218–19.
14. H. W. Turnbull (ed.), *The Correspondence of Isaac Newton*, vol. 2, 1676–87 (Cambridge, 1960), p. 329, Newton to Burnet, January 1681; p. 475, Halley to Wallis, 9 April 1687.

15. Oldroyd, 'Geological controversy in the seventeenth century', pp. 219–20.
16. Hooke, *Posthumous Works*, p. 546. P. J. Pugliese, 'The scientific achievement of Robert Hooke: method and mechanics' (Harvard University Ph.D. thesis, 1982), p. 563.
17. R. S. Westfall, *Never at Rest* (Cambridge, 1980), p. 459.
18. Birch, *History of the Royal Society*, vol. iv, pp. 544–8. R. Waller, *Life of Robert Hooke*, in Gunther, *Early Science in Oxford*, vol. 6 (Oxford, 1930), p. 59.
19. Hooke, *Posthumous Works*, pp. 411–12.

24. A Revolution and Old Battles

1. R. Waller (ed.), *The Posthumous Works of Robert Hooke* (London, 1705, reprinted with an introduction by R. S. Westfall (London, 1968)), pp. 428–33. Mercers' Company, Gresham Repertory, 1678–1722, p. 195, 29 June 1688. Royal Society Journal Book, vol. vii, 18 July 1688.
2. Hooke's Diary, 16 May, 8 November 1689.
3. Hooke, *Posthumous Works*, pp. 433–6.
4. Royal Society Classified Papers, 1680–1740, vol. xx (Hooke Papers), f. 77.
5. Royal Society Classified Papers, 1680–1740, vol. xx (Hooke Papers), f. 78.
6. Royal Society Classified Papers, 1680–1740, vol. xx (Hooke Papers), f. 78.
7. L. D. Patterson, 'Hooke's gravitation theory and its influence on Newton' (part II), *Isis*, 41 (1950), pp. 32–45. Also in R. T. Gunther, *Early Science in Oxford*, vol. 7 (Oxford, 1930), pp. 714–16.
8. Royal Society Journal Book, vol. 7, pp. 244–5, 4, 11 December 1689.
9. Derham, *Philosophical Experiments and Observations*, pp. 210–12.
10. Royal Society Journal Book, vol. vii, pp. 247–50. R. Burton, *The Anatomy of Melancholy* (ed. W. H. Gass, New York, 2001), Part Two, pp. 247, 251.
11. Hooke's Diary, 14, 18, 20, 23 December 1689; 8, 21, 29 January 1690. Royal Society Journal Book, vol. 7, pp. 250, 256–7.
12. Hooke, *Posthumous Works*, pp. 563–4.
13. R. S. Westfall, 'Robert Hooke, mechanical technology and scientific investigation', pp. 97–9, in J. G. Burke, *The Uses of Science in the Age of Newton* (Los Angeles, 1983), pp. 85–110.
14. A. R. Hall, 'Two unpublished lectures of Robert Hooke', *Isis*, 42 (1951), pp. 219–30. P. J. Pugliese, 'The scientific achievement of Robert Hooke:

method and mechanics' (Harvard University Ph.D. thesis, 1982), pp. 532–4.
15. R. S. Westfall, *Never at Rest* (Cambridge, 1980), pp. 511–12.

25. The Fear of Being Forgotten

1. M. Hunter, *Establishing the New Science* (Woodbridge, Suffolk, 1989), p. 326.
2. British Museum Sloane mss 1039, f. 131, Hooke's memo to Levett, 20 June 1691.
3. N. Pevsner, B. Cherry, *The Buildings of England. Devon* (2nd edn, 1989), pp. 83, 651.
4. R. Waller (ed.), *The Posthumous Works of Robert Hooke* (London, 1705, reprinted with an introduction by R. S. Westfall (London, 1968)), pp. 203–7, 'Of Dr Dee's Book of Spirits'.
5. J. Henry, 'Robert Hooke, the incongruous mechanist', in M. Hunter, S. Schaffer (eds), *Robert Hooke: New Studies* (Woodbridge, Suffolk, 1989), pp. 149–80. P. Gouk, *Music, Science and Natural Magic in Seventeenth-Century England* (New Haven and London, 1999), pp. 190–223.
6. Hooke, *Posthumous Works*, pp. 553–62.
7. R. T. Gunther, *Early Science in Oxford*, vol. 8 (Cutler Lectures) (Oxford, 1931), p. vii, addenda to vols 6 and 7.
8. W. Derham (ed.), *Philosophical Experiments and Observations of the Late Eminent Dr Robert Hooke* (1726; reprinted 1967), pp. 225–48.
9. Royal Society Journal Book, vol. 8, p. 99, 24 February 1692.
10. Derham, *Philosophical Experiments and Observations*, pp. 257–68.
11. Royal Society Journal Book, vol. 8, 25 February 1691, 23 March 1692. Hooke's Diary, Guildhall Library, 14 March 1682.
12. R. Waller, *The Life of Dr Robert Hooke*, pp. 60–1, in R. T. Gunther, *Early Science in Oxford*, vol. vi (Oxford, 1930).
13. I. A. Archer, *The History of the Haberdashers' Company* (Chichester, 1991), pp. 104–8, 185.
14. Hooke's Diary, 28 April 1693.
15. Hooke's Diary, 18, 19, 26 April; 9, 12, 19, 20 May; 7 June; 5, 6, 18, 31 July. Hunter, *Establishing the New Science*, pp. 326–7.
16. Hooke's Diary, 23 December 1692; 18 January, 13 April, 12 May, 19, 20 July, 3, 5 August 1693.
17. Mercers' Company Gresham Repertory, 17 May 1682.
18. Hooke's Diary, 11 January, 6 March 1693.

26. Hooke's Last Years

1. M. A. R. Cooper, 'Robert Hooke's work as a Surveyor for the City of London', part two, p. 220, *Notes and Records of the Royal Society of London*, 52 ii (1998), pp. 205–20. M. A. R. Cooper, 'Robert Hooke, City Surveyor. An assessment of the importance of his work as a Surveyor for the City of London in the aftermath (1667–74) of the Great Fire', City University (London) Ph.D. thesis (1999), p. 239, n. 597.
2. Hooke's Diary, 23 December 1692; 18 January, 13 April, 12 May, 19, 20 July, 3, 5 August 1693. Royal Society Journal Book, vol. 8, 1, 29 November 1693; 3, 10 January 1694; 7, 14, 21 November 1694; 5 December 1694.
3. K. Thomas, *Man and the Natural World* (Harmondsworth, 1984), pp. 17–24.
4. R. Waller (ed.), *The Posthumous Works of Robert Hooke* (London, 1705, reprinted with an introduction by R. S. Westfall (London, 1968)), pp. 446–50. R. Rappaport, 'Hooke on earthquakes: lectures, strategy and audience', *British Journal for the History of Science* 19 (1986), pp. 129–46, has a good discussion of Hooke's attempts to persuade the Royal Society of his views, and also dates his series of geological lectures.
5. Royal Society Journal Book, vol. 8, pp. 264–72; Derham, *Philosophical Experiments and Observations*, pp. 292–6.
6. Derham, *Philosophical Experiments and Observations*, pp. 296–9.
7. Hunter, *Establishing the New Science*, pp. 327–8.
8. Hunter, *Establishing the New Science*, pp. 328–9.
9. Hooke, *Posthumous Works*, pp. 534–6.
10. Royal Society Journal Book, vol. 8, pp. 264, 319.
11. R. T. Gunther, *Early Science in Oxford*, vol. 7 (Oxford, 1930), p. 759, an extract from Hooke's Diary. Royal Society Journal Book, vol. 8, 30 May, 5 June 1695; 22 July 1696.
12. Royal Society Journal Book, vol. 8, pp. 311–13, 24 and 31 July 1695.
13. Hunter, *Establishing the New Science*, pp. 329–30. R. Waller, *The Life of Dr Robert Hooke*, pp. 61–2, in R. T. Gunther, *Early Science in Oxford*, vol. 6 (Oxford, 1930).
14. Derham, *Philosophical Experiments and Observations*, pp. 388–91.
15. Derham, *Philosophical Experiments and Observations*, pp. 304–14.
16. Derham, *Philosophical Experiments and Observations*, pp. 315–38.
17. Hooke, *Posthumous Works*, pp. 438–41. Royal Society Journal Book, vol. 9, 2 June 1697.
18. Waller, *The Life of Dr Robert Hooke*, p. 63.

19. Mercers' Company, Acts of Court, 1693–1700, p. 103: Court of Assistants approves Godfrey's burial in the Mercers' Chapel.
20. British Museum Sloane mss 1039, f. 108. Royal Society meetings had been on Wednesdays since January 1681.
21. Royal Society Classified Papers, volume xx, ff. 91a, 91b, 92.
22. Hooke, *Posthumous Works*, pp. 444–5.
23. Ned Ward, *The London Spy*, ed. A. L. Hayward (London, 1927), pp. 48–9.
24. R. Chartres, D. Vermont, *A Brief History of Gresham College, 1597–1997* (London, 1998), pp. 40–1. Mercers' Company, Gresham Repository, 1678–1722, minutes for August 1699 to March 1702.
25. Royal Society Classified Papers, vol. xx, f. 93.
26. Royal Society Journal Book, vol. 9, 6 and 13 December 1699.
27. *Philosophical Transactions* (1701), vol. 22, p. 791.
28. Royal Society, Southwell Papers, manuscript 248, ff. 14, 18. British Museum Sloane mss, f. 101, Southwell to Hooke, 6 January 1702.
29. Copley is quoted in M. Hunter and S. Schaffer (eds), *Robert Hooke: New Studies* (Woodbridge, Suffolk, 1989) p. 290. Somers' remark is quoted in C. Jones and G. Holmes (eds), *The London Diaries of William Nicholson, Bishop of Carlisle, 1702–1718* (Oxford, 1985), p. 358. Robert Knox, *Concerning Severall Remarkable Passages of my Life that hath Hapnd since my Deliverance out of my Captivity* (Glasgow, 1911), pp. 382–3.
30. Waller, *The Life of Dr Robert Hooke*, p. 63.
31. Knox, *Concerning Severall Remarkable Passages*, pp. 382–3.
32. M. A. R. Cooper, 'The Legacy of Robert Hooke: Science, Surveying and the City', a talk given at Gresham College on 4 April 2001, and available at www.royalsoc.ac.uk./events/hooke.

Bibliography

PRIMARY SOURCES

1. Manuscript sources

Royal Society

Classified Papers of the Royal Society, 1667–1740, vol. XX contains nearly a hundred manuscript accounts of Hooke's experiments and observations. Other volumes have letters sent by Hooke as the Secretary of the Royal Society, and some miscellaneous papers.

The Journal Book of the Royal Society (copy), vols 7–10, contains the minutes of meetings after 1687, which were not printed in Birch's *History of the Royal Society*.

The British Library

Sloane Mss 4024: Hooke's 1688–93 diary.

Several volumes containing letters or papers written by Hooke, especially Add. Mss 4441, which includes Oldenburg's papers relating to Hooke's attack on him in *Lampas*, and Sloane Mss 1039, which has Hooke correspondence and papers, including material on Pitt's atlas.

Add. Mss 5238, a volume containing architectural drawings by Hooke.

The Guildhall Library

Hooke's 1672–83 diary, which includes entries for March to July 1672 and for 1681–3 missing from the printed version.

A collection of papers on foreign countries and exploration (written by others), and a few of Hooke's lectures.

Bodleian Library, Oxford

The Aubrey papers have several letters from Aubrey to Hooke, and one from Hooke to Aubrey, and the Rawlinson papers, B. 363, have an account, partly by Hooke, of work done on the Monument.

Trinity College Library, Cambridge

Has letters (all published) from Hooke to Newton, and scientific papers by Hooke dealing with longitude, gravity, motion, music, memory and colours. Some of these have been published, and are listed in the bibliography under Kassler, Oldroyd, Gouk and Wright.

Mercers' Company Archives

Papers relating to the administration of Gresham College: the Acts of Court, 1693–1700, and the Gresham Repertory, 1660–1703.

2. Robert Hooke's published works

The fullest listing of Hooke's works is Geoffrey Keynes' *Bibliography of Robert Hooke* (Oxford, 1966).

The biggest collection of Hooke's writings is in vols 6–8, 10 and 13 of Gunther, R. T., *Early Science in Oxford: The Life and Work of Robert Hooke* (1930–8). Vols 6–7 reprint Hooke's contributions to Royal Society meetings up to 1687, selected letters (especially Hooke's letters to Boyle in the 1660s), and assorted scientific papers and brief diary entries. It also contains:

An Attempt for the Explication of the Phænomena Observable in an Experiment Published by the Honourable Robert Boyle (1661), in vol. 10 (1935), pp. 1–50.

Micrographia (1665), in vol. 13 (1938).

Lectiones Cutlerianae, or a Collection of Lectures, Physical, Mechanical, Geographical & Astronomical (1679), in vol. 8 (1931). This includes all the published Cutler lectures: *An Attempt to Prove the Motion of the Earth from Observations* (1674), *Animadversions on the First Part of the Machina Coelestis of Johannes Hevelius* (1674), *A Description of Helioscopes* (1676), *Lampas* (1677), *Cometa* (1678), *Microscopium* (1678) and *Lectures de Potentia Restituva or of Spring* (1678).

The Diary of Robert Hooke, 1688–1693, in vol. 10 (1935), pp. 69–294.

Hooke, R., *Philosophical Experiments and Observations of the Late Eminent*

Dr Robert Hooke (ed. W. Derham, 1726, repr. 1967). Miscellaneous scientific papers, including some collected by Hooke but written by others.

Hooke, R., *Posthumous Works* (ed. R. Waller, 1705; R. S. Westfall, New York, 1969), contains Hooke's lectures on earthquakes, light, navigation, comets and 'A general scheme of the present state of natural philosophy'.

Robinson, H. W., Adams, W. (eds), *The Diary of Robert Hooke, 1672–1680* (1935).

Philosophical Transactions contains many papers by Hooke, especially in vols 1–6 (1665–71), 9 (1674), 16 (1688), and 17 (1693).

Philosophical Collections, numbers 1–7 (1679–82), were edited by Hooke, but only number 33 (1681) has articles signed by him.

Knox, R., *An Historical Relation of the Island Ceylon* (1911) has an introduction by Hooke.

Pitt, M., *The English Atlas*, vol. 1 (1680), has an introduction by Hooke, and Hooke's 'map of the North Pole and the parts adjoining'.

3. Other printed primary sources

Aubrey, J., *Brief Lives* (ed. O. Lawson Dick, 1962).

Bannerman, W. B. (ed.), *The Registers of St Helen, Bishopsgate* (1904).

Birch, T., *The History of the Royal Society of London* (4 vols., 1756–7).

———, *The Life and Work of the Honourable Robert Boyle* (6 vols., 1772).

Evelyn, J., *The Diary and Correspondence of John Evelyn* (ed. W. Bray, n.d.).

Forbes, E. G., Murdin, L., Willmoth, F. (eds), *Correspondence of John Flamsteed* (2 vols., 1995).

Hall, A. R., Hall, M. B., *The Correspondence of Henry Oldenburg* (13 vols., 1965–86).

Hunter, M., Clericuzo, A., Principe, L. M. (eds), *The Correspondence of Robert Boyle* (6 vols, 2001)

Knox, R., *An Historical Relation of the Island Ceylon* (1681; 1911).

———, *Concerning Severall Remarkable Passages of my Life that hath Hapnd since my Deliverance out of my Captivity* (included in the 1911 ed. of *An Historical Relation of the Island Ceylon.*

MacPike, E. F., *Correspondence and Papers of Edmond Halley* (1932).

Pepys, S., *The Diary of Samuel Pepys* (ed. Latham, R., Matthews W., 11 vols, 1983).

Shadwell, T., *The Virtuoso: a Comedy* (ed. Nicolson, M. H., Rodes, D. S., 1966).

Sprat, T., *A History of the Royal Society* (ed. Cope, J. I., Jones, H. W., 1959).

Turnbull, H. W. (ed.), *The Correspondence of Isaac Newton*, vols 1 and 2 (1959).
Waller, R., *The Life of Dr Robert Hooke*, in Gunther, op. cit., vol. 6 (1930).
Ward, N., *The London Spy* (1927).

SECONDARY SOURCES

Agassi, J., 'Who discovered Boyle's Law?', *Studies in the History and Philosophy of Science*, 8 (1977), 189–250.
Anderson, R. G. W., Bennett, J. A., Ryan, W. F., *Making Instruments Count. Essays for G. l'E Turner* (1993).
Andrade, E. N. da C., 'Robert Hooke', *Proceedings of the Royal Society* (1950), series A, number 201, 439–73.
Andrewes W. J. H. (ed.), *The Quest for Longitude* (1996).
Andrews, J., Briggs, A., et al., *The History of Bethlem* (1997).
Archer, I. A., *The History of the Haberdashers' Company* (1991).

Barker F., and Jackson, P., *The History of London in Maps* (1990).
Batten, M. I., 'The Architecture of Dr Robert Hooke, F.R.S.', *Walpole Society*, 25 (1936–7), 83–113.
Beier, L. M., 'Experience and experiment: Robert Hooke, illness and medicine', in Hunter and Schaffer, *Robert Hooke, New Studies*, op. cit., 235–52.
Bennett, J. A., 'A study of *Parentalia*, with two unpublished letters of Sir Christopher Wren', *Annals of Science*, 30 (1973), 129–47.
———, 'Hooke and Wren and the system of the world', *British Journal for the History of Science*, 8 (1975), 32–61.
———, 'Robert Hooke as mechanic and natural philosopher', *Notes and Records of the Royal Society*, 35 (1980–1), 33–48.
———, *The Mathematical Science of Christopher Wren* (1983).
———, 'The mechanics' philosophy and the mechanical philosophy', *History of Science*, 24 (1986), 1–28.
———, *The Divided Circle* (1987).
———, 'Hooke's instrument for Astronomy and Navigation', in Hunter and Schaffer, *Robert Hooke, New Studies*, op. cit., 21–32.
Birrell, T. A., *The Cultural Background of Two Scientific Revolutions: Robert Hooke's London and James Logan's Philadelphia* (1963).
Bradley, S., Pevsner, N., *London: the City Churches* (1998).

Bud, R., Warner, D. J. (eds), *Instruments of Science* (1998).
Burke, J. G. (ed.), *The Uses of Science in the Age of Newton* (1983).

Centore, F. F., *Robert Hooke's Contributions to Mechanics* (1970).
Chapman, A., 'England's Leonardo: Robert Hooke (1635–1703) and the art of experiment in Restoration England', *Proceedings of the Royal Institution of Great Britain*, 67 (1996), 239–275.
Chartres, R., Vermont, D., *A Brief History of Gresham College, 1597–1997* (1998).
Clark, Sir G., *A History of the Royal College of Physicians*, vol. 1 (1964).
Clifton, G., Turner, G. l'E. (eds), *Directory of British Instrument Makers, 1550–1851* (1995).
Colvin, H., M., *A Biographical Dictionary of British Architects, 1600–1840* (3rd edn, 1995).
———, 'Robert Hooke and Ramsbury Manor', in Colvin, H. M., *Essays in English Architectural History* (1999).
Cooper, M. A. R., 'Robert Hooke's work as a surveyor for the City of London in the aftermath of the Great Fire', parts 1–3, *Notes and Records of the Royal Society*, 51 (1997), 161–74; 52 (1998), 25–38; 52 (1998), 205–20.
———, 'Robert Hooke, City Surveyor. An assessment of the importance of his work as a Surveyor for the City of London in the aftermath (1667–74) of the Great Fire'. City University Ph.D. thesis, May 1999.
———, 'The Legacy of Robert Hooke: Science, Surveying and the City', a talk given at Gresham College on 4 April 2001, http://www.royalsoc.ac.uk./events/hooke.
Crook J. M., *The British Museum* (1972).

Deacon, M., 'Founders of marine science in Britain: the work of the early fellows of the Royal Society', *Notes and Records of the Royal Society*, 20 (1965), 28–50.
Dennis, M. A., 'Graphic Understanding: Instruments and Interpretation in Robert Hooke's *Micrographia*', *Science in Context*, 3 (1989), 309–64.
Derry, T. K., Williams, T. I., *A Short History of Technology* (1960).
Dictionary of National Biography, ed. Stephen, L., Lee, S. (1975).
Downes, K., *The Architecture of Sir Christopher Wren* (1982).
Dowson, D., *History of Tribology* (1979).
Drake, E. T., *Restless Genius. Robert Hooke and his Earthly Thoughts* (1996).

Earle, P., *The Making of the English Middle Class* (1989).
Edwardes, E. L., *The Story of the Pendulum Clock* (1977).

Ehrlich, M. E., 'Mechanism and activity in the Scientific Revolution: the case of Robert Hooke', *Annals of Science*, 52 (1995), 127–51.
Ellis, A., *The Penny Universities* (1956).
Erlichson, H., 'Newton and Hooke on centrifugal force motion', *Centaurus*, 35 (1992), 46–63.
Espinasse, M., *Robert Hooke* (1956).
———, 'The Decline and Fall of Restoration Science', *Past and Present*, 14 (1958), 71–89.

Feingold, M., 'Astronomy and Strife: John Flamsteed and the Royal Society', in Willmoth, F. (ed.), *Flamsteed's Stars*, op. cit.
Feisenberger, H. A., 'The libraries of Newton, Hooke and Boyle', *Notes and Records of the Royal Society*, 38 (1966), 42–55.
Ford, B. J., *Single Lens. The Story of the Simple Microscope* (1985).
Fournier, M., *The Fabric of Life* (1996).
Frank, R. G., *Harvey and the Oxford Physiologists. A Study of Scientific Ideas and Social Interaction* (1980).

Gal, O., 'Producing knowledge in the workshop: Hooke's "inflection" from optics to planetary motion', *Studies in the History and Philosophy of Science* (1996), 27, 181–205.
Geraghty, A., 'Nicholas Hawksmoor and the Wren City church steeples', *The Georgian Group Journal*, 10 (2000), 1–14.
Gorden, J. E., *The New Science of Strong Materials* (2nd ed., 1976).
Gouk, P.,'The role of acoustics and music theory in the scientific work of Robert Hooke', *Annals of Science*, 37 (1980), 575–605.
———, *Music, Science and Natural Magic in Seventeenth-Century England* (1999).
Greenberg, *The Problem of the Earth's Shape from Newton to Clairant* (1995).

Hall, A. R., 'Robert Hooke and Horology', *Notes and Records of the Royal Society*, 8 (1951), 167–77.
———, 'Two unpublished lectures of Robert Hooke', *Isis* 42 (1951), 219–30.
———, *Hooke's Micrographia, 1665–1965* (1966).
———, 'Horology and Criticism: Robert Hooke', *Studia Copernicana*, 16 (1978), 261–81.
———, *The Revolution in Science, 1500–1700* (1983).
Hall, A. R., Hall, M. B., 'Why blame Oldenburg?', *Isis* 53 (1962), 482–91.
Hall, M. B., 'Oldenburg, the Philosophical Transactions, and Technology', in J. G. Burke (ed.), *The Uses of Science in the Age of Newton*, op. cit., 21–47.

———, *Promoting Experimental Learning. Experiment and the Royal Society, 1660–1727* (1991).

Harris, H., *The Birth of the Cell* (1999).

Harwood, J. T., 'Rhetoric and graphics in *Micrographia*', in Hunter, M., Schaffer, S. (eds), *Robert Hooke, New Studies*, op. cit., 119–148.

Helden, A. Van, 'The birth of the modern scientific instrument, 1550–1700', in Burke, *The Uses of Science in the Age of Newton*, op. cit.

Henry, J., 'Occult qualities and the experimental philosophy', *History of Science*, 24 (1986), 335–81.

———, 'Robert Hooke, the incongruous mechanist', in Hunter, M., Schaffer, S. (eds), *Robert Hooke, New Studies*, op. cit.

Hesse, M. B., 'Hooke's philosophical algebra', *Isis*, 57 (1966), 67–83.

———, 'Hooke's vibration theory and the isochrony of springs', *Isis*, 57 (1966), 433–41.

Hoppen, K. T., 'The nature of the early Royal Society', *British Journal for the History of Science*, 9 (1976), 1–24, 243–73.

Howse, D., *Greenwich Observatory*. Vol. 3, *The Buildings and Instruments* (1975).

Hunter, M., *Science and Society in Restoration England* (Cambridge, 1981).

———, *The Royal Society and its Fellows, 1660–1700: the Morphology of an Early Scientific Institution* (1982).

———, *Establishing the New Science: the Experience of the Early Royal Society* (1989).

———, *Science and the Shape of Orthodoxy: Intellectual Change in Late Seventeenth-Century Britain* (1995).

———, *Robert Boyle, 1627–1691. Scrupulosity and Science* (2000).

———, 'The life and thought of Robert Boyle', http://www.bbk.ac.uk/Boyle/biog.html.

———(ed.), *Robert Boyle by Himself and his Friends* (1994).

Hunter, M., Schaffer, S. (eds), *Robert Hooke, New Studies* (1989).

Hunter, M. et al., 'Hooke's Possessions at his Death: a hitherto unknown inventory', in Hunter, M., Schaffer, S. (eds), *Robert Hooke: New Studies*, op. cit.

Hutchison, K., 'What happened to occult qualities in the scientific revolution?', *Isis*, 73 (1982), 233–53.

Iliffe, R., 'Material doubts: Hooke, artisan culture and the exchange of information in 1670s London', *British Journal for the History of Science*, 28 (1995), 285–318.

———, '"In the Warehouse": Privacy, Property and Priority in the Early Royal Society', *History of Science*, 30 (1992), 29–68.

Inwood, S., *A History of London* (1998).

Jardine, L., *Ingenious Pursuits: Building the Scientific Revolution* (1999).

———, 'Monuments and microscopes: scientific thinking on a grand scale in the early Royal Society', *Notes and Records of the Royal Society*, 55 (2001), 289–308.

Jeffery, P., *The Parish Churches of Sir Christopher Wren* (1996).

Kassler, J. C., Oldroyd, D. R., 'Robert Hooke's Trinity College "Musick Scripts"', *Annals of Science*, 40 (1983), 559–95.

Keynes, M., 'The Personality of Isaac Newton', *Notes and Records of the Royal Society*, 49 (1995), 1–56.

King, H. C., *A History of the Telescope* (1955).

Koyré, A., 'A note on Robert Hooke', *Isis*, 41 (1950), 195–6.

———, *Newtonian Studies* (1965).

Landes, D., *Revolution in Time: Clocks and the Making of the Modern World* (1983).

Lillywhite, B., *London Coffee Houses* (1963).

Little, B., *Sir Christopher Wren. A Historical Biography* (1975).

Lloyd, C., 'Shadwell and the virtuosi', *Publications of the Modern Language Association of America*, 44 (1929), 472–94.

Lohne, J., 'Hooke versus Newton. An analysis of the documents in the case on free fall and planetary motion', *Centaurus*, 7 (1960), 6–52.

London Topographical Society, *The A to Z of Restoration London* (1992), introduced by R. Hyde.

Louw, H., 'The origin of the sash-window', *Architectural History*, 26 (1983), 49–72.

Louw, H., Crayford, R., 'A constructional history of the sash-window c. 1670–1725, parts 1 and 2', *Architectural History*, 41 (1998), 82–130; 42 (1999), 173–239.

Macintosh., J. J., 'Perception and imagination in Descartes, Boyle and Hooke', *Canadian Journal of Philosophy*, 13 (1983), 327–52.

McKellar, E., *The Birth of Modern London* (1999).

McKie, D., 'Fire and the *Flamma Vitalis*: Boyle, Hooke and Mayow', in Underwood, E. A., *Science, Medicine and History* (1953), 469–88.

Merton, R. K., *Science, Technology and Society in seventeenth-century England* (1970).

Metcalf, P., *The Halls of the Fishmongers' Company* (1977).

Middleton, W. E. K., *The Invention of the Meteorological Instruments* (1969).

Middleton, W. S., 'The medical aspects of Robert Hooke', *Annals of Medical History*, 9 (1927), 227–43.

Montagu, M. F. A., *Edward Tyson, M.D., F.R.S., 1650–1708* (1943).
Moyer, A. E., 'Robert Hooke's ambiguous presentation of "Hooke's Law"', *Isis*, 68 (1977), 266–75.
Mulligan, L., 'Robert Hooke and certain knowledge', *Seventeenth Century*, 7 (1992), 151–69.
———, 'Robert Hooke's "memoranda": memory and natural history', *Annals of Science*, 49 (1992), 47–61.
———, 'Self-scrutiny and the study of nature: Robert Hooke's diary as natural history', *Journal of British Studies*, 35 (1996), 311–42.

Nakajima, H., 'Robert Hooke's family and his youth', *Notes and Records of the Royal Society*, 48 (1994), 11–16.
Nichols, R., *The Diaries of Robert Hooke, the Leonardo of London* (1994).
Nicolson, M. H., Mohler, N. M., 'The scientific background of Swift's voyage to Laputa', *Annals of Science*, 2 (1937), 299–334.

Ochs, K. H., 'The Royal Society of London's History of Trades programme: an early episode in applied science', *Notes and Records of the Royal Society*, 39 (1985), 129–58.
Oldroyd, D. R., 'Robert Hooke's methodology of science as exemplified in his Discourse of Earthquakes', *British Journal for the History of Science*, 6 (1972), 109–30.
———, 'Some "Philosophicall Scribbles" attributed to Robert Hooke', in *Notes and Records of the Royal Society*, vol. 35 (1980), 17–32.
———, 'Some writings of Robert Hooke on procedures for the prosecution of scientific enquiry', *Notes and Records of the Royal Society*, 41 (1987), 145–67.
———, 'Geological controversy in the seventeenth century: "Hooke vs Wallis" and its aftermath', in Hunter, M., Schaffer, S. (eds), *Robert Hooke: New Studies* op. cit., 207–37.
———, *Thinking About the Earth* (1996).

Park, D., *The Fire Within the Eye* (1997).
Palter, R., 'Early measurement of magnetic force', *Isis* 63 (1972), 544–58.
Patterson, L. D., 'Hooke's gravitation theory and its influence on Newton' (parts I and II), *Isis*, 40 (1949), 327–41, and 41 (1950), 32–45.
Pevsner, N., Cherry, B., *The Buildings of England. Devon* (2nd ed., 1989).
Powell, A., *John Aubrey and his Friends* (1963).
Pugliese, P. J., 'The scientific achievement of Robert Hooke: method and mechanics' (Harvard University Ph.D. thesis, 1982).

——, 'Robert Hooke and the dynamics of motion in a curved path', in Hunter, M., Schaffer, S. (eds), *Robert Hooke: New Studies*, op. cit., 181–206.

Pumfrey, S., 'Ideas above his station: a social study of Hooke's curatorship of experiments', *History of Science*, 29 (1991), 1–44.

Rappaport, R., 'Hooke on earthquakes: lectures, strategy and audience', *British Journal for the History of Science*, 19 (1986), 129–146.

——, *When Geologists were Historians, 1665–1750* (1997).

Reddaway, T. F., *The Rebuilding of London After the Great Fire* (1940).

Robinson, E. F., *The Early History of Coffee Houses in England* (1963).

Robinson, H. W., 'Robert Hooke as a surveyor and architect', *Notes and Records of the Royal Society*, 6 (1949), 48–55.

Rossi, P., *The Dark Abyss of Time. The History of the Earth and the History of Nations from Hooke to Vico* (1984).

Rostenberg, L., *The Library of Robert Hooke: the Scientific Book Trade of Restoration England* (1989).

Sabra, A. I., *Theories of Light from Descartes to Newton* (1967).

Saunders, A., *The Royal Exchange* (1997).

Schaffer, S., 'Natural Philosophy as a Public Spectacle', *History of Science*, 21 (1983), 1–43.

——, 'Godly men and mechanical philosophers: souls and spirits in Restoration natural philosophy', *Science in Context*, 1 (1987), 55–85.

Shapin, S., 'The house of experiment in seventeenth-century England', *Isis*, 79 (1988), 373–404.

——, 'Who was Robert Hooke?', in Hunter, M., Schaffer, S. (eds), *Robert Hooke, New Studies*, op. cit., 253–86.

——, *A Social History of Truth* (1994).

Shapin, S., Schaffer, S., *Leviathan and the Air-Pump: Hobbes, Boyle and the Experimental Life* (1985).

Shapiro, A. E., *Fits, Passions and Paroxysms* (1993).

Shapiro, B. J., *John Wilkins, 1614–1672: an Intellectual Biography* (1969).

——, *Probability and Certainty in Seventeenth-Century England* (1983).

Simpson, A. D. C., 'Robert Hooke and practical optics', in Hunter, M., Schaffer, S. (eds), *Robert Hooke, New Studies*, op. cit., 33–61.

Singer, B. R., 'Robert Hooke on memory, association and time perception', *Notes and Records of the Royal Society*, 31 (1979), 115–131.

Singer, C. (ed.), *The History of Technology* (1957).

Stewart, L., *The Rise of Public Science. Rhetoric, Technology and Natural Philosophy in Newtonian Britain, 1660–1750* (1992).

Straus, R., *Carriages and Coaches* (1912).
Summerson, Sir J., *Architecture in Britain, 1530–1830* (1953; 1970).
Symonds, R. W., *Thomas Tompion. His Life and Work* (1951).

Taylor, E. G. R., 'The Geographical Ideas of Robert Hooke', *Geographical Journal*, 89 (1937), 525–38.
———, 'Robert Hooke and the Cartographical Projects of the Late Seventeenth Century (1666–1696)', *Geographical Journal*, 90 (1937), 529–40.
———, 'The English Atlas of Moses Pitt, 1680–1683', *Geographical Journal*, 95 (1940), 292–9.
Thomas, K., *Man and the Natural World* (1984).
Thornbury, W., Walford, E., *Old and New London* (6 vols, 1883–5).
Turner, A. J., 'Hooke's theory of the Earth's axial displacement: some contemporary opinion', *British Journal for the History of Science*, 7 (1974), 166–70.
Turner, G. l'E., *Essays on the History of the Microscope* (1980).
Turner, H. D., 'Robert Hooke and theories of combustion', *Centaurus* 4 (1956), 297–310.

Ward, J., *Lives of the Professors of Gresham College* (1740).
Webster, C., *The Great Instauration* (1975).
Westfall, R. S., 'Hooke and the law of universal gravitation', *British Journal for the History of Science*, 3 (1967), 245–61.
———, 'Introduction' in Waller (ed.), *Posthumous Works of Robert Hooke* (1969).
———, 'Robert Hooke', in *Dictionary of Scientific Biography* (1972).
———, *Never at Rest. A Biography of Isaac Newton* (1980).
———, 'Robert Hooke, Mechanical Technology and Scientific Investigation', in J. G. Burke (ed.), *The Uses of Science in the Age of Newton*, op. cit., 85–110.
Willmoth, F. (ed.), *Flamsteed's Stars. New Perspectives on the Life and Work of the First Astronomer-Royal* (1997).
Wilson, C., *The Invisible World. Early Modern Philosophy and the Invention of the Microscope* (1995).
Woodbury, R. S., *History of the Gear-Cutting Machine* (1958).
Wren, C., *Parentalia* (1750).
Wren Society, vol. x, *The Parochial Churches of Sir Christopher Wren, 1666–1728* (1933), and vol. xix, *The City Churches* (1942).
Wright, M., 'Robert Hooke's Longitude Timekeeper', in M. Hunter, S. Schaffer (eds), *Robert Hooke. New Studies*, op. cit., 63–118.

Index

The index covers the introductory quotations and Chapters 1 to 26, but not 'A Chronology of the Life of Robert Hooke', the Preface, the Notes or the Illustrations, a guide to which can be found on p. xv. Fellowship of the Royal Society is indicated – by the abbreviation '*FRS*' – because of its importance to the scientific activity of the time and to Hooke's life in particular. Other abbreviations used are 'H' for Robert Hooke, 'RS' for Royal Society and 'C17th' for seventeenth century. The alphabetical arrangement is word-by-word. Numbers are arranged as though spelled out. Book titles are given in *italic* whilst the titles of lectures and articles are enclosed in single quotation marks. Locations given in bold face indicate a definition of a term or an explanation of a concept *eg* mechanics **13–14**.